What Is Adaptive about Adaptive Memory?

What Is Adaptive about Adaptive Memory?

EDITED BY

BENNETT L. SCHWARTZ

MARK L. HOWE

MICHAEL P. TOGLIA

HENRY OTGAAR

OXFORD
UNIVERSITY PRESS

OXFORD
UNIVERSITY PRESS

Oxford University Press is a department of the University of Oxford.
It furthers the University's objective of excellence in research, scholarship, and education by publishing worldwide.

Oxford New York
Auckland Cape Town Dar es Salaam Hong Kong Karachi
Kuala Lumpur Madrid Melbourne Mexico City Nairobi
New Delhi Shanghai Taipei Toronto

With offices in
Argentina Austria Brazil Chile Czech Republic France Greece
Guatemala Hungary Italy Japan Poland Portugal Singapore
South Korea Switzerland Thailand Turkey Ukraine Vietnam

Published in the United States of America by
Oxford University Press
198 Madison Avenue, New York, NY 10016

Library of Congress Cataloging-in-Publication Data
What is adaptive about adaptive memory? / edited by Bennett L. Schwartz,
Mark L. Howe, Michael P. Toglia, Henry Otgaar.
pages cm
Includes bibliographical references and index.
ISBN 978-0-19-992805-7
1. Memory. 2. Evolutionary psychology. 3. Human evolution. I. Schwartz, Bennett L. editor of compilation.
BF371.W46 2014
153.1′2—dc23
2013017237

9 8 7 6 5 4 3 2 1
Printed in the United States of America
on acid-free paper

TABLE OF CONTENTS

CONTRIBUTORS

Jeanette Altarriba, University at Albany, State University of New York, USA
Catherine M. Baker, University of North Florida, USA
Raoul Bell, Heinrich-Heine-Universität, Düsseldorf, Germany
David F. Bjorklund, Florida Atlantic University, USA
Brock R. Brothers, Oakland University, USA
Axel Buchner, Heinrich-Heine-Universität, Düsseldorf, Germany
Daniel J. Burns, Union College, USA
Anne M. Cleary, Colorado State University, USA
Eileen M. Dacey, University of North Florida, USA
Mary H. Derbish, Lancaster University, UK
Edgar Erdfelder, University of Mannheim, Germany
Joshua Hart, Union College, USA
Mark L. Howe, City University, London, UK
Stephanie Kazanas, University at Albany, State University of New York, USA
Stanley B. Klein, University of California, Santa Barbara, USA
Meike Kroneisen, University of Mannheim, Germany
Aaron D. Leedy, University of North Florida, USA
Corrina Maguinness, Trinity College Dublin, Ireland
Kelly W. Mosteller, Springhill College, USA
James S. Nairne, Purdue University, USA
Fiona N. Newell, Trinity College Dublin, Ireland
Henry Otgaar, Maastricht University, The Netherlands
Maarten J. V. Peters, Maastricht University, The Netherlands
Linsey Raymaekers, Maastricht University, The Netherlands
Bennett L. Schwartz, Florida International University, USA
Patrick Douglas Sellers II, Florida Atlantic University, USA
Tom Smeets, Maastricht University, The Netherlands
Nicholas C. Soderstrom, University of California, Los Angeles, USA
Michael P. Toglia, University of North Florida, USA
Johan van Beers, Maastricht University, The Netherlands
Jennifer Vonk, Oakland University, USA
Adam M. Wilde, University of North Florida, USA

1

What Is Adaptive Memory?

HENRY OTGAAR, MARK L. HOWE, BENNETT L. SCHWARTZ, AND MICHAEL P. TOGLIA

Clearly, our memory plays a decisive role in our daily life. It gives us striking abilities that allow us both to recollect the past and predict the future. For example, our memory allows us to think back about past episodes, such as witnessing a crime, but also to mentally plan prospective contingencies that can aid us in our future plans, such as avoiding the location where the crime took place. As memory scientists, we should attend both to the functions of memory as well as to the processes that permit optimal performance. Our interest in both function and process, for example, has led to discoveries that we possess conscious and unconscious memories and that they can be subdivided into even more categories (episodic, semantic, procedural; Tulving, 1983, 2002, 2005). However, function does not necessarily inform us about form. An issue that has been receiving increasing attention is the question about how and why our memory evolved. Although surely function follows from evolution, it may not always be the case that current function stems directly from natural selection. This makes the issue of addressing evolutionary questions in human memory fraught with difficulties. It is our contention that the chapters presented in this book deal with these difficulties with scientific acumen and methodological rigor.

Pinpointing the exact evolutionary paths of memory is nearly impossible, if for no other reason than because records of fossilized memory traces simply do not exist (Nairne, Pandeirada, & Thompson, 2008). However, many researchers agree that our memory was shaped by natural selection and that it is possible to examine the origin of human memory using an evolutionary framework (e.g., Klein, Cosmides, Tooby, & Chance, 2002). In particular, this will require a detailed examination of the nature of current function and its possible correlation with evolutionary origin (see Howe & Otgaar, 2013; Klein, Chapter 2 in this volume). In the present volume, we are concerned with the evolution of memory, in particular with what has been termed adaptive memory. We and other contributors to this volume demonstrate which mechanisms and functions may reveal distinct facets of memory evolution. Insight into these aspects of the evolution of memory is imperative for understanding memory specifically, as well as for psychology in general. In this chapter, we will first briefly introduce the basic developmental patterns of human and animal memory. Then, we will focus on a special type of memory processing that may be relevant in the context of memory evolution. Finally, we will provide an overview of the chapters that together provide the cornerstone of this volume.

THE DEVELOPMENT OF MEMORY

When we experience an event, our memory is able to encode and store the contents of this event as an abstract mental representation. When needed in the future, we retrieve part or all of this stored representation in order to guide our decision-making. Obviously, memory is beneficial to survival and reproductive success in many species, and it is likely that complex human memory abilities also evolved via natural selection. Imagine someone who is unable to encode or retrieve. Such a person will be unlikely to engage in appropriate social activities, will be unable to remember and therefore avoid hazardous situations, and will not be able to plan for future contingencies. Thus, that memory in humans evolved through natural selection should not be a controversial statement.

Amnesic patients such as H.M. do not possess either normal encoding or retrieval processes. Because of their amnesia, they lack the capacity to successfully execute tasks, such as decision-making, that are based in other people on being able to recall the past. Although these patients are of particular interest for our understanding about memory functioning, it is obvious that such patients have extreme difficulties managing their daily life activities without the help of sufficient working memory processes and the help of other persons (Milner, 1964). Although in modern society we care for such individuals, amnesia would have been a major barrier to survival in prehistoric times.

Some have argued that implicit memory may be an evolutionarily old memory system (Squire, 2004). Moreover, implicit memory already exists quite early in development, whereas the development of episodic memory emerges gradually through infancy and into later childhood. It is likely, therefore, that early development correlates with older evolutionary roots. Thus, looking at development for clues to evolution may be relevant, even if evolutionary scientists are no longer committed to the mantra of "ontogeny recapitulates phylogeny." With respect to episodic memory, the gradual developmental trajectory of episodic memory is particularly relevant for autobiographical memory that starts to develop when a sense of self evolves (Howe, 2011). Indeed, the developmental stages of autobiographical memory coincide with the offset of infantile amnesia, the inability to retrieve autobiographical memories from infancy and early childhood (Rubin, 2000). These point to a later evolutionary introduction of the processes that contribute to episodic memory (e.g., see Hampton & Schwartz, 2004).

Thus, we think it is clear that the development of both implicit and explicit memories is crucial for our survival. Although humans may be the only species to have such a sophisticated memory system, many animals are capable of developing representations of earlier experiences (Nelson, 2005). It is also easier to demonstrate the relation between function and evolution when one considers how memory works differently in closely related species. Therefore, to understand the course of memory evolution, one should also consider how memory evolved in nonhumans.

COMPARATIVE APPROACHES TO MEMORY EVOLUTION

That evolution of memory systems has occurred is clear from examining the comparative literature (Shettleworth, 2013; Vonk & Mosteller, Chapter 14 in this volume). Examining memory in nonhuman animals allows us to compare memory behaviors across closely related species that differ in terms of their ecology. This research has revealed important differences in the way memory functions in response to ecological demands, such as foraging ecology and reproductive strategies. It should not be a stretch to consider that human memory is also adapted to our particular (and peculiar) ecological demands.

The ecological demands of an herbivore, an insectivore, or a carnivore are quite different. An ungulate that eats grass does not need to engage memory processes for feeding. It simply puts its mouth down and chews—the stomach does the rest. One might suspect that this foraging ecology puts low demands on memory systems. However, predators, such as a wolf or an eagle, would do well to remember the spatial locations in which prey species may be found. For an animal such as a hummingbird with high energy needs, it would be advantageous to remember not only where sources of nectar can be found but which flowers have recently been visited so as to avoid returning to used resources. Animals with specific nutritional needs, such as koalas, may be able to remember the location of specific feeding resources, such as eucalyptus trees. Indeed, when one looks across the comparative literature, one finds that foraging ecology influences memory performance in these species.

Foraging ecology and its relation to memory have been best studied in birds (see Sherry & Schacter, 1987; Shettleworth, 2013). For example, Balda and Kamil (2006) examined spatial memory across four species of American corvids (jays and nutcrackers). The four species varied to the extent to which they store food for later use, with the nutcracker being the biggest food storer, the Mexican jay not storing food at all, and pinyon jays and Western scrub jays of intermediate dependence on stored food. The extent to which the birds store food in the wild predicted their performance in a lab-based spatial memory task; that is, the nutcracker showed the best spatial memory performance and the Mexican jay the worst. However, their feeding ecology did not predict their ability to perform in a color-based delayed matching task, in which performance was equivalent across species Thus, it is likely that spatial memory evolved along with food-storing, allowing the birds to better recall locations of stored food, thus supporting the view that memory is shaped by natural selection.

Hummingbirds are nectar-eaters with high energy demands. Flowers are quickly depleted of nectar and will not rebuild their nectar supply for at least 24 hours. Thus, it is potentially advantageous for hummingbirds to have good working memory, which allows them to avoid recently visited flowers. This exactly what research has found (see Flores-Abreu, Hurly, & Healy, 2012). Sulikowski and Burke (2012) found a similar pattern in an Australian nectar feeder. When regent honeyeaters were brought into the lab, they learned win-shift tasks more easily than win-stay tasks, consistent with their nectar-feeding ecology. Scrub jays—one of the intermediate performers in the Balda and Kamil (2006) study—store both perishable (e.g., worms) and nonperishable food (nuts). Thus, it is potentially adaptive for them to be able to remember both what food is stored in what location and how long ago it was stored. Indeed, research shows that scrub jays remember the what, the where, and the when of a food-storing event (Clayton & Dickinson, 1998). Thus, across a wide range of bird species, memory ability tracks feeding ecology.

Foraging is not the only ecological factor that can affect memory. In a series of studies with closely related species of voles (a kind of small rodent), Gaulin and his colleagues showed that reproductive strategies could affect memory as well (see Jacobs, Gaulin, Sherry, & Hoffman, 1990). In species of voles in which males mate with several females in different geographic areas, spatial memory is stronger than in those species of voles that are monogamous. Moreover, in the polygynous species, males outperform females, but spatial memory is equivalent in the monogamous species. This suggests that male voles in the polygynous species may have been under selective pressure to develop better spatial memory in order to remember where the dens of their female mates are located.

One can see the relation between foraging ecology and memory in primates as well. Platt, Brannon, Briese, and French (1996) examined performance in a spatial delayed matching-to-sample task and a radial maze task in two species of New World monkeys, golden lion tamarins (*Leontopithecus rosalia*) and Weid's marmosets (*Callithrix kuhli*). Golden lion tamarins are omnivores, eating both fruit and insects, so they must forage widely. Marmosets eat tree sap, which means once the monkeys have penetrated a tree's bark, they do not need to move to feed. That is, a marmoset returns to the same food source, whereas tamarins look for food in new locations on subsequent days. These ecological features predicted how the two monkey species performed in the maze. Marmosets performed better at shorter retention intervals (5 minutes), but tamarins performed better at the longer retention intervals (24 hours). Therefore, similar to the studies with birds, foraging ecology predicted memory performance.

In theory, it would be possible to apply such logic to human memory. Consider our closest genetic neighbors, the great apes. Clearly, our "ecology," even our ancestral "ecology," differs greatly among humans and the great apes. Could we use ecological distinctions to predict memory performance? Would nonsocial orangutans perform less well in memory tasks for social information than humans and the African great apes? Are humans and chimpanzees (as meat-eaters) better at remembering the movements of animal in a simulated environment or an experimental analog with moving computer stimuli than are gorillas and orangutans (as herbivores)? Unfortunately, the logic of ecological niches has not been applied to studies among these species. Usually, the goal is to determine if a particular ape species can accomplish a task similar to ones humans can without regard to foraging ecology (see Hampton & Schwartz, 2004). But we think this would be a productive area for future memory research (see Vonk & Mosteller, Chapter 14 in this volume).

A "SPECIAL" ASPECT OF MEMORY: SURVIVAL PROCESSING?

Researchers have started to examine a number of ways in which evolution may have influenced the development of human memory. These include but are not limited to face recognition, cheater detection, planning for the future, and spatial memory (see Schwartz & Brothers, Chapter 9 in this volume). However, all of these hypotheses are retrospective in nature. We know that humans are good at face recognition and that it has great current utility, and therefore we conclude that it must have evolved.

The goal of good evolutionary science is to develop testable hypotheses from evolutionary principles and then determine if they can predict current performance. Although research is beginning to apply this principle to the above functions, there has been one very influential program of research that started from this first principle and then began to do the research.

Nairne and his colleagues reasoned that the most fundamental aspect of natural selection is survival. An individual who does not survive cannot contribute to the next generation. Therefore, variables that influence the ability of an individual to survive will likely influence the evolution of that species. Taking this view directly, Nairne argued processing information in terms of its survival value might lead to stronger memory, as such stronger memory might influence natural selection (Nairne, Pandeirada, & Thompson, 2008). Thus, an intriguing way to examine the evolution of memory is to examine whether nature "tuned" memory to be sensitive to survival-related information.

Nairne et al. (2008) were the first who examined this idea. They led participants to imagine being stranded on the grasslands of a foreign land and highlighted the potential danger of predators. Meanwhile, participants received a list of unrelated words that had to be rated for the relevance to the imagined situation. Nairne et al. showed, in a series of experiments, that processing information for survival value results in superior retention and performance in free recall compared to a host of control conditions known to enhance memory performance (for an overview, see Nairne & Pandeirada, 2008). This effect has come to be known as the *survival processing advantage.*

In the current volume, a number of the chapters are devoted to experiments using this survival processing paradigm. These chapters investigate the robustness of the survival processing advantage, when it occurs and when it does not, and which proximate mechanisms are likely to cause this effect. The authors themselves take varied positions. Some see the survival processing effect as a route to understand the evolution of memory, whereas others see it as an interesting feature of elaborative encoding, but not likely to be based in an evolved module. Regardless of the eventual fate of survival processing, we see it as an important first step in generating hypotheses about human memory that derive from the theory of natural selection.

THIS VOLUME

This volume focuses on the evolution of memory. It is divided into three sections, roughly corresponding to some of the important distinctions in the field of adaptive memory. We have selected chapters that provide a critical overview of the evolutionary constraints of memory, yet this volume also contains chapters describing experimental research in the field of adaptive memory. In Section 1, the emphasis is on the factors that drove the natural selection of memory. These chapters draw on some of the likely candidates for natural selection to have influenced memory, including cheater selection, memory for the future, and avoiding dangerous situations. Specifically, in Chapter 2, Klein discusses whether planning and imagining the future might be regarded as an adapted function of memory. Furthermore, he argues that increased reliance on the self might explain part of the superior memory performance during survival tasks. In Chapter 3, Bell and Buchner provide an intriguing overview about how and why memory for the faces of cheaters is enhanced relative to other types of faces. They argue that this finding could be regarded as evidence for a highly specific cheater detection module that has co-developed with human cooperation. In Chapter 4, Burns and Hart explain that retention advantages due to reliance on survival situations are likely caused by differential processing of item-specific and relational processing. Also, they argue that thinking of one's mortality might be regarded as a specialized function of memory.

Section 2 focuses on the survival processing effect. Our authors approach this topic from a number of perspectives, testing its generalizability, its relevance to natural selection, and a topic we consider to be increasingly important: the proximate mechanisms that mediate the survival processing advantage. In Chapter 5 Howe and Derbish present evidence that the survival processing effect is likely the result of enhanced elaborative processing in the survival instruction. They show that the survival processing effect vanishes when scenarios are equated on elaborative processing. In Chapter 6 Soderstrom and Cleary investigate whether survival processing effects in memory can be considered domain-specific or domain-general. In Chapter 7 Altarriba and Kazanas examine whether survival processing effects in memory are caused by changes in attentional processing. More experimental studies on critical tests of the survival processing advantage are discussed in Chapter 8 (by Toglia, Leedy, Wilde, Baker, and Dacey) and Chapter 9 (by Schwartz and Brothers), which discover some boundary conditions of the adaptive memory.

Section 3 discusses the evolution of memory in other distinctive populations. This is defined rather broadly in this volume and includes children and older adults, nonhuman species, and victims of crime. The goal is to expand the search for clues to natural selection in human memory.

Collectively, this volume consists of the latest findings in the area of the evolution of memory. We hope that these chapters will stimulate novel and exciting perspectives in the field of adaptive memory, ones that have the potential to unravel why and how our memory evolved. Only then will we know more about what is adaptive about adaptive memory.

REFERENCES

Balda, R. P., & Kamil, A. C. (2006). Linking life zones, life history traits, ecology, and spatial cognition in four allopatric southwestern seed caching corvids. In M. F. Brown and R. G. Cook (Eds.), *Animal spatial cognition: Comparative, neural, and computational approaches*. Available at http://www.pigeon.psy.tufts.edu/asc/Balda/Default.htm

Clayton N. S., & Dickinson, A. (1998). Episodic-like memory during cache recovery by scrub jays. *Nature, 395*, 272–274.

Flores-Abreu, I. N., Hurley, T. A., & Healy, S. D. (2012). One-trial spatial learning: wild hummingbirds relocate a reward after a single visit. *Animal Cognition, 15*, 631–637.

Hampton, R. R., & Schwartz, B. L. (2004). Episodic memory in nonhumans: what, and where, is when? *Current Opinion in Neurobiology, 14*, 192–197.

Howe, M. L. (2011). *The nature of early memory: An adaptive theory of the genesis and development of memory*. New York: Oxford University Press.

Howe, M. L., & Otgaar, H. (2013). Proximate mechanisms and the development of adaptive memory. *Current Directions in Psychological Science, 22*, 16–22.

Jacobs, L. F., Gaulin, S. J. C., Sherry, D. F., & Hoffman, G. E. (1990). Evolution of spatial cognition: Sex-specific patterns of spatial behavior predict hippocampal size. *Proceedings of the National Academy of Science USA, 87*, 6349–6352.

Klein, S. B., Cosmides, L., Tooby, J., & Chance, S. (2002). Decisions and the evolution of memory: Multiple systems, multiple functions. *Psychological Review, 109*, 306–329.

Milner, B. (1964). Some effects of frontal lobectomy in man. In J. M. Warren & K. Akert (Eds.), *The frontal granular cortex and behavior* (pp. 313–334). New York: McGraw-Hill.

Nairne, J. S., & Pandeirada, J. N. S. (2008). Adaptive memory: Remembering with a stone-age brain. *Current Directions in Psychological Science, 17*, 239–243.

Nairne, J. S., Pandeirada, J. N. S., & Thompson, S. R. (2008). Adaptive memory: the comparative value of survival processing. *Psychological Science, 19*, 176–180.

Nelson, K. (2005). Evolution and development of human memory systems. In B. J. Ellis & D. F. Bjorklund (Eds.), *Origins of the social mind: Evolutionary psychology and child development* (pp. 354–382). New York: Guilford.

Platt, M. L., Brannon, E. M., Briese, T. L., & French, J. A. (1996). Differences in feeding ecology predict differences in performance between golden lion tamarins (*Leontopithecus rosalia*) and Wied's marmosets (*Callithrix kuhli*) on spatial and visual memory tasks. *Animal Learning and Behavior, 24*, 384–393.

Rubin, D. C. (2000). The distribution of early childhood memories. *Memory, 8*, 265–269.

Sherry, D. F., & Schacter, D. L. (1987). The evolution of multiple memory systems. *Psychological Review, 94*, 439–454.

Shettleworth, S. J. (2013). *Fundamentals of comparative cognition*. New York: Oxford University Press.

Squire, L. R. (2004). Memory systems of the brain: A brief history and current perspective. *Neurobiology of Learning and Memory, 82*, 171–177.

Sulikowski, D., & Burke, D. (2012). Win shifting in nectarivorous birds: selective inhibition of the learned win-stay response. *Animal Behaviour, 83*, 519–524.

Tulving, E. (1983). *Elements of episodic memory*. New York: Oxford University Press.

Tulving, E. (2002). Episodic memory: From mind to brain. *Annual Review of Psychology, 53*, 1–25.

Tulving, E. (2005). Episodic memory and autonoesis: Uniquely human? In H. S. Terrace & J. Metcalfe (Eds.), *The missing link in cognition: Origins of self-reflective consciousness* (pp. 3–56). New York: Oxford University Press.

SECTION 1

Adaptive Memory: What Are the Factors that Drove the Natural Selection of Memory?

2

Evolution, Memory, and the Role of Self-Referent Recall in Planning for the Future

STANLEY B. KLEIN

> "The adaptive function of information storage is intrinsically prospective: It is used to support future decisions and judgments, which cannot be known in advance with certainty. To the extent that the character of subsequent decisions and judgments can be predicted, the memory system can be tailored to flag relevant information and precompute variables that are required to make them." (Klein, Cosmides, Tooby, & Chance, 2002, p. 313)

The goal in this chapter is to make a case for evolution as one of the core concepts in the scientific study of memory (e.g., Klein, Cosmides, Tooby, & Chance, 2002; Sherry & Schacter, 1987). At its most general level, the task seems straightforward: Few scientists seriously question the relevance of evolutionary principles to understanding the function of biological systems. Such general considerations do not, however, warrant the conclusion that evolutionary principles are essential to understanding memory *per se*. Whereas biological systems all share a basic set of features, each system exhibits a functional organization specialized to address a particular set of adaptive problems recurrent in the organism's evolutionary past. To argue for evolution as a core concept for memory research requires the case be made that consideration of the problems memory was designed to address offers a richer understanding of memory's origin, function, and scope.

EVOLUTION AND MEMORY: GENERAL CONSIDERATIONS

Evolutionary Biology Provides Theories of the Various Adaptive Problems Memory Systems Evolved to Solve, Thereby Guiding Research on Their Design

Anatomists dissect organs of the body. Dissection does not imply random cutting; it is a theoretically driven attempt to divide the body's parts into functional units.

Psychologists rarely dissect the brain physically, but we dissect it conceptually. The goal of my research is to characterize the information processing architecture of the brain: to dissect the mind into *functional units*. This requires theories of function.

Yet most modern psychologists define the function and scope of memory in a rather general way: memory as designed to enable the encoding, storage, and retrieval of information. This is undoubtedly true, but too general to be a helpful guide to research. As a result, psychologists have spent many years studying the processes that encode, store, and retrieve information as if they were the same across all domains—as if it makes no difference whether the words you are hearing are unrelated items on a list or your spouse explaining that s/he's fallen in love with someone else. If all memory processes are the same, regardless of content, then it doesn't matter what content a subject is remembering: Word lists are as good as anything else and afford more experimental control. This assumption led to an exploration of what memory is *capable* of doing, without allowing any further dissection of memory into *functional* units.

Starting with evolutionary theories of adaptive function results in a different research program. Behavioral ecologists studying the memory of nonhuman animals always start with the adaptive problem: A bird may need to learn its species' song and also remember the random locations of hundreds of food items it cached for another season. By starting with a crisp definition of each adaptive problem the animal needs to solve, these researchers quickly became aware that different adaptive problems have different computational requirements when it comes to memory (e.g., Klein, Cosmides, Tooby, & Chance, 2002; Sherry & Schacter, 1987).

For instance, song learning requires the neural development of a species-typical template, selective attention to songs produced only by its own species while ignoring other songs, sound analysis of these samples, encoding of ambient song variations, and retrieval of stored songs in the breeding season rather than at other times. This requires specialized memory systems designed to interact with specialized systems of attention and learning. Underlining this point, in some species the brain nuclei involved in song learning change size depending on whether it is time to serve their function: They are large in the breeding season and shrink at other times (e.g., Nottebohm, 1981).

Consider the differences, however, between the functional requirements of a memory system that is well designed for song learning versus one that is well designed for seed caching. A system well designed for song learning does not need the huge memory storage capacity of a system designed for caching hundreds of seeds; retrieval of songs would have to be activated by the presence of potential mates, not by habitat features; location information is irrelevant to memory for songs but crucial for remembering where seeds are cached.

Each problem—song learning and cache memory—has quite specific and different computational requirements, and different memory systems have evolved to solve each. Notice also that, on this view, memory processes are not just in the animal's head. Memory systems have procedures that are *designed* to use specific cues that are reliably present in the animal's environment: certain songs and not others, cues to breeding season, presence of mates, and so on (e.g., Healy & Hurly, 2004; Henderson, Hurly, & Healy, 2006). Some of these cues activate the learning and memory system (breeding season cues), some determine what information is encoded and stored (particular song templates/song features), and some govern the precise timing of retrieval (presence of mates or competitors).

This means that the design of memory can only be understood in the context of the animal's ecology, because natural selection shaped its procedures to exploit some features of that ecology and not others. Just as crucially, each encoding, storage, and retrieval feature has co-evolved not just with the environment, but with attention and learning mechanisms that are just as problem-specialized as the memory features themselves, and without which these memory components would be inert.

This type of research program, focused on the identification and analysis of adaptive problems, has seldom been tried in the study of human memory (for discussion, see Klein, Cosmides, Tooby & Chance, 2002). Instead, most psychologists have been cataloging samples of the seemingly inexhaustible set of things a memory system is capable of doing, without asking what it was designed to do. This is an inefficient method for honing in on the small set of highly ordered, interlocking elements that embody the functional design of a biological system.

Starting with Adaptive Problems Leads to a Focus on Specialized Problem-Solving Systems, Each of Which Has Proprietary Memory Processes

An evolutionary perspective provides a principled way of dividing a system into component parts that explains why those parts take the form they do. It does this by focusing on adaptive problems and looking for systems that form functional units by virtue of the fact that they exist in their present arrangement because that form solved certain recurrent problems the organism faced in its evolutionary history. In many cases, memory processes—processes dedicated to the encoding, storage, and retrieval of data—will be one component of a functional (that is, problem-solving) system (for discussion, see, Klein, Cosmides, Gangi, Jackson, Tooby, & Costabile, 2009; Klein, Cosmides, Tooby, & Chance, 2002; Klein, German, Cosmides, & Gabriel, 2004). This puts the concept of a memory system in a new light: From this perspective, memory can be conceptualized as the *interplay* of a set of computational processes, only some of which involve encoding, storage, and retrieval (e.g., Klein et al., 2004). Many psychological processes will interact and through their interaction will transform stored information into the subjective experiences we call memory. Similarly, it is through the interaction of these co-evolved processes that learning occurs—that past experiences influence which objective behaviors we produce.

In what follows, I draw out some implications of this argument with respect to (a) memory systems at the biological level of analysis and (b) memory systems more broadly construed as the conjoining of neurological, biological, and ecological components in the service of functional adaptations for survival and reproduction.

Systems of Memory at the Biological Level of Analysis

A biological system, defined with respect to functionality, consists in those component processes—and their coordinated interactions—that are necessary to accomplish a specific adaptive task. As an example, the circulatory system consists of many components—for instance, a heart, a vascular network, valves, the lungs—that interact in a precise and coordinated way to accomplish an adaptive task: transporting energetically important material to and from the host in which the system resides. The functional identity of each part is defined with respect to the job it performs within the system that evolved to perform this adaptive function. Thus, despite the ability to rhythmically contract, the heart is not a pump unless there is fluid to transport and a mechanism for its delivery. It becomes a pump in the context of a system whose function is to ensure the movement of blood through the body.

Memory, from this perspective, can be viewed as a co-adapted relation between component processes whose joint efforts are required to perform certain adaptive functions. These processes include the usual suspects—encoding, storage, and retrieval—but are not limited to those capabilities. For an organism to behave "more appropriately" (that is, more adaptively) at a later time because of experiences at an earlier time, it must be equipped not only with mechanisms that retrieve ontogenetically acquired information but also with mechanisms that use this information to make decisions and drive behavior. Moreover, because decision mechanisms or rules often differ in what information they require, different sets of decision rules will activate different search engines (retrieval processes) and may access different data systems (storage systems). Without engines that can search for and retrieve the right information, supplying it to the right decision rule at the right time, an organ designed to store ontogenetically acquired information would be a pointless appendage. Seen in this way, there may be no such thing as "memory." Instead, there may be a large number of memory-related systems, each associated with a particular set of problem-solving machinery.

A RESEARCH EXAMPLE: PLANNING AS AN ADAPTED FUNCTION OF MEMORY

Researchers adopting an evolutionary perspective take as a starting assumption that systems exist in their present arrangement because that form solved certain recurrent problems the organism faced in its evolutionary history, and attempt to use knowledge of those problems to reverse-engineer the design of the systems (e.g., Barkow, Cosmides, & Tooby, 1992: Dawkins, 1976; Glenberg, 1997; Klein, Cosmides, Tooby, & Chance, 2002; Nairne, 2005; Sherry & Schacter, 1987; Suddendorf & Corballis, 1997; Williams, 1966). From this perspective, memory can be viewed as the result of the complex interplay of a set of processes that enable the organism to draw on past experiences to guide current behavior and plan for future contingencies (for discussion, see, Adam, 2004; Boyer, 2007; Ingvar, 1985; Klein, 2007; Klein, Cosmides, Tooby & Chance, 2002; Klein et al., 2009; Klein et al., 2010; Llinas, 2001; Suddendorf & Corballis, 1997; for a differing opinion, see Jones & Paschler, 2007). To the extent that these abilities affect the organism's reproductive potential, they are acted on by natural selection.

In a recent series of papers (e.g., Klein, Robertson, & Delton, 2010, 2011, Klein, Robertson, Delton, & Lax, 2012) my colleagues and I tested the notion that memory, from an evolutionary point of view, is inherently prospective: It enables its possessor to behave more appropriately at a later date by using information acquired in the past. Accordingly, the longstanding focus in memory research on one end of the temporal spectrum (an organism's ability to recall the past) unintentionally downplays the future-orientation of memory for dealing with current circumstances and anticipating future contingencies that cannot be known in advance with certainty. Since events that transpired in the past have already had their effect on our ability to survive and reproduce, the mere capacity to recollect personal life events is of limited benefit to the organism's reproductive fitness (although, under some circumstances, such recollection has important fitness consequences—for review and discussion, see Klein, Cosmides, Tooby, & Chance, 2002; Klein et al., 2009; Klein, Robertson, & Delton, 2010; Suddendorf & Corballis, 1997). By contrast, there would be fitness benefits from a memory system designed to take past events and extract information from them that increases our ability to deal with the current situation and aids planning for future contingencies (i.e., the "now and the next").

We first outlined and tested this hypothesis in Klein et al. (2010). In that paper we argued that when memory is engaged in monitoring, storing, and accessing information in the service of planning for personal future contingencies, its performance should be especially efficient (for discussion and additional relevant evidence, see Klein et al., 2010). This prediction trades on a conceptual distinction proposed by Klein and colleagues between a memory system's capabilities and its functions (e.g., Klein et al., 2009; Klein, Cosmides, Tooby, & Chance, 2002; Klein et al., 2010).

As is well known, the capabilities of memory systems are vast (for reviews, see Schacter & Tulving, 1994, and Tulving & Craik, 2000). These systems can, with varying degrees of success, encode, store, and retrieve an enormous variety of information, including much that is adaptively arbitrary and evolutionarily novel—from nonsense syllables, jingles, and batting averages to chess moves, equations, and the triple lutz.

However, there is an important difference between what a system can do and its evolved function. Memory is a system whose parts exist in their present form because of the functional contributions they made to adaptive behavior in the past (e.g., Klein, Cosmides, Tooby, & Chance, 2002). When such a system is engaged by tasks that maximally utilize its evolved machinery—for example, tasks that encourage the memory system to use information acquired in the past to plan and coordinate the organism's behavior in the future—especially efficient performance is likely to be found.[1]

In our study, we compared memorial performance for a set of encoding tasks that differed with respect to the temporal orientation they encouraged participants to take while they thought about a list of experimentally presented stimuli. All participants were asked to imagine being in the woods. In one condition—the past-oriented encoding condition—the participants were instructed to recall a specific time in their past when they went camping in the woods and then to determine whether each item on a list was part of their memory of the recalled experience. In the future-oriented encoding condition, the participants were asked to imagine that they were planning to go camping and then to decide whether each of the items on the list was relevant to planning their trip. In the atemporal encoding condition, the participants were asked to use their generic (i.e., nonpersonal, semantic) knowledge of camping to form an image of a camping trip and then to decide whether each item was part of their representation; no mention was made of the temporal placement (i.e., past, present, or future) of the camping context.

Following encoding, the participants received a surprise recall test in which they were asked to remember as many list words as possible. As anticipated, encoding that asked participants to plan what items to take on a camping trip (i.e., the future condition) gave rise to much better recall of those items than did encoding that asked participants to remember what they had taken on a previous camping trip (i.e., the past condition) or encoding that requested participants form an image of a typical camping trip and report whether the to-be-recalled stimuli formed part of their image (i.e., the atemporal condition). The latter two conditions did not differ in the recall they produced.

A Possible Confound: Survival Versus Planning

However, our future-oriented task asked participants to plan a camping trip in the woods. Thus, it could be argued that it was *survival* aspects of our future-oriented task, rather than future orientation *per se* that accounted for our recall findings. Such reasoning is favored by Nairne and his colleagues (e.g., Nairne & Pandeirada, 2008a, 2008b; Nairne, Pandeirada, Gregory, & Van Arsdall, 2009; Nairne, Pandeirada, & Thompson, 2007; Nairne, Thompson, & Pandeirada, 2007), who have conducted a large series of studies examining the memorial consequences of processing information for its relevance to a person's survival. Although these authors acknowledge that "survival processing" likely is not a unified domain (and therefore unlikely to be handled by a single cognitive mechanism; e.g., Nairne & Pandeirada, 2008b), they argue there is a strong *a priori* likelihood such processing will activate at least some specialized motivational machinery that interfaces with memory and give rise to superior recall and recognition. Consistent with this perspective, Nairne and colleagues consistently have demonstrated recall advantages for material judged for survival relevance when compared with material judged with respect to tasks (including planning tasks) that do not entail obvious relevance to survival.

However, it was our belief (supported by empirical evidence presented in the new experiment reported in this chapter [see the section just prior to the chapter summary], as well as in Klein, 2012a) that survival processing, as operationalized in studies by Nairne and colleagues, confounds survival with planning (e.g., Klein et al., 2011). Decisions about an item's relevance to one's survival often entail a future-orientation and planning (e.g., what items would you take with you to survive if you were stranded on the savannah?). But it is not clear that it has to be this way: There appears to be no principled reason why reference to survival requires a causal connection to planning. For example, in certain situations, a person can focus on his or her present circumstances when making survival-relevant decisions (e.g., deciding to take immediate evasive action as a truck careens toward you out of control). And, of course, planning need not logically entail considerations of survival: It is perfectly possible to plan a party, a picnic, or even a psychological experiment without such plans having relevance to one's survival.

Thus, although planning and survival are correlated in many life experiences, their relation is contingent rather than one of logical necessity. Moreover, planning seems a more circumscribed candidate for the effects of natural selection than does the more general concept of survival. Accordingly, we sought to examine independently the contributions of one (survival processing) to subsequent recall while controlling for the contributions of the other (planning processing).

Specifically, Klein et al. (2011) compared two survival encoding conditions, both of which took place on the savannah. They differed, however, in the extent to which they entailed future-oriented processing. In one condition participants judged which of a list of items (half of which were edible and half inedible) would be best to bring in anticipation of being stranded without food on the savannah. This condition combined both an obvious survival scenario (i.e., the need to eat) with a clear future orientation (i.e., planning what food items to take). This we termed our "Survival with Planning" condition.

A second condition also highlighted food-related survival concerns but minimized the planning component of the task: Participants were told that they were stranded on the savannah and had come upon a cache of items, some of which were edible, some of which were not. Their job was to decide which of the items they could eat to survive. We predicted, and subsequent pretesting confirmed (see below), that under these circumstances, participants' concern with the need for planning would be greatly minimized. We labeled this the "Survival without Planning" condition.

A third condition emphasized future-oriented planning but minimized survival concerns: Participants were asked to imagine they were going to shop for food items for a dinner party they were planning. Since this condition was not survival relevant (see pretesting below) nor did it take place on the savannah (i.e., the environment of evolutionary adaptation [EEA]—see discussion in the next paragraph), it is a conservative test of the future-orientation hypothesis—that is, that planning, rather than survival concerns, or location of the encoding context, mediates the good recall typically found for survival encoding. We labeled this task "Planning without Survival."

Our survival conditions were modeled closely after that used by Nairne and colleagues (e.g., Nairne et al., 2007). Participants were asked to imagine that they were stranded without provisions on the savannah. The employment of the savannah as the context of encoding deserves explanation. Extensive research by Nairne has suggested that the environmental context of survival processing is of primary importance for obtaining a memorial benefit. Given the likelihood that our memory-based survival abilities evolved over the extended period during which our hominid ancestors roamed the African savannah (i.e., the EEA), Nairne reasons that survival effects in recall are most likely to be evidenced when the encoding context matches the conditions present during their selection—that is, the savannah.

One could, of course, argue that the future-oriented planning condition used by Klein et al (2010)—that is, being on a camping trip—also entailed survival concerns, thereby confounding the hypothesized effects of future orientation with survival. However, pretesting suggested this was not the case: Participants consistently rated the future-oriented planning task as highly unlikely to evoke thoughts of personal survival.

The results of Klein et al. (2011) provided further evidence that planning is a key variable in the superior recall performance found with survival processing. Specifically, "Planning without Survival" produced recall equal to "Survival with Planning," and both tasks reliably exceeded recall associated with the "Survival without Planning" task. Thus, by systematically varying whether or not survival encoding included planning, we found that survival without a planning component was inferior to survival that also entailed planning.

The equivalence between "Survival with Planning" and "Planning without Survival" was obtained despite the latter task (a) having virtually no survival relevance (confirmed by pretest results and subsequent manipulation checks) and (b) having no reference to the EEA—the savannah. In combination with our previous study (Klein et al., 2010) these findings suggest that the planning component of survival processing may be a key factor in the excellent recall that our future-orientation tasks yield.

This is not to say that survival processing does not have an important role to play in memory. There are numerous studies documenting its beneficial effects to recall and recognition (e.g., Nairne, 2005; Nairne & Pandeirada, 2008a, 2008b), and it is doubtful that any one factor (even planning) can explain all those results. It simply is our contention that *one* part of the beneficial effects of survival processing to subsequent recall can be explained by the future orientation it engenders—that is, the planning component. Although planning may be only one of several factors relevant to the excellent recall achieved with survival processing, it provides memorial benefits not present when it is absent.

A Problem with Planning Tasks: Previous Failures to Enhance Recall Relative to Survival Tasks

If one accepts my arguments about planning and its memorial potency, however, one is confronted by a puzzling result: Tasks that entail planning often have been used as control tasks in studies demonstrating the superior recall associated with survival processing. Consistent with Nairne and colleagues' proposal about the evolutionary memorial advantage of survival encoding, researchers typically have found planning tasks to yield recall performance inferior to survival tasks (e.g., Kang, McDermott, & Cohen, 2008; Nairne et al., 2007; Nairne & Pandeirada, 2008a, 2008b; Nairne et al, 2009; Nairne, Pandeirada, & Thompson, 2007; Weinstein, Bugg, & Roediger, 2008; but for conflicting findings, see Butler, Kang, & Roediger, 2009; Klein et al., 2010, 2011). These results seem to weaken our argument about the memorial potency of planning and its potential role in survival processing memory advantage.

However, the planning and survival tasks used in studies finding survival encoding superior to planning in its memory enhancing effects differed along multiple dimensions, including, but not limited to, the context of encoding (e.g., survival in the wild vs. planning an extended vacation, planning a move to a foreign country or being on a spaceship) and the plausibility/familiarity of the planning scenario (planning a bank robbery, planning to go on a hunting trip, planning an extended stay in a luxury hotel, planning a move to a new home in a foreign land, evading an assassin in a strange city).

Moreover, participants in virtually all of the planning conditions were likely to have few, if any, personally relevant experiences regarding the planning being encouraged (e.g., moving to a foreign land, fighting attackers in the streets of a strange country, robbing a bank, taking a hunting trip on the savannah). Accordingly, these tasks often lacked a personal, real-world relevance that connected them in a meaningful way to participants' actual experience—what Klein, Loftus, and Kihlstrom (2002) labeled *lived time*.

Lived time is personal, self-referential, and based largely on episodic memory (see also Tulving, 2002). When it is future-oriented, lived temporality recruits self-referential episodic recollections, which now are well known to serve as the foundation for imagined scenarios involving the self (e.g., Atance & O'Neill, 2005; Buckner & Carroll, 2007; Klein, Loftus, & Kihlstrom, 2002; Schacter, Addis, & Buckner, 2008; Suddendorf & Corballis, 1997, 2007; Szpunar & McDermott, 2008; Wheeler, Stuss, & Tulving, 1997; Weiler & Daum, 2008). Thus, a connection with lived time may be important to the future-oriented function of memory for constructing personally relevant scenarios (for discussion, see Klein et al., 2010, 2011; Suddendorf & Corballis, 2007). Given these considerations, our previous studies used planning tasks (e.g., a camping trip, a dinner party) for which participants reported considerable personal experience.

It was our contention that this aspect of planning (i.e., the ability to recruit episodic personal recollections pertaining to the planning scenario) is fundamental to planning a personal future (e.g., Atance & O'Neill, 2005; Klein, Loftus, & Kihlstrom, 2002; Suddendorf & Corballis, 2007; Tulving, 1985) and thus necessary for the superior recall found in the planning scenarios my colleagues and I have used (Klein et al., 2010, 2011). By contrast, planning scenarios serving as control encoding tasks in other studies of survival processing typically did not reflect situations of planning in participants' real-world experience (and hence were likely to lack personally relevant representation in episodic memory). Such scenarios were thus unlikely to draw on personal recollections—a form of memory that repeatedly has been shown to enable its owner to plan future actions and anticipate future events (e.g., Klein, Loftus, & Kihlstrom, 2002; Schacter, Addis & Buckner, 2008; Suddendorf & Corballis, 2007; Szpunar & McDermott, 2008; Tulving & Lepage, 2000; Wheeler, Stuss, & Tulving, 1997). Klein, Loftus, and Kihlstrom (2002) have labeled this form of planning *known time.*

Known time also is a form of mental time travel, but it is associated with a type of semantic-based memory that allows a person to orient to imagined future events *without* experiencing them as a future part of his or her personal history (Klein, Loftus, & Kihlstrom, 2002). *Known time* thus enables its owner to know about, but not re-experience, previous states of the world and to draw on this generic knowledge to construct *impersonal* scenarios for the future. Given the close connection observed between episodic memory and an imagined personal future, tasks that rely on *known time* encoding should not exhibit the memorial potency of tasks involving *lived time* when one's goal is personal planning.

The Importance of *Lived Time* in Personal Planning

In Klein, Robertson, Delton, and Lax (2012), we selected four encoding conditions, all of which involved planning. They differed, however, in the degree to which the planning scenario was one for which participants had personal experience: Half the tasks were ones for which most participants had self-relevant personal experience (planning a dinner party and planning a picnic) and half involved planning that few if any participants could reasonably be expected to have first-person familiarity (planning a meal to feed the animals at a local zoo and planning an Arctic trek). Our intuitions about the personal relevance of the individual planning scenarios were substantiated by pretesting as well as manipulation checks for each of the four planning scenarios. Manipulation checks also revealed that, as expected, all four scenarios entailed approximately equal degrees of planning on the part of participants, though they varied in the degree to which they engendered thoughts of personal survival.

As predicted, the dramatic differences in personal experience mapped reliably onto differences in recall. The "high personal experience" groups (dinner and picnic planning) produced reliably greater recall than did groups for whom personal experience/familiarity with the planning scenario was severely limited (zoo meal and Arctic trek planning).

The variable of personal experience also helped explain why planning scenarios in previous studies often yielded inferior recall performance relative to comparison tasks (e.g., survival on the savannah). These planning scenarios typically required participants to plan for events such as "assembling things one would need to rob a bank", "planning to avoid a potential assassin", or "planning a move to a foreign country"—events outside the scope of most undergraduates' personal experience. For most participants, such scenarios lack personal, scenario-relevant experience. This would likely require participants to rely on *known time* to formulate their answers to questions of item relevance, drawing on generic knowledge rather than knowledge derived from retrieval of personally relevant episodic memories.

Accordingly, the type of planning involved in the "high personal experience" groups of the present study (dinner and picnic planning) was more likely than previously adopted planning tasks—which typically lacked connection to participants' personal experience—to involve lived temporality. Similar considerations explain the comparatively poor performance found for our zoo meal and Arctic trek planning scenarios (i.e., "low experience" groups), which lacked a clear connection to *lived time*—a factor that, we believe, plays a crucial role in future-oriented functions of memory.

A question that immediately arises is why the savannah-related survival processing combined with a planning component should yield good recall in light of my earlier arguments that planning for personal contingencies works optimally when it is based on personal recollections. After all, it is unlikely many of our participants have spent time on the African savannah. Indeed, because of frequent requests for explication of the term "savannah" during previous studies, prior to the start of the current study we informed all participants that "savannah" could be likened to "grasslands." It is my contention that many of our participants (upon hearing that savannah is a grassland area) simply searched episodic memory for a personally relevant grassland-like context.

To explore this conjecture, Klein et al. (2011) randomly queried participants ($N = 57$) in survival conditions in their studies about (a) how they conceptualized the "savannah" as they performed their respective tasks and (b) whether that conceptualization (if any) was based on personal experience. Since our instructions to participants likened the savannah to grasslands (to enhance the clarity of a potentially unfamiliar term; as per Nairne, Thompson, & Pandeirada, 2007), it was not surprising that the majority of responses pertained to images of fields and plains—both of which were also were rated as highly likely to be based on personal experience (for participant responses, see Klein et al., 2011, Appendix 3). However, more work clearly needs to be done to obtain a better understanding of the episodic/planning connection in the savannah survival paradigm.

ARE SURVIVAL AND PLANNING TASKS ULTIMATELY REDUCIBLE TO SELF-REFERENTIAL ENCODING?

A recent explanation of the memorial potency of survival encoding notes its similarities to self-referential encoding (Burns, Hwang, & Burns, 2011). Certainly, few things are more self-relevant than one's personal survival. However, as Nairne and colleagues (e.g., Nairne & Pandeirada, 2008b; Nairne, Thompson, & Pandeirada, 2007) have shown, survival on the savannah promotes better recall than does self-referential processing (participants in the self-reference condition were asked to rate "how easily the word brings to mind a personal experience" (Nairne & Pandeirada, 2008b, p. 269)). The list words, all of which were nouns, included such items as "river," "snake," "liver," "dirt," and "fabric."

Given the nature of the self-referential task used by Nairne and colleagues, it is quite reasonable to assume that it entailed processing nouns with respect to episodic memory. However, as Klein, Loftus, and Burton (1989) have shown, not all self-reference tasks, particularly those using noun stimuli, are equivalent in their memorial potency. In particular, although some studies have replicated the self-reference effect (SRE) using nouns (e.g., Brown, Keenan, & Potts, 1986; Klein & Kihlstrom, 1986; Klein & Loftus, 1988; Warren, Chattin, Thompson, & Tomsky, 1983), a number have not (e.g. Karylowski & Buczek, 1984; Lord, 1980, 1987; Maki & McCaul, 1985; Pressley, Levin, Kuiper, Bryant, & Michner, 1982).

One key factor underlying this variability is the familiarity of the encoding context (the importance of context familiarity is also evidenced by its broad applicability in the SRE paradigm—e.g., it is known to influence memory for both nouns and trait stimuli; e.g., Klein et al., 1989). Another factor influencing self-referential recall of nouns is the use of an autobiographical context for encoding: SREs with nouns typically are obtained with self-reference tasks promoting encoding with respect to the retrieval of episodic memories of life events.

Consistent with the documented variability of SRE tasks when noun stimuli serve as stimulus items, it is of note that in some studies in which Nairne and colleagues compared self-referential processing to survival processing (e.g., Nairne & Pandeirada, 2008b), the self-reference task failed to produce a recall advantage over control tasks that normally produce recall *inferior* to self-reference (for a comprehensive review of the relevant data, see Klein et al., 1989). This may be due (as I argue below) to the specific wording of the self-referential task Nairne employed.

Another line of evidence suggesting a self-referential component for the survival processing tasks comes from an analysis of participants' ratings of self-referential thoughts during performance of planning and survival scenarios (reported in Klein et al., 2012). Statistical analyses revealed that participants' number of self-thoughts in the dinner party, picnic, and Arctic trek planning conditions (Ms = 3.17, 2.73, and 2.90, respectively) did not differ reliably. All three, however, reliably exceeded self-thoughts produced in the zoo meal planning condition ($M = 2.03$).

These findings are consistent with my intuitions about the self-referential nature of different planning tasks—dinner parties, picnics, and Arctic treks all would seem to entail thoughts directed toward oneself, whereas zoo meal planning, though somewhat self-referential (after all, one is doing the planning), is more outwardly directed toward the needs of the animals being fed. While "outward directedness" might also seem applicable to the dinner and picnic planning, certain considerations lessen the impact of this concern. Specifically, dinner and picnic conditions have a social/personal component (e.g., sensitivity to social expectancies, evaluation of self in the eyes of others, etc.; e.g., Goffman, 1959; Meade, 1934) that does not seem applicable when the evaluators of one's planning efforts are nonverbal beasts.

Analysis of self-referential thoughts among participants in African savannah "survival plus planning" condition (data collected, but not reported, by Klein et al., 2011) yielded a mean value (M = 3.00), a score comparable to that found with the "high experience" planning tasks. The finding of a high frequency of self-referential thought in the savannah planning conditions adds weight to the proposal that survival processing ultimately may be reducible to a form of self-referential processing in which participants attempt to relate noun stimuli to personal experience retrieved from episodic memory (a personal experience found lacking in the Arctic trek condition) in an effort to plan for future contingencies. As a side note, recent philosophical treatments of the self have maintained that it is the agential, planning aspect of the person—the need to plan one's acts in anticipation of future contingencies—that helps create a sense of self in the first place (e.g., Gertler, 2011; Moran, 2001; O'Brien, 2007).

Further support for the proposed "survival/planning/self" relation comes from a recent study of survival processing by Burns et al. (2011), who found that survival-based processing promotes the encoding of *both* item-specific and relational information (e.g., Hunt & Einstein, 1981). The *only* other task known to promote both of these processes at encoding is the self-reference task employed by Klein and Loftus (1988; see also a meta-analytic review by Symons & Johnson, 1997).

Nairne's survival processing is similar in many respects to the type of self-referential encoding conditions employed by Klein and Loftus (1988)—for example, both used scenario-based encoding contexts (i.e., "does this word relate to personal recollections of a day at the beach?," "does this word relate to personal recollections of a party?" [both from Klein & Loftus, 1988], "does this word relate to survival on the savannah?" [from Nairne and colleagues]); both ask participants to encode the nouns with regard to personally familiar scenarios (for this claim to be warranted, one must accept our assumption, and the supporting data, that participants transformed an unfamiliar context—the African savannah—into a context for which personal experience was available in memory; e.g., recollections of time spent on a grassy field—see above); and (a) in both, only a subset of the noun stimuli were relevant to any particular scenario, (b) both required encoding of nouns in the context of episodic memory retrieval, and (c) both found evidence for item-specific and relational processing.

Although a proposed relation between survival processing and self-referential encoding clearly is speculative, the pattern of findings pertaining to self-referential processing, episodic memory, personal familiarity, and planning, combined with what is about known about survival processing, suggests an interesting degree of overlap. I believe these considerations provide some intriguing clues that might help tie together three empirically separable, yet logically, as well as contingently, related areas of interest: the parts played by time, self, and memory in human experience in general and memory performance in particular.

A Study of the Role of Self-Reference in Survival Processing

In the present study I selected four encoding conditions. One was the standard survival encoding task used by Nairne and colleagues and one was a replication of the picnic planning task used by Klein et al. (2011). In addition, I included two self-referential encoding tasks. One, was identical in its instructions to those used by Nairne and Pandeirada (2008b) and one was identical to that used by Klein et al (1989). Specifically, the Nairne version of the task requested that for each of a list of noun stimuli, participants rate how easily each word brings to mind an important personal experience. By contrast, the Klein et al. (1989) version of the self-reference task requested that participants try to (a) bring to mind an important personal experience involving each of the items presented on the next page and (b) rate how easily the item brought to mind an important experience.

Although differences in wording between these two self-referential tasks may seem trivial, in actuality research suggests they can be a crucial factor in whether one obtains a SRE. Specifically, Klein et al. (1989) have shown that the *actual* retrieval of episodic memories is necessary for the recall enhancement found with the episodic-based self-reference tasks using nouns as stimuli.

A priori, it might seem that the mere involvement of a self-referential question, rather than episodic retrieval *per se*, is responsible for the high levels of recall found with self-reference tasks. However, Klein et al. (1989) compared what they termed the "modified self-reference task" with the standard version of the task. The former task simply requested that participants rated how difficult it would be to recall an experience in which they manifested the trait, but were not asked explicitly to attempt to retrieve a relevant episodic memory. This task was modeled after the well-documented finding that people can accurately predict the likelihood that they will be able to remember a specified bit of information in memory without having first to retrieve it (e.g., Hart, 1965, 1967a, 1967b; Schacter, 1983). Indeed, Reder (1987) has shown that people can make "feeling of knowing" judgments faster than they could retrieve the item. By contrast, in the latter task (i.e., the standard episodic self-reference task), participants were first asked to recall a specific personal experience in which they manifested the trait and then rate the difficulty of retrieving the memory. Our results, as expected, found that, compared to a control task (semantic encoding of the stimulus words), the standard but not the modified self-reference task produced recall reliably greater than that found with semantic processing of the same material.

It is thus possible that Nairne's failure to find equivalence between survival and self-referential processing resulted from his adoption of a version of Klein et al.'s (1989) "modified self-reference task." If retrieval of a personal episode is essential for the recall enhancement found with self-referential, noun-based tasks, then the modified version of the self-reference task should be less beneficial to recall than would be the standard version of the task.

With regard to the present study, I predicted that a self-referential task that promotes explicit retrieval of episodic recollections would yield better recall of stimulus items than would a self-reference task that only requests participants rate the ease with which such retrieval *could* occur.

Method

Seventy-two undergraduates in an introductory psychology class participated in the study. The experiment was conducted in a single, mass-testing session lasting approximately 25 min.

Materials

Participants were randomly assigned to one of four encoding conditions (N = 18 per condition): the modified self-reference condition, the standard self-reference condition, the picnic planning condition, and the survival encoding condition. All participants viewed the same list of 30 stimulus words. The words all represented small, movable, potentially edible objects (e.g., sugar, carrots, glue, kerosene) that previously had been tested and found relevant to food-related planning decisions (see Klein et al., 2011, for documentation of the relevance of the to-be-remembered items to food-relevant planning decisions). Half the participants in each experimental condition were randomly assigned to receive one of two random orderings of the list words.

Design and Procedure

At the start of the study, each participant received a booklet containing the experimental material appropriate to his or her condition. Participants were instructed to remain on the page on which they currently were working and not to turn from that page until explicitly requested to do so (the word "Stop" was prominently displayed at the bottom-center of each booklet page). Participants also were instructed not to refer back to a previous page in the booklet once work on that page had been completed. To ensure compliance, two research assistants monitored participant performance. There were no reports of any participants who failed to comply with these instructions.

The first page of the booklet contained instructions describing one of the four encoding tasks used in the study. Below the task instructions were the 30 list words, printed one per line. Participants were asked to judge each word in accordance with their instructions from the previous page (i.e., how easy *would it be* for you to recall a memory involving the item? [i.e., the modified self-reference condition]; how easy *was it* for you to recall a memory involving this item? [i.e., the standard self-reference condition]; how likely are you to select this item for your planned picnic? [i.e., the picnic planning condition]; how likely are you to select this item to help you survive on the savannah? [i.e., the survival condition]). Each word was accompanied by a 5-point scale relevant to their task (e.g., 1 = "very easy" or "very likely" or 5 = "very difficult" or "very likely"). Four minutes were allotted to complete this portion of the study (pretesting indicated this interval was sufficient for participants to read instructions and make ratings at a comfortable pace; see Klein et al., 2011, for discussion). Following the rating/encoding task, participants were instructed to turn to page 2, which contained a series of anagram completions and served as a 3-minute distracter task. Participants were asked to complete as many anagrams as they could in the time provided.

Participants then were asked to continue to page 3, which contained instructions for the recall portion of the study. The instructions read: "I now would like you to try to recall the words you rated in the first part of the study. Please write the words, one per line, in the spaces provided below. You may recall the words in any order they come to mind." The page contained 30 blank lines. Four minutes were provided for recall.

On completion of recall, participants were instructed to turn to the final page of the booklet. This page contained questions designed as manipulation checks of my assumptions about the extent to which participants had survival-relevant thoughts when rating stimulus items, the extent to which stimulus ratings evoked thoughts of planning, and the degree to which self-relevant thoughts occurred during performance of the assigned rating tasks. Questions were answered on a 4-point scale with values ranging from 1 = "Not at All" to 4 = "Always." Two minutes were allowed to complete the ratings, after which all the booklets were collected and participants debriefed.

Results[2]

Of the hundreds of items remembered, only 5 were extra-list intrusions. These were not counted as part of a participant's recall score. A one-way ANOVA conducted on the mean recall scores yielded a reliable effect of task, $F(3, 68) = 11.41$, $p < .01$, MSE = 5.44. Consistent with predictions, participants in the survival and picnic planning conditions recalled more words (*M*s = 14.50 and 14.61, respectively) than did participants in the modified self-reference condition (*M* = 10.66). However, participants in the standard self-reference condition recalled on average 12.94 words, a performance that did not reliably differ from either the survival or the picnic planning conditions. All three conditions, however, produced significantly greater recall than did the modified self-referential encoding task (Tukey testing: $p < .05$). (Note that the finding of self-referential thought absent explicit requests for episodic retrieval in our modified self-reference condition was not sufficient for optimal recall fits well with our previous finding that Arctic planning, though high in self-referential thought, lacked episodic familiarity—and thus was unaccompanied by personal recollections and yielded comparatively poor recall.)

Analysis of participants' ratings of self-thought during encoding found no differences in thoughts of self associated with the four tasks (*M*s = 2.95, 2.89, 3.00, and 3.00 for the survival, picnic, standard self, and modified self tasks, respectively). Additional manipulation checks examined participants' ratings of the extent to which they had thoughts of (a) planning and (b) survival while judging stimuli. For the "survival thoughts" question, the picnic, modified self, and standard self groups produced means of 1.34, 1.39, and 1.33 respectively. By contrast, the survival group's mean score was 3.50. Thus, as we consistently have shown, high levels of recall with encoding conditions absent survival considerations can be obtained, provided the task under consideration entails either self-referential episodic recollection or planning (which, as I hope this chapter makes clear, are inextricably intertwined).

With regard to the planning question, as expected, the two self-referential conditions exhibited little if any such thought during encoding (*M*s = 1.16 and 1.55 for the modified and standard self tasks, respectively). Also, as anticipated, the picnic planning and survival planning tasks reported comparable levels of thoughts of planning during encoding (Ms = 2.90 and 3.28). Once again, we find that recall comparable to that achieved by survival processing can be obtained with tasks that entail either self-reference or planning.

In sum, although these findings do not demonstrate conclusively that self-referential memory retrieval is a component of the survival processing (comparable recall does not logically warrant the conclusion of comparable memorial processes), they do suggest that it may play a role in the high levels of recall typically found with survival processing. At a minimum, these findings open the door to such speculation. Clearly, additional research is necessary.[3]

SUMMARY

Planning as an Evolved Adaptation of the Human Mind

Over the course of evolution, modifications in the design of human memory were incorporated to the extent that they improved the functional operation of the system—that is, they increased the ability of neurocognitive machinery to successfully solve adaptive information processing problems (e.g., Cosmides & Tooby, 1987; Sherry & Schacter, 1987). Planning and a personal orientation toward the future, two highly interrelated skills, greatly enhance their owner's chances for survival. Accordingly, planning for future contingencies is, in our view, a specific, evolved set of mechanisms designed to help solve a general problem—how to remain alive long enough to reproduce and care for one's offspring.

What separates humans from our closest evolutionary relatives is not planning *per se* but the extent and complexity with which we are able to plan future contingencies—that is, imagine our actions and their consequences (e.g., Bar, 2010; Brandimonte, Einstein, & McDaniel, 1996; Cosmides & Tooby, 1987; Donald, 1991; Klein et al., 2010, 2011; Passingham, 1982). Although all organisms capable of long-term memory are of necessity oriented toward the future, what is unique to human long-term memory is the extent to which our plans entail complex, flexible, and temporally extensive responses to the environment (e.g., Donald, 1991; Lombardo, 2008; Klein et al., 2010; Suddendorf & Corballis, 1997).

By virtue of this sophistication, humans, unlike most other mammals, can anticipate and plan for personal future contingencies (e.g., Klein, Loftus, & Kihlstrom, 2002) that transcend their immediate needs—that is, planning that is neither a species-typical fixed action pattern nor tied directly to current motivational states (e.g., Bischof-Koehler, 1985; Suddendorf & Corballis, 1997). Such versatility confers an enormous selective advantage on human planning ability (e.g., Klein et al., 2010, 2011; Lombardo, 2008).

The Self

This chapter (and related research; Klein, 2012a) also makes clear that it is a mistake to ignore self-referential memory—an aspect of information processing associated with all of the key constructs that play a role in survival processing (e.g., planning, familiarity, episodic retrieval, etc.). No matter how one chooses to unpack the findings to date, it is impossible to ignore the conclusion that mnemonic effectiveness of various tasks under consideration entail a role for the neural instantiation of self (e.g., Klein, in press b), a set of brain-based systems known to be heavily influential in a great number of mnemonic contexts (for review, see Klein & Gangi, 2010; Klein, Robertson, Gangi, & Loftus, 2008).

Survival as an Evolved Adaptation of the Human Mind

I have great admiration for Nairne's program of adopting evolutionary theory as a heuristic for promoting the discovery of functional mechanisms of mind. His research has certainly sparked interest among a substantial component of the community of memory researchers and theorists and has helped place evolutionary considerations into the mindset of those interested not simply in what memory can do, but in what functions it might serve. As noted earlier in this chapter, a focus on function (which ultimately is an evolutionary question) can bring to light aspects of an organ's design that would be missed if one were exclusively focused—as memory research traditionally has been—on questions concerning an organ's capabilities (e.g., questions concerning how much, how long, what type of representational format, etc.).

However, although survival processing certainly appears to capture an important function of episodic recollective abilities, from a conceptual standpoint survival may not be the optimal starting point for exploring the effects of natural selection on an organism. As previously argued (Klein et al., 2010), it is not easy to enumerate a set of minimally necessary or sufficient behaviors that entail survival. Indeed, as developed in biological theory, the concept of survival encompasses everything that is not growth or reproduction—clearly an extremely heterogeneous set of activities and processes. So although natural selection can design distinct adaptations for distinct aspects of survival, it would be difficult for it to build a single adaptation for survival *per se* (see also Nairne & Pandeirada, 2008b; Soderstrom & McCabe, 2011).

Planning is not a conceptual primitive either; rather, it consists in a set of evolved component processes, each of which has its own evolved function (for traditional treatments, see Miller, Galanter, & Pribram, 1960; Schank & Abelson, 1977; for a recent summary, see Lombardo, 2008). For example, Klein et al. (2010) argued that to plan for one's future, one must, at a minimum, be able to (a) imagine possible futures, (b) set long-term personal goals, (c) construct hypothetical scenarios whose enactment will result in the realization of an intended goal, and (d) evaluate the likelihood an imagined act will achieve the goal. However, when compared with survival, it is my contention that planning offers a more circumscribed set of behaviors on which natural selection can act.

Survival is Too Broad to Be Targeted by Natural Selection

In this chapter I have been guided by the assumption that survival *per se* is too amorphous a concept to be a focus of natural selection. I have argued instead that rather than being taken as a conceptual primitive, the concept of survival is best seen as consisting in a large collection of fitness-relevant variables, some of which would clearly benefit from enhanced memorial abilities (e.g., planning), whereas others would not (e.g., incest avoidance).

In strong support for this proposal, a recent set of studies by Sandry, Trafimow, Marks, and Rice (2012) found that although Nairne's survival processing task produced high levels of recall, other EEA-relevant tasks with clear fitness relevance (e.g., mate selection, incest avoidance, and cheater detection [e.g., Barkow, Cosmides, & Tooby, 1992]) failed to produce recall substantially better memory than that found for control conditions. These findings reinforce my argument that survival processing *per se* is too broad a construct to be worked on by natural selection.

Sandry et al.'s (2012) findings also reinforce my contention that the memorial consequences of survival processing tasks typically employed in research on adaptive memory result primarily from the memorial demands of *correlated* factors—such as episodically based (a) self-referential processing and (b) personal planning. In short, survival-relevant tasks that can be performed in a proficient manner absent a need for enhanced memorial capabilities should not be—and, as the Sandry et al. (2012) results suggest, have not been—selected by evolution to interface with mechanisms that enable superior memory. For example, it likely would not take much in the way of enhanced memory capability to successfully keep track of potential mates—particularly during the EEA, when the size of one's group numbered, at a maximum, 120 individuals (e.g., Dunbar, 1996), an even smaller subset of whom would be relevant to one's mating opportunities (due to considerations of age and gender).

FINAL THOUGHTS

The memorial potency of the survival processing task now is well established. One response to this robust finding has been to explore conditions under which survival processing can be attenuated or enhanced (e.g., Butler et al., 2009; Klein et al., 2011; Weinstein et al., 2008). Another strategy has been to enumerate variables and conditions that have a an influence on or are affected by survival processing (e.g., developmental stage of participants, true and false memory decisions, stereotype activation, implicit memory, face processing [e.g., Howe & Derbish, 2010; Nairne et al., 2007; Otgaar & Smeets, 2010; Otgaar, Smeets, & van Bergen, 2010; Otgaar et al., 2011; Savine, Scullin, & Roediger, in press; Tse & Altarriba, 2010). Clearly, these two research agendas are closely related.

Such knowledge obviously is valuable. There is, however, a relative dearth of research focused *specifically* on explicating the mechanisms underlying the survival processing recall advantage. This is not to say such work does not exist—clearly it does (e.g., Burns et al., 2011; Kang et al., 2008; Klein et al., 2011, 2012; Kroneisen & Erdfelder, 2011; Nairne, 2010)! While I applaud what has been accomplished to date, my goal in this section is to call attention to the need for more work specifically devoted to clarifying the underlying mechanism. Demonstrations of the reliability of an effect and exploring its parameters are an essential initial component of a well-grounded research program. But once that program is well underway, the next, crucially important step is to better understand the cognitive mechanisms that mediate the effect (e.g., Craver, 2007).

While the functional origins of our ability to manifest superior recall of survival-relevant information may be traced to conditions located in the EEA (for discussion see Klein et al., 2010; Nairne, 2005, 2010; Soderstrom & McCabe, 2011), the neural and cognitive underpinnings of the effect, as currently observed, are based on mechanisms embedded in the brain of modern *Homo sapiens*. As such, direct attention to theoretical explication of the mechanism is the next essential step to understanding this remarkably robust phenomenon. Otherwise, we run the risk of carefully exploring the "edges" of the phenomenon (e.g., the variables that influence survival processing and the variables affected by it) but fail to satisfactory explain what survival processing *per se* consists in.

NOTES

1. In a recent communication Endel Tulving inquired, "What exactly is the logic that takes you and Jim Nairne from the role that Darwinian evolution plays in the survival of a species, and its mechanism as per theory, to enhanced memory of survival-related WORDS in a free recall experiment in a psychology lab tens of thousands of years later?" (August 23, 2011).

   My response is that our evolved machinery works best when the task for which it was selected is instantiated by the present context. Planning for future contingencies is of key importance for human reproductive success (for a recent review, see Bar, 2010) and thus likely to have been a target for natural selection. Accordingly, memory as an evolved neural resource should function optimally when it is engaged in doing that for which it has been selectively designed.

   The fact we use "words" as stimuli in most of our experiments is largely incidental. What matters is what participants *do* with those words. When words are employed in the context of a task calling for personal planning, they serve as a means for invoking from memory that which *really matters* (as I try to show in this chapter)—specifically, recollection of past items and events in the service of future planning for self-relevant contingencies. Thus, when list words cue our planning system to make available episodic recollections that might potentially play a role in personal planning, the memories evoked are well matched to the function of the system activated, and the information these recollections impart is consciously retained in the service of deliberations for a successful planning outcome.
2. I did not include within the present analyses measures of effect size and variability (e.g., confidence intervals). The interested reader is referred to Klein (2012a) for a full presentation of these data. As seen in that publication, effects sizes for all relevant comparisons were in the moderate range. Moreover, standard deviations and 95 percent confidence intervals demonstrated that both reliable differences and failures to obtain differences were not attributable to the absence of suitable sample sizes or to excessive variability in the data. Given these analyses, I did not include similar measures—along with discussion—in the present chapter in an effort to (a) not repeat what already exists elsewhere and (b) minimize what turned out to be a chapter greatly exceeding the editors' specified word limits.
3. It is worth noting that the results (greatly expanded) of the present study have been replicated and generalized in a separate set of studies by Klein (2012a).

REFERENCES

Adam, B. (2004). Memory of futures. *KronoScope*, *4*, 297–315.

Atance, C. M., & O'Neill, D. K. (2005). The emergence of episodic future thinking in humans. *Learning and Motivation*, *36*, 126–144.

Bar, M. (2010, Ed.). *Predictions and the brain: Using our past to generate a future.* New York: Oxford University Press.

Barkow, J. H., Cosmides, L., & Tooby, J. (Eds.). (1992). *The adapted mind: Evolutionary psychology and the generation of culture.* New York: Oxford University Press.

Bischof-Koehler, D. (1985). On the phylogeny of human motivation. In L. H. Eckensberger & E. D. Lantermann (Eds.), *Emotion and reflexivity* (pp. 3–47). Vienna: Urban & Schwarzenberg.

Boyer, P. (2007). Why do we have episodic memories? Mental time-travel as a strategic incentive mechanism. Unpublished manuscript.

Brandimonte, M., Einstein, G. O., & McDaniel, M. A. (Eds.). (1996). *Prospective memory: Theory and applications.* Mahwah, NJ: Erlbaum.

Brown, P., Keenan, J. M., & Potts, G. R. (1986). The self-reference effect with imagery encoding. *Journal of Personality and Social Psychology, 51,* 897–906.

Buckner, R. L., & Carroll D. C. (2007). Self-projection and the brain. *Trends in Cognitive Sciences, 11,* 49–57.

Burns, D. J., Hwang, A. J., & Burns, S. A. (2011). Adaptive memory: Determining the proximate mechanisms responsible for the memorial advantage of survival processing. *Journal of Experimental Psychology: Learning, Memory, and Cognition, 37,* 206–218.

Butler, A. C., Kang, S. H. K., & Roediger, H. L. III. (2009). Congruity effects between materials and processing tasks in the survival processing paradigm. *Journal of Experimental Psychology: Learning, Memory, and Cognition, 35,* 1477–1486.

Cosmides, L., & Tooby, J. (1987). From evolution to behavior: Evolutionary psychology as the missing link. In J. Dupre (Ed.), *The latest on the best: Essays on evolution and optimality* (pp. 277–306). Cambridge, MA: MIT Press.

Craver, C. F. (2007). *Explaining the brain: Mechanisms and the mosaic unity of neuroscience.* Oxford, UK: Oxford University Press.

Dawkins. R. (1976). *The selfish gene.* Oxford, UK: Oxford University Press.

Donald, M. (1991). *Origins of the modern mind: Three stages in the evolution of culture and cognition.* Cambridge, MA: Harvard University Press.

Dunbar, R. (1996). *Grooming, gossip, and the evolution of language.* Cambridge, MA: Harvard University Press.

Gertler, B. (2011). *Self-knowledge.* New York: Routledge.

Glenberg, A. M. (1997) What memory is for. *Behavioral and Brain Sciences, 20,* 1–55.

Goffman, E. (1959). *The presentation of self in everyday life.* Garden City, NY: Doubleday Anchor Books.

Hart, J. T. (1965). Memory and the feeling-of-knowing experience. *Journal of Educational Psychology, 56,* 208–216.

Hart, J. T. (1967a). Memory and the memory-monitoring process. *Journal of Verbal Learning and Verbal Behavior, 6,* 685–691.

Hart, J. T. (1967b). Second-try recall, recognition, and the memory monitoring process. *Journal of Experimental Psychology, 58,* 193–197.

Healy, S. D., & Hurly, T. A. (2004). Spatial learning and memory in birds. *Brain Behavior and Evolution, 63,* 211–220.

Henderson, J., Hurly, T. A., & Healy, S.D. (2006). Spatial relational learning in rufous hummingbirds (*Selasphorus rufus*). *Animal Cognition, 9,* 201–205.

Howe, M. L., & Derbish, M. H. (2010). On the susceptibility of adaptive memory to false memory illusions. *Cognition, 115,* 252–267.

Hunt, R. R., & Einstein, G. O. (1981). Relational and item-specific information in memory. *Journal of Verbal Learning and Verbal Behavior, 20,* 497–514.

Ingvar, D. H. (1985). "Memory for the future": An essay on the temporal organization of conscious awareness. *Human Neurobiology, 4*, 127–136.

Jones, J., & Paschler, H. (2007). Is the mind inherently forward looking? Comparing prediction and retrodiction. *Psychonomic Bulletin & Review, 14*, 295–300.

Kang, S. H. K., McDermott, K. B., & Cohen, S. M. (2008). The mnemonic advantage of processing fitness-relevant information. *Memory & Cognition, 36*, 1151–1156.

Karylowski, J., & Buczek, M. (1984, August). *Cognitive representation of self and others: Words and images.* Paper presented at the 92nd Annual Convention of the American Psychological Association, Toronto.

Klein, S. B. (2007). Phylogeny and evolution: Implications for understanding the nature of a memory system. In H. L. Roediger, Y. Dudai, & S. M. Fitzpatrick (Eds.), *Science of memory: Concepts* (pp. 377–381). Oxford, UK: Oxford University Press.

Klein, S. B. (2012a). A role for self-referential processing in tasks requiring participants to imagine survival on the savannah. *Journal of Experimental Psychology: Learning, Memory & Cognition, 38*, 1234–1242.

Klein, S. B. (2012b). The self and its brain. *Social Cognition, 30*, 474–518.

Klein, S. B., Cosmides, L., Gangi, C. E., Jackson, B., Tooby, J., & Costabile, K. A. (2009). Evolution and episodic memory: An analysis and demonstration of a social function of episodic memory. *Social Cognition, 27*, 283–319.

Klein, S. B., Cosmides, L., Tooby, J., & Chance, S. (2002). Decisions and the evolution of memory: Multiple systems, multiple functions. *Psychological Review, 109*, 306–329.

Klein, S. B., & Gangi, C. E. (2010). The multiplicity of self: Neuropsychological evidence and its implications for the self as a construct in psychological research. *The Year in Cognitive Neuroscience 2010: Annals of the New York Academy of Sciences, 1191*, 1–15.

Klein, S. B., German, T. P., Cosmides, L., & Gabriel, R. (2004). A theory of autobiographical memory: Necessary components and disorders resulting from their loss. *Social Cognition, 22*, 460-490.

Klein, S. B., & Kihlstrom, J. F. (1986). Elaboration, organization, and the self-reference effect in memory. *Journal of Experimental Psychology: General, 115*, 26–38.

Klein, S. B., & Loftus, J. (1988). The nature of self-referent encoding: The contributions of elaborative and organizational processes. *Journal of Personality and Social Psychology, 55*, 5–11.

Klein, S. B., Loftus, J., & Burton, H. A. (1989). Two self-reference effects: The importance of distinguishing between self-descriptiveness judgments and autobiographical retrieval in self-referent encoding. *Journal of Personality and Social Psychology, 56*, 853–865.

Klein, S. B., Loftus, J., & Kihlstrom, J. F. (2002). Memory and temporal experience: The effects of episodic memory loss on an amnesic patient's ability to remember the past and imagine the future. *Social Cognition, 20*, 353–379.

Klein, S. B., Robertson, T. E., & Delton, A. W. (2010). Facing the future: Memory as an evolved system for planning future acts. *Memory & Cognition, 38*, 13–22.

Klein, S. B., Robertson, T. E., & Delton, A. W. (2011). The future-orientation of memory: Planning as a key component mediating the high levels of recall found with survival processing. *Memory, 19*, 121–139.

Klein, S. B., Robertson, T. E., Delton, A. W., & Lax, M. L. (2012). Familiarity and personal experience as mediators of recall when planning for future contingencies. *Journal of Experimental Psychology: Learning, Memory, and Cognition, 38*, 240–245.

Klein, S. B., Robertson, T. E., Gangi, C. E., & Loftus, J. (2008). The functional independence of trait self-knowledge: Commentary on Sakaki (2007). *Memory, 16*, 556–565.

Kroneisen, M., & Erdfelder, E. (2011). On the plasticity of the survival processing effect. *Journal of Experimental Psychology: Learning, Memory, & Cognition*, *37*, 1553–1562.

Llinas, R. R. (2001). *I of the vortex: From neurons to self.* Cambridge, MA: MIT Press.

Lombardo, T. (2008). *The evolution of future consciousness.* Bloomington, IN: AuthorHouse.

Lord, C. G. (1980). Schemas and images as memory aids: Two modes of processing social information. *Journal of Personality and Social Psychology*, *38*, 257–269.

Lord, C. G. (1987). Imaging the self and others. Reply to Brown, Keegan, and Potts. *Journal of Personality and Social Psychology*, *53*, 445–450. Chicago Press.

Maki, R. H., & McCaul, K. D. (1985). The effects of self-reference versus other reference on the recall of traits and nouns. *Bulletin of the Psychonomic Society*, *23*, 169–172.

Mead, G. H. (1934). *Mind, self & society.* Chicago, IL: The University of Chicago Press.

Miller, G. A., Galanter, E., & Pribram, K. H. (1960). *Plans and the structure of behavior.* New York: Holt, Rinehart and Winston, Inc.

Moran, R. (2001). *Authority and estrangement: As essay on self-knowledge.* Princeton, NJ: Princeton University Press.

Nairne, J. S. (2005). The functionalist agenda in memory research. In A. F. Healy (Ed.), *Experimental cognitive psychology and its applications: A Festschrift in honor of Lyle Bourne, Walter Kintsch and Thomas Landaur* (pp. 115–126). Washington DC: American Psychological Association.

Nairne, J. S. (2010). Adaptive memory: Evolutionary constraints on remembering. In B. H. Ross (Ed.), *The psychology of learning and motivation, Vol. 53* (pp. 1–32). London, UK: Academic Press.

Nairne, J. S., & Pandeirada, J. N. S. (2008a). Adaptive memory: Is survival processing Special? *Journal of Memory and Language*, *59*, 377–385.

Nairne, J. S., & Pandeirada, J. N. S. (2008b). Adaptive memory: Remembering with a stone-age brain. *Current Directions on Psychological Science*, *17*, 239–243.

Nairne, J. S., Pandeirada, J. N. S., Gregory, K. J., & Van Arsdall, J. E. (2009). Adaptive memory: Fitness relevance and the hunter-gatherer mind. *Psychological Science*, *20*, 740–746.

Nairne, J. S., Pandeirada, J. N. S., & Thompson, S. R. (2007). Adaptive memory: The comparative value of survival processing. *Psychological Science*, *19*, 176–180.

Nairne, J. S., Thompson, S. R., & Pandeirada, J. N. S. (2007). Adaptive memory: Survival processing enhances retention. *Journal of Experimental Psychology: Learning, Memory, and Cognition*, *33*, 263–273.

Nottebohm, F. (1981). A brain for all seasons: Cyclical anatomical changes in song control nuclei of the canary brain. *Science*, *214*, 1368–1370.

O'Brien, L. (2007). *Self-knowing agents.* New York: Oxford University Press.(2003).

Otgaar, H., & Smeets, T. (2010). Adaptive memory: Survival processing increases both true and false memory in adults and children. *Journal of Experimental Psychology: Learning, Memory, & Cognition*, *36*, 1010–1016.

Otgaar, H., Smeets, T., Merckelbach, H., Jelicic, M., Verschuere, B., Galliot, A., & van Riel, L. (2011). Adaptive memory: Stereotype activation is not enough. *Memory & Cognition*, *39*, 1033–1041.

Otgaar, H., Smeets, T., & van Bergen, S. (2010). Picturing survival memories: Enhanced memory after fitness-relevant processing occurs for both verbal and visual stimuli. *Memory & Cognition*, *38*, 23–28.

Passingham, R. E. (1982). *The human primate.* San Francisco, CA: W. H. Freeman and Company.

Pressley, M., Levin, J. R., Kuiper, N. A., Bryant, S. L., & Michener, S. (1982). Mnemonic versus non-mnemonic vocabulary-learning strategies: Additional comparisons. *Journal of Educational Psychology, 74*, 693–707.

Reder, L. M. (1987). Strategy selection in question answering. *Cognitive Psychology, 19*, 90–138.

Sandry, J., Trafimow, D., Marks, M., & Rice, S. (2012). Adaptive memory: Evaluating alternative forms of fitness-relevant processing in the survival processing paradigm. Unpublished manuscript.

Savine, A. C., Scullin, M. K., & Roediger, H. L. (2011). Survival processing of faces. *Memory & Cognition, 39*, 1359–1373.

Schank, R. C., & Abelson, R. P. (1977). *Scripts, plans, goals and understanding: An Inquiry into human knowledge structures*. Hillsdale, NJ: Lawrence Erlbaum Associates, Publishers.

Schacter, D. L. (1983). Feeling of knowing in episodic memory. *Journal of Experimental Psychology: Learning, Memory, and Cognition, 9*, 39–54.

Schacter, D. L., Addis, D. R., & Buckner, R. L. (2008). Episodic simulation of future Events: Concepts, data, and applications. *Year in Cognitive Neuroscience 2008: Annals of the New York Academy of Sciences, 1124*, 39–60.

Schacter, D., & Tulving, E. (Eds.) (1994). *Memory systems*. Cambridge, MA: MIT Press.

Schank, R. C., & Abelson, R. P. (1977). *Scripts, plans, goals and understanding: An inquiry into human knowledge structures*. Hillsdale, NJ: Lawrence Erlbaum Associates, Publishers.

Sherry, D. F., & Schacter, D. L. (1987). The evolution of multiple memory systems. *Psychological Review, 94*, 439–454.

Soderstrom, N. C., & McCabe, D. P. (2011). Are survival processing memory advantages based on ancestral priorities? *Psychonomic Bulletin & Review, 18*, 564–569.

Suddendorf, T., & Corballis, M. C. (1997). Mental time travel and the evolution of the human mind. *Genetic, Social, and General Psychology Monographs, 123*, 133–167.

Suddendorf, T., & Corballis, M.C. (2007). The evolution of foresight: What is mental time travel, and is it unique to humans? *Behavioral and Brain Sciences, 30*, 299–313.

Symons, C. S., & Johnson, B. T. (1997). The self-reference effect in memory: A meta-analysis. *Psychological Bulletin, 121*, 371–394.

Szpunar, K. K., & McDermott, K. B. (2008). Episodic future thought and its relation to remembering: Evidence from ratings of subjective experience. *Consciousness and Cognition, 17*, 330–334.

Tse, C.-S., & Altarriba, J. (2010). Does survival processing enhance implicit memory? *Memory & Cognition, 38*, 1110–1121.

Tulving, E. (1985). Memory and consciousness. *Canadian Psychology/ Psychologie Canadienne, 26*, 1–12.

Tulving, E. (2002). Chronesthesia: Awareness of subjective time. In D. T. Stuss & R. C. Knight (Eds.), *Principles of frontal lobe function* (pp. 311–325). New York: Oxford University Press.

Tulving, E., & Craik, F. I. M. (Eds.). (2000). *The Oxford handbook of memory*. New York: Oxford University Press.

Tulving, E., & Lepage, M. (2000). Where in the brain is awareness of one's past? In D. L. Schacter & E. Scarry (Eds.), *Memory, brain, and belief* (pp. 208–228). Cambridge, MA: Harvard University Press.

Warren, M. W., Chattin, D., Thompson, D. D., & Thomsky, M. T. (1983). The effects of autobiographical elaboration on noun recall. *Memory & Cognition, 11*, 445–455.

Weiler, J. A., & Daum, I. (2008). Mental time travel: The neurocognitive basis of future thinking. *Fortschritte Der Neurologie Psychiatrie, 76*, 539–548.

Weinstein, Y., Bugg, J. M., & Roediger, H. L. (2008). Can the survival recall advantage be explained by basic memory processes? *Memory & Cognition, 36*, 913–919.

Wheeler, M. A., Stuss, D. T., & Tulving, E. (1997). Toward a theory of episodic memory: The frontal lobes and autonoetic consciousness. *Psychological Bulletin, 121*, 331–354.

Williams, G. C. (1966). *Adaptation and natural selection: A critique of some current evolutionary thought.* Princeton: Princeton University Press.

3

Remembering Cheaters

RAOUL BELL AND AXEL BUCHNER

INTRODUCTION

When cognitive psychologists are interested in how associations between different memoranda are formed, they often use highly artificial stimulus materials. For instance, when interested in source memory (that is, memory for the context in which an item was encountered), researchers may examine how their participants succeed in remembering the font color of randomly selected words. The question of *why* participants should remember this type of material in the first place is rarely addressed. This type of research is considered important even with such artificial material because it could provide basic knowledge about the functioning of memory. However, this approach to the understanding of human cognition was criticized by evolutionary psychologists who deny the existence of general-purpose learning mechanisms that apply across different domains of biological function. Evolutionary Psychology offers a fundamentally different approach to understanding human cognition that is based on the assumption that the human nervous system and its information processing mechanisms have evolved to solve adaptive problems faced by our ancestors during long periods of human evolution (Cosmides & Tooby, 1997; Tooby & Cosmides, 1992, 2005).

A core assumption of Evolutionary Psychology is the accumulative modularity of the human mind (Tooby & Cosmides, 2005). Each mental module represents the solution of a well-defined adaptive problem and is thus shaped by natural selection to fulfill a specific purpose, just as body parts do (e.g., just as a heart pumps blood). From this point of view human memory is adaptive in that a novel evolutionary challenge leads to the selection of individuals equipped with additional cognitive modules that support coping successfully with this challenge. Given that the short time since industrialization is insignificant in evolutionary terms, Evolutionary Psychology assumes that our "modern skulls house stone-age minds" (Cosmides & Tooby, 1997, p. 84) (which, by the way, seems an interesting proposal given that the evolutionary history of skulls cannot be shorter than that of the mind). An implication of this view is that information processing mechanisms are content-specific

(Cosmides & Tooby, 1992). Therefore, an analysis of human memory that is based on research focusing on the processing of stimuli without biological relevance reveals little about the basic architecture of the human mind.

This evolutionary approach, in turn, has been criticized for failing to consider alternatives to adaptationist explanations (Gould & Lewontin, 1979), for neglecting relevant neurobiological and neuropsychological evidence that questions the modularity assumption (e.g., Panksepp & Panksepp, 2000), and for lacking convincing empirical support (e.g., Buller, 2005). According to its harshest critics, the whole field of Evolutionary Psychology is based on "just so" stories instead of empirically founded theories. The gist of this criticism is that it is easy to come up with a plausible *post hoc* "explanation" of why a certain behavior might have been adaptive in a stone-age environment while at the same time it is notoriously difficult to make scientifically valid assumptions about the driving forces of human evolution during the Pleistocene era. Many mechanisms that are believed to be adaptations might in fact be byproducts or epiphenomena of evolution that have not been directly selected for their adaptive value (Gould & Lewontin, 1979).

Cartwright (2008) traced the discussion back to an old controversy between sociobiology and behaviorism. Behaviorism was based on the assumption that it is possible to identify a few universal general-purpose learning principles that generalize readily across species and stimuli. One implication of the equipotentiality assumption is that any two stimuli can be associated in the brain through associative learning, regardless of the nature of these stimuli (see Öhman, Fredrikson, Hugdahl, & Rimmö, 1976, for a critical discussion of the concept). As is well known, the equipotentiality assumption was challenged by the observation that some associations between stimulus and consequences (e.g., a taste and nausea) are much more easily acquired than others (e.g., a taste and an electric shock, or a bright light and nausea), which was interpreted as evidence for an innate preparedness to learn these associations that results from adaptations to species-specific problems of survival (e.g., food poisoning) (Garcia & Koelling, 1966). Although these findings have entered psychology textbooks and common wisdom about memory, the idea of biological preparedness is often met with skepticism when it comes to human learning and memory. For instance, there is no consensus about whether the learning of associations between pictures of poisonous animals (snakes, spiders) and negative events (electric shocks) is mediated by an ancient fear module that serves to protect against poisonous animals (Flykt, Esteves, & Öhman, 2007; Hugdahl & Johnsen, 1989; Öhman & Mineka, 2003). Furthermore, there is a lively debate about the recent observation that rating words for their survival relevance in the grasslands leads to exceptionally good memory for this material (Nairne, Pandeirada, & Thompson, 2008). This finding was used to speculate that human episodic memory might only be properly explained by referring to "survival" modules that have developed during the evolution of human cognition (e.g., Nairne, 2010; Nairne & Pandeirada, 2010). However, this evolutionary explanation is challenged by studies showing that the result may be explained by well-known, domain-general memory mechanisms (e.g., Erdfelder & Kroneisen, Chapter 10 in this volume; Kroneisen & Erdfelder, 2011). The present chapter focuses on another aspect of adaptive memory, memory for cheaters and cooperators in social exchange situations.

RECIPROCAL ALTRUISM AND CHEATER DETECTION

Given that the social environment has exerted important selection pressures during human evolution (Dunbar, 2003), researchers committed to the Evolutionary Psychology paradigm have claimed that diverse modules have evolved as psychological adaptations to group living (Geary, 2007). Most prominently, it is assumed within social contract theory (Cosmides & Tooby, 1989, 1992, 2005) that a cheater detection module has evolved as a component of an adaptation for social cooperation. Cooperation among non-kin is fascinating for evolutionists because it seems to be inconsistent with the struggle for survival postulated by Darwin. At first glance, it seems obvious that members of a group can benefit from mutual cooperation. A great example for the benefits of cooperation at group level was given by Darwin (1871, p. 113):

> When two tribes of primeval man, living in the same country, came into competition, if the one tribe included (other circumstances being equal) a greater number of courageous, sympathetic, and faithful members, who were always ready to warn each other of danger, to aid and defend each other, this tribe would without doubt succeed best and conquer the other.

However, the dilemma arises because cooperation is often costly for the individual providing support for other group members. Therefore, each member of the group can increase his or her fitness by shying away from the costs of reciprocating. Given that cheaters benefit at the expense of cooperators, the egoistic strategy seems to be more successful in evolutionary terms. Hence, natural selection would work against individuals who sacrifice themselves or their reproductive success in favor of other members of their groups (Cosmides & Tooby, 2005; Trivers, 1971). Darwin (1871, p. 113f) recognized this problem:

> It is extremely doubtful whether the offspring of the more sympathetic and benevolent parents, or of those which were the most faithful to their comrades, would be reared in greater number than the children of selfish and treacherous parents of the same tribe. He who was ready to sacrifice his life [...] rather than betray his comrades, would often leave no offspring to inherit his noble nature. The bravest men, who were always willing to come to front in war, and who freely risked their lives for others, would on an average perish in larger numbers than other men. Therefore, it hardly seems probable, that the number of men gifted with such virtues, or that the standard of their excellence, could be increased through natural selection.

Cooperation can only evolve when the associated fitness costs are compensated by benefits. This is the case when the cooperators provide benefits to others *conditional* on the recipient's providing benefits in return. This solution to the problem of human cooperation was also anticipated by Darwin (1871, p. 114):

> As the reasoning powers and foresight of the members became improved, each man would soon learn from experience that if he aided his fellow-men, he would

commonly receive aid in return. From this low motive he might acquire the habit of aiding his fellows.

The idea of reciprocal altruism was refined and formalized by more recent evolutionists (Axelrod & Hamilton, 1981; Trivers, 1971). Reciprocal altruism is based on the idea that cooperation can be beneficial when it is restricted to other cooperators who return the favors they have been receiving. Given that cheating is attractive because cheaters receive support from others but have to bear no costs, cooperative groups are always in danger of being invaded by cheaters. Therefore, cooperation can only evolve together with mechanisms of cheater detection and avoidance. From these premises, Cosmides and Tooby (1989, 1992, 2005) have derived the prediction that humans, as members of a highly cooperative species, have developed a cheater detection module.

Empirical tests of social contract theory (Cosmides & Tooby, 2005) have concentrated on the Wason Selection Task (Wason, 1968). The original version of the task is widely used to examine human reasoning. Participants are shown a set of four cards, each of which has a number on one side and a letter on the other. The visible sides of the cards show *A*, *B*, *4*, and *7* (see upper panel of Fig. 3.1). The participants are asked which cards they have to turn over to test the truth of the proposition "*if a card shows a consonant on one side, then there is an odd number on the other*" ("*if p then q*" in the abstract version of the task). The logically correct answer to this problem is to select the cards that allow for a falsification of the rule (i.e., p and $\neg q$, or *B* and *4* in the example presented in Fig. 3.1).

The task is interesting for many cognitive psychologists because most participants fail to select the logically correct answer. However, when the task is reformulated so that people have to detect violations of social contract rules such as "*if you take the benefit, then you have to pay the cost*" (lower panel of Fig. 3.1), most people solve the task correctly (i.e., they know that they have to check those who have accepted the benefit [p] and those who have failed to pay the cost [$\neg q$]). This pattern of finding has been interpreted as evidence of an evolved reasoning module that is specialized

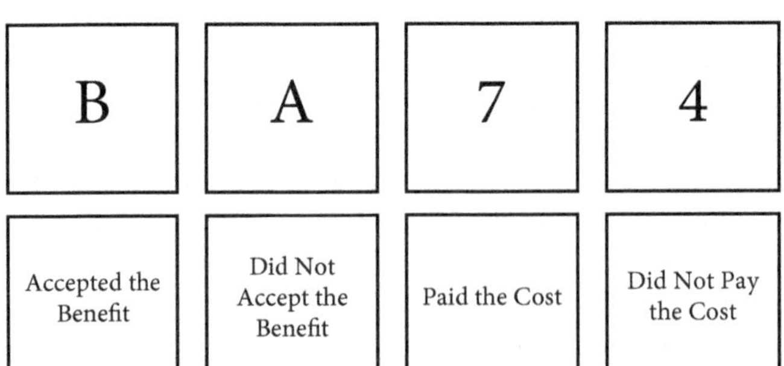

Figure 3.1 Examples for the Wason Selection Task. In its original version (upper panel), participants see four cards and are asked, "Which cards do you have to turn over to test the truth of the rule 'If a card shows a consonant on one side, then there is an odd number on the other'?" In the deontic version of the task (lower panel), participants are asked, "Which of the cards do you have to turn over to see whether someone breaks the rule 'If you take the benefit, then you have to pay the cost'?"

in the detection of cheaters (Cosmides & Tooby, 1992, 2005). This interpretation of the findings is, however, highly controversial, and the results have been attributed, for instance, to confounds in the stimulus material or to demand characteristics of the task (e.g., Carlisle & Shafir, 2005; Fodor, 2000). An alternative theory, relevance theory, argues that participants simply choose the options that are most relevant, relevance being defined as the degree to which the information leads to new, or to the abandonment of old, beliefs in a given context (Sperber, Cara, & Girotto, 1995). Given that the research using the Wason Selection Paradigm does not seem to lead to a consensus, it may be necessary to use other paradigms to validate or invalidate the hypothesis of an evolved cheater detection module.

ENHANCED MEMORY FOR FACES OF CHEATERS?

Several researchers (Chiappe et al., 2004; Mealey, Daood, & Krage, 1996; Oda, 1997) have argued that the cheater detection module has to be complemented by memory mechanisms that enable the individual to learn particularly well from previous negative experiences with cheaters. Reciprocal strategies require the ability to remember and recognize cheaters "in order to know when to cooperate, when to defect, and when to punish defection" (Cosmides & Tooby, 1992, p. 177). For instance, consider *tit for tat*, which is one of the simplest forms of reciprocity and has been considered to be one of the most successful strategies in repeated social exchange (Axelrod & Hamilton, 1981). This strategy consists of always cooperating unless being provoked by a cheater, and to retaliate against cheaters in the next encounter. Obviously, an individual following this strategy would need to have good memory for cheaters to know when to refuse cooperation.

Mealey, Daood, and Krage (1996) were the first to examine whether people have enhanced memory for faces of cheaters. In their experiment, participants saw faces presented together with short behavior descriptions in which the persons shown were described as cheaters, as cooperators, or as neutral with respect to the cooperator–cheater dimension. In a surprise old–new recognition test, the faces were presented together with new faces, and the participants were asked to indicate whether the face was old or new. Old–new face recognition was significantly influenced by the behavior descriptions. However, recognition of cheater faces was only descriptively enhanced when these faces were associated with professions that confer low social status. When the faces were associated with high social status, the data pattern was descriptively reversed. Furthermore, the significant effect of the behavior descriptions on old–new face recognition was not due to a recognition *advantage* for faces of cheaters. Rather, faces of cooperative persons were *less well* recognized than faces of neutral persons and faces of cheaters, which makes no sense from an evolutionary perspective. Nevertheless, the results were interpreted as evidence for a cheater recognition mechanism. However, not only are the results counterintuitive and inconsistent, they also could not be replicated in several follow-up studies, all of which found that old–new recognition was not affected by the behavior descriptions at all (e.g., Barclay & Lalumière, 2006; Buchner, Bell, Mehl, & Musch, 2009; Mehl & Buchner, 2008). The cheater recognition advantage found by Mealey et al. might have been due either to chance or to cofounds in the stimulus material (see Mehl & Buchner, 2008, for a more thorough discussion).

However, this null finding is not necessarily inconsistent with the predictions of an evolutionary account, because knowing that a face is familiar (item memory) does not help to avoid cheaters in social exchange. For instance, if you meet a person who has cheated on you in the past, it will not help you if this person's face seems familiar to you. What is worse, familiarity of the face may even increase the chances that you will like the person (Bornstein, 1989; Zajonc, 1968), which may make you vulnerable to social exploitation. To be able to avoid and punish cheaters, it is necessary to remember that the face belongs to a cheater. In other words, you have to remember the cheating context in which the face was encountered (source memory). Therefore, Buchner, Bell, Mehl, and Musch (2009) examined memory for faces of cheaters using a source memory test. In a learning phase, participants saw the faces together with descriptions of cheating, cooperative, and neutral behavior (Table 3.1). In the test phase, participants had to complete a source memory test in which they first classified the faces as old or new. If a face was classified as old, participants were additionally asked to classify the person as a cheater, a cooperator, or a neutral control person.

When comparing source memory between different types of sources, it is important to use a source memory measure that is not affected by differences in old–new recognition and guessing. This problem can be best illustrated using an example. Let us assume that a hypothetical participant classified 90 percent of the cooperator faces correctly as cooperators. From this finding, one may be tempted to conclude that the participant had good source memory for cooperators. However, this conclusion may not be valid. For instance, the participant may have labeled 90 percent of all faces as cooperators, regardless of whether the faces actually belonged to cooperators or not. In this case a valid conclusion would have been that the good classification performance for cooperator faces was due to a guessing bias rather than to a memory advantage for cooperators. Thus, it is necessary to use a measurement model that allows us to disentangle source memory and guessing. The multinomial source monitoring model of Bayen, Murnane, and Erdfelder (1996) has the advantage of yielding a source memory parameter d that has empirically been shown to be unaffected by manipulations of old–new recognition and guessing, and to be sensitive to manipulations of source memory (Bayen et al., 1996; Bayen, Nakamura, Dupuis, & Yang, 2000).

When source memory was measured in this way, a very robust data pattern was obtained (Fig. 3.2). Source memory for faces associated with cheating has been found

Table 3.1. Examples of the Behavior Descriptions Used in Buchner et al. (2009)

Cheater	R. K. is a second-hand car dealer. Regularly, he sells restored crash cars as supposedly accident-free and conceals serious defects from the customers.
Control	P. M. is a dyer. He lives in an old windmill and is specialized in the fabrication of medieval-looking clothes.
Cooperator	T. S. is a pool attendant. Once on his way home he risked his own life to save a child who was about to drown in a frozen lake.

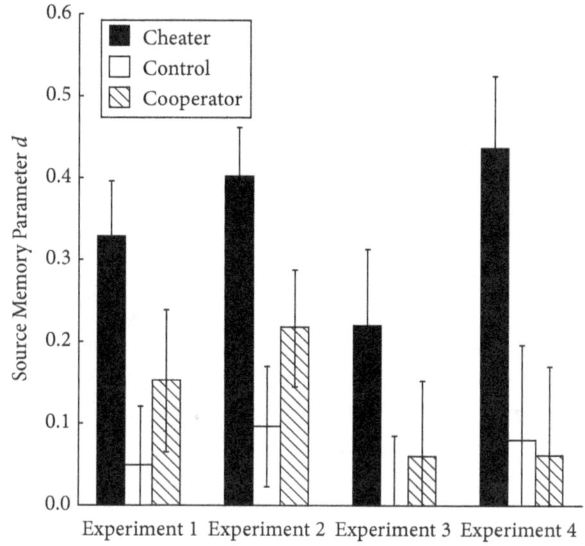

Figure 3.2 Social source memory for cheaters, cooperators, and neutral control persons in Buchner et al.'s (2009) Experiments 1–4. Source memory was better for faces of cheaters than for faces of cooperators or neutral control persons. The source memory parameters were obtained using Bayen et al.'s (1996) source monitoring model and represent the conditional probability of remembering the source of an item given that the item was correctly recognized as old. The error bars represent the 95 percent confidence intervals. Source memory for control persons in Experiment 3 was estimated to be zero.

to be superior to source memory for faces associated with neutral or trustworthy reputational information immediately after the encoding phase (Experiments 1, 2, and 4) and after a 1-week retention interval (Experiment 3). The effect was independent of the exceptionality of the behavior descriptions and the social status of the stimulus persons (Buchner et al., 2009). The source memory advantage has been subsequently replicated under a variety of different conditions using a variety of carefully normed stimulus materials (Bell & Buchner, 2010a; Bell, Buchner, Erdfelder, et al., 2012).

HOW DO WE REMEMBER SOCIAL-EXCHANGE PARTNERS?

Follow-up studies examined whether participants are more likely to remember details of the cheating episodes in comparison to episodes of trustworthy or neutral behaviors. Source memory is often equated with a vivid, detailed recollection of the learning episode. However, it is known that correct source judgments are sometimes based on the retrieval of partial source information (Dodson, Holland, & Shimamura, 1998; Klauer & Wegener, 1998). With respect to source memory for cheaters, the source memory advantage for faces of cheaters may be due to (a) enhanced retrieval of the specific details of the cheating episodes, (b) a retrieval of a "cheater" tag (i.e., memory for the emotional category without a detailed recollection of the specific details of the cheating episode), or (c) a change in valence of the previously neutral face due to the prior association to a negative context. To examine the specificity of source memory for cheaters, we (Bell, Buchner, Erdfelder, et al., 2012) showed participants faces together with descriptions of cheating, trustworthy, or neutral behavior. At test, participants were required to remember the type of behavior that accompanied the faces at encoding (i.e., "cheating," "trustworthy," or "neutral"). Unlike previous studies, participants were also required to recall the behavior episodes in as much detail as possible. This cued-recall test served to examine whether the participants were more likely to remember details of the cheating contexts compared to the trustworthy or neutral contexts. A multinomial model of partial and specific source memory (Dodson et al., 1998; Klauer & Wegener, 1998) was used to tease apart specific source memory for the individual cheating episodes and partial source memory in the sense of only a rough classification of the behavior as belonging to a particular emotional category—cheating, trustworthy, or neither of these. The results revealed that participants were able to remember that a face was associated with cheating but were unable to retrieve more specific details of the cheating episodes, consistent with the assumption that the source memory advantage for faces of cheaters is due to emotional tagging. Nevertheless, participants succeeded in discriminating between two different types of negative context information (cheating and disgusting behavior) at test (Bell, Giang, & Buchner, 2012), which shows that the emotional tags that were associated with the faces are specific for the particular type of negative experience at encoding (e.g., cheating or disgust).

This data pattern fits with theories about how we form and store impressions about other people in general. These theories assume that people spontaneously derive very robust trait inferences (e.g., *kind, tolerant, selfish, aggressive,* etc.) from other people's behavior (Carlston & Skowronski, 1994; Todorov & Uleman, 2004). This trait knowledge continues to influence the behavior toward that person even when the episode leading to the trait inference is forgotten (Carlston & Skowronski, 1994; Carlston, Skowronski, & Sparks, 1995; Todorov & Uleman, 2002). Neuropsychological studies suggest that this type of evaluative learning does not depend on hippocampal structures that are typically involved in the learning of nonemotional associations but instead relies on emotion-processing brain circuits, including the amygdala (Johnson, Kim, & Risse, 1985; Todorov & Olson, 2008). Consistent with the results summarized above, it has been suggested that negative social behaviors have more weight in the impression formation process than positive social behaviors (Peeters & Czapinski, 1990) because they may be more diagnostic of the person's true disposition (Rothbart & Park, 1986; Skowronski & Carlston, 1989; Tausch, Kenworthy, & Hewstone, 2007).

IS MEMORY FOR CHEATERS SPECIAL?

The question arises whether the mechanism responsible for the source memory advantage for faces of cheaters is indeed specific to the processing of exchange-related information, as assumed by evolutionary theorists. Most research in Evolutionary Psychology tends to focus on very narrow topics. Often, findings that fit with evolutionary hypotheses are explained by invoking highly specific mental modules designed by natural selection to solve well-defined adaptive problems. However, the specificity of the effect in question should be empirically investigated before drawing conclusions about content-specific cognitive adaptations, and alternative hypotheses should be considered. Cheating elicits strong negative emotions (*anger, disgust, fear*). Therefore, it seems possible that the source memory advantage for faces of cheaters is just an example of the general phenomenon that negative information is especially well retained. In an influential literature review, Baumeister, Bratslavsky, Finkenauer, and Vohs (2001) examined the effects of emotion on cognitive processing and arrived at the conclusion that "it appears to be a basic, pervasive fact of psychology that bad is stronger than good" (p. 360). There are well-documented negativity effects on memory (Kensinger, 2007), and several studies found enhanced source memory for negative contexts (Kensinger & Corkin, 2003; Kensinger & Schacter, 2006). Based on this literature, one may assume that (everything else being equal) negative contexts may be generally better remembered than positive or neutral contexts. This hypothesis was tested by examining whether a source memory advantage for faces associated with negative contexts could be obtained using disgusting behavior descriptions.

In an encoding phase, faces were paired with descriptions of disgusting, neutral, or pleasant behavior (Table 3.2). In a source memory test, participants were required to indicate whether faces were old or new, and to select the type of context these faces had been presented in. In two experiments, source memory for faces associated with disgusting behaviors was enhanced in comparison to source memory for faces associated with neutral or pleasant behaviors (Fig. 3.3). Thus, the source memory advantage generalizes to faces associated with other types of negative contexts and is not tied to cheating (Bell & Buchner, 2010b).

What is more, disgusting contexts were remembered just as well as cheating when the qualitative characteristics of source memory for these two types of contexts was compared directly within a single experiment (Bell, Giang, et al., 2012). Furthermore, individual differences in the tendency to respond with strong negative emotions to injustice (Schmitt, Gollwitzer, Maes, & Arbach, 2005) moderated the source memory advantage for faces of cheaters (Bell & Buchner, 2010a). Participants with a high sensitivity to injustice were particularly likely to remember when a face was associated with cheating. This finding suggests that the source memory advantage may be mediated by the strong negative feelings elicited by descriptions of unjust behavior. An fMRI study (Singer, Kiebel, Winston, Dolan, & Frith, 2004) showed that, relative to neutral faces, both faces of cooperators and faces of cheaters were associated with enhanced activity in a wide network of emotion-processing brain regions, including the prefrontal cortex and the amygdala. The emotional mediation hypothesis was further supported by the finding that a reduction in the negative emotionality of cheating behavior leads to a drastic reduction of the source memory advantage for faces of cheaters (Bell & Buchner, 2011).

Table 3.2. Examples of the Behavior Descriptions Used in Bell and Buchner (2010b)

Disgust	H. M. is a concrete worker. He suffered a burn on one leg, which does not heal and which secretes a foul smell.
Control	J. B. is a glazier. He buys his clothes from a young saleswoman at the fashion store in the city center.
Pleasantness	O. D. is a cheesemonger. He is such an expert in the art of cooking that his friends get already overwhelmed by the delicious scent of his exotic culinary creations.

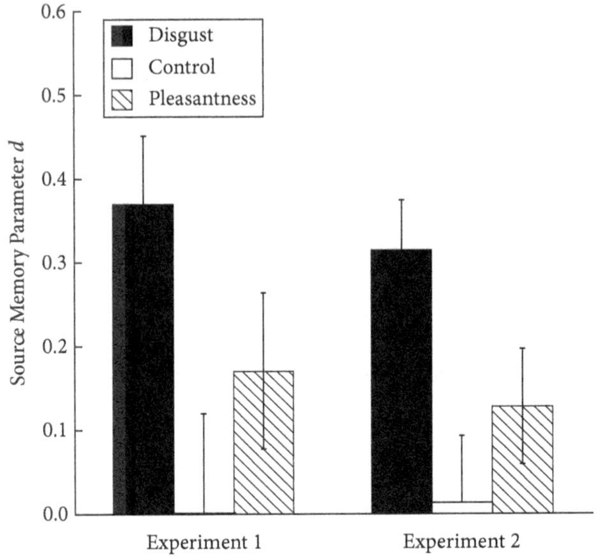

Figure 3.3 Source memory for faces associated with disgust, pleasantness, and control contexts in Bell and Buchner's (2010b) Experiments 1 and 2. Source memory was better for faces associated with disgust than for the other types of faces. The source memory parameters were obtained using Bayen et al.'s (1996) source monitoring model and represent the conditional probability of remembering the source of an item given that the item was correctly recognized as old. The error bars represent the 95 percent confidence intervals. Source memory for control persons was estimated to be zero in Experiment 1.

The most general theory that could explain these results is that negative valence or arousal enhances source memory. It is thus tempting to speculate that memory simply improves as a function of absolute (positive or negative) valence or arousal. Unfortunately, this simple hypothesis is not supported by recent evidence. For instance, memory for aggressive behavior shows an advantage that is equivalent to the original cheater advantage, but only if the aggressive behavior is other-relevant (i.e., threatening to an observer) and not if it is self-relevant (i.e., only negative for the person displaying this behavior), although *a priori* norming has shown that the valence of both types of behavior is equivalent (Bell, Giang, et al., 2012). Furthermore, the cheating information *per se* does not lead to an advantage in source memory because faces of victims of cheating are remembered less well than faces of perpetrators despite the fact that essentially the same context descriptions were used in the two conditions, and the *a priori* valence of the context information was identical for victims and perpetrators (Bell & Buchner, 2011). This shows that negative valence and arousal do not automatically lead to enhanced memory. Rather, human memory seems to be biased toward processing social information that helps to avoid harmful encounters with dangerous persons, consistent with the assumption that source memory plays a functional role in threat avoidance.

HOW ADAPTIVE IS MEMORY FOR CHEATERS?

The research reported above suggests that the source memory advantage for cheaters may be just another example of a general threat bias in information processing

(Baumeister et al., 2001; Öhman & Mineka, 2001; Peeters & Czapinski, 1990; Pratto & John, 1991). Such threat biases in information processing are often interpreted within an evolutionary functional framework. A focus on negative, threatening information may be adaptive because negative encounters with the environments have greater effects on the organism's survival and reproduction than positive ones. For instance, there are many negative events that lead to death, but there are no positive events that can guarantee survival over long periods of time (Baumeister et al., 2001; Peeters & Czapinski, 1990). Hence, it may be adaptive to put more weight on negative information.

In direct contrast to these hypotheses, Rothermund, Voss, and Wentura (2008) have argued that a rigid focus on negative events would be maladaptive because such an inflexible mechanism would make it difficult to escape from a negative motivational state. The emotional system should allow the individual to flexibly adapt to the changing demands of a variety of situations. In social exchange, it may depend on the social environment whether it is more beneficial to remember cheaters or cooperators (e.g., Aktipis, 2006). When cooperation is the rule, we may routinely engage in social exchange and do not expect to be cheated. These positive expectations toward the social environment may facilitate cooperation between strangers, but they have to be balanced by a cheater recognition mechanism that helps to protect against exploitation of these cooperative tendencies. However, when cheaters represent the majority, a rigid focus on cheating may ultimately be suboptimal because the benefits of cooperation are missed out. Thus, success in social exchange requires not only skills in cheater detection, but also the ability to revise unjustified negative expectations toward others. In an aversive social situation, a focus on potential cooperation partners could increase the possibility of forming successful alliances against the hostile social environment. Thus, it may not always be adaptive to prioritize negative over positive information.

Consistent with these assumptions, several studies (Barclay, 2008; Bell, Buchner, & Musch, 2010; Volstorf, Rieskamp, & Stevens, 2011) have shown that source memory for cheaters and cooperators depends on the relative proportion of cheaters and cooperators in the learning phase. When participants played cooperation games in which most interaction partners were cooperative, cheating was better remembered than cooperation. However, when the ratio of cheaters to cooperators was changed so that most interaction partners were cheaters, this pattern reversed, and source memory for cooperators was enhanced. Furthermore, Chang and Sanfey (2009) showed that source memory varies as a function of the participants' expectations about the future behavior of the interaction partners prior to the experiment. Memory for cooperators and cheaters was accompanied by activation in brain regions such as the striatum, the anterior insula, and the anterior cingulate cortex that can be viewed as parts of a general system that is specialized in the processing of expectancy violation, conflict monitoring, and the correction of inappropriate behavioral responses.

Suzuki and Suga (2010) showed that cheaters who looked trustworthy were better remembered than untrustworthy-looking cheaters. Bell, Buchner, Kroneisen, and Giang (2012) replicated this finding (Fig. 3.4) and showed that smiling cheaters were better remembered than cheaters with an angry facial expression. Kroneisen and Bell (2013) showed that the expectancy violation effect generalizes to source memory for information that is not directly relevant to social exchange, which is consistent with

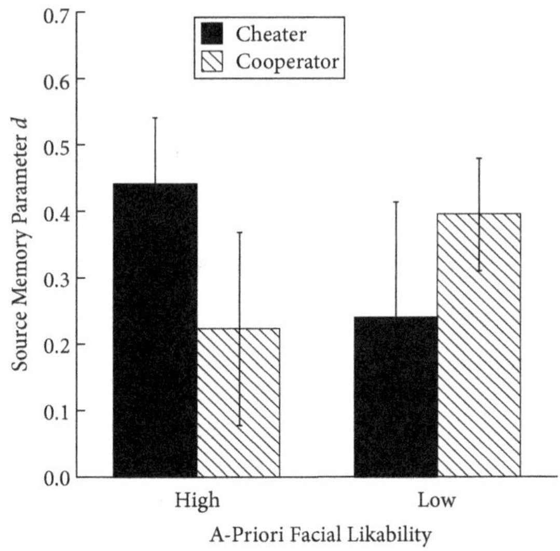

Figure 3.4 Source memory for faces of cheaters and cooperators as a function of the pre-experimental likableness of the faces in Bell, Buchner, Kroneisen, et al.'s (2012) Experiment 1. Source memory was enhanced for emotionally incongruent information. The source memory parameters were obtained using Bayen et al.'s (1996) source monitoring model and represent the conditional probability of remembering the source of an item given that the item was correctly recognized as old. The error bars represent the 95 percent confidence intervals.

a general expectancy violation account. These findings reinforce the interpretation that people remember events particularly well when they are emotionally incongruent to currently activated approach or avoidance tendencies, and therefore require a change of behavior. Given that many findings suggest that (source) memory for unexpected events is generally enhanced (Cook, Marsh, & Hicks, 2003; Ranganath & Rainer, 2003; van Kesteren, Ruiter, Fernandez, & Henson, 2012), the expectancy violation effects on source memory for cheaters and cooperators can be viewed as byproducts of general mechanisms serving to prioritize the processing of unanticipated information. These mechanisms may serve to promote efficient information processing because cognitive resources are only invested when the appropriate approach-or-avoidance behavior cannot be generated by less resource-demanding mechanisms (expectancy-based guessing). Therefore, focusing on unexpected information may represent a simple and efficient way to optimize information processing in a wide range of situations.

CONCLUSION

We found across a large number of experiments that source memory for faces of cheaters was enhanced in comparison to other types of faces (Bell & Buchner, 2010a; Bell et al., 2012; Buchner et al., 2009). These findings are consistent with an evolutionary analysis of the cognitive prerequisites of cooperation. If seen in isolation, these findings could be viewed as evidence for a highly specific cheater recognition module that has co-evolved with human cooperation. However, before drawing such strong conclusions, it has to be empirically tested whether the effect is specific for cheating. Indeed, subsequent studies showed that a similar source memory advantage can be obtained with other types of threatening contexts (Bell & Buchner, 2010b; Bell, Giang, et al., 2012). Another series of experiments showed that people do not focus exclusively on cheating, but instead focus on cheating and cooperation depending on whether cooperation or cheating is expected (Barclay, 2008; Bell, Buchner, et al., 2012; Bell et al., 2010; Suzuki & Suga, 2010; Volstorf et al., 2011). As discussed above, flexibility in social information processing is necessary to establish successful cooperation in a variety of situations. Taken together, the results suggest that memory can be conceived of as adaptive in the general sense that information processing is focused on events that are relevant because they require behavioral adjustment. More precisely, participants focus on information that is (a) threatening and/or (b) emotionally incongruent to currently activated approach or avoidance tendencies. These general mechanisms can explain not only why source memory for cheaters and cooperators is enhanced, but also why these findings generalize to other types of emotional information. Explaining parallel effects with a single general mechanism (e.g., a threat advantage) seems to be more parsimonious than postulating fresh modules to explain the identical pattern of results obtained with different types of emotionally charged information (e.g., cheating, disgusting, aggressive behaviors, etc.). From an evolutionary perspective, a single mechanism that leads to appropriate information processing in a wide variety of situations seems to be more efficient than a large collection of independent but functionally equivalent modules.

Nevertheless, a functional analysis of the requirements of social information processing has proven useful in predicting that source memory for social interaction partners varies as a function of its approach or avoidance relevance. Source memory was not generally enhanced for negative information. Instead, people preferentially remembered information that can be considered instrumental in avoiding negative encounters with dangerous persons (Bell & Buchner, 2011; Bell, Giang, et al., 2012). These findings suggest that a functional analysis can be useful in specifying how the cognitive system helps us to solve biologically relevant problems (e.g., threat avoidance).

Conversely, it is possible to draw conclusions about the plausibility of biological reciprocity models from the findings reported here. A strategy such as *tit for tat* (that is superior to other strategies in simulation studies; Axelrod & Hamilton, 1981) implies that the reciprocal cooperators remember the most recent interaction with each partner. Assuming that each individual engages in social exchange with a large number of partners over longer periods of time, it seems questionable whether such a model is compatible with our knowledge about the limitations of the cognitive systems (cf. Stevens, Volstorf, Schooler, & Rieskamp, 2010). Given that *tit for tat* relies on memory for the single last interaction in a series of similar interactions with the same individual, the strategy is highly susceptible to source memory errors. Volstorf et al. (2011) argue that a solution to this problem might be to use strategies that do not rely on the most recent behavior but instead rely on a categorization of the partner that represents a judgment of the sum of previous interactions. Such a categorization strategy would be consistent with our results showing that people fail to remember the specific aspects of unique behavior episodes, but instead use emotional tags to classify other people's behavior (Bell et al., 2012). Thus, research addressing the limitations of human social cognition can inform models of reciprocal altruism.

Our conclusion with respect to the field of adaptive memory as a whole is that it is probably unwise to rely too much on the concept of domain-specific modules to explain novel findings without first considering our rich knowledge about the more general mechanisms that determine memory in a wide variety of situations. Specifically, by focusing too much on fine-grained modularity, one may very easily miss the point that a general mechanism is capable of solving not only a single problem, but very large classes of evolutionary problems simultaneously in an extremely parsimonious way. A mind equipped with such general memory mechanisms is thus much more adaptive than what we have been made to believe by the modularist variety of Evolutionary Psychology.

REFERENCES

Aktipis, C. A. (2006). Recognition memory and the evolution of cooperation: How simple strategies succeed in an agent-based world. *Adaptive Behavior, 14*, 239–247.

Axelrod, R., & Hamilton, W. D. (1981). The evolution of cooperation. *Science, 211*, 1390–1396.

Barclay, P. (2008). Enhanced recognition of defectors depends on their rarity. *Cognition, 107*, 817–828.

Barclay, P., & Lalumière, M. L. (2006). Do people differentially remember cheaters? *Human Nature, 17*, 98–113.

Baumeister, R. F., Bratslavsky, E., Finkenauer, C., & Vohs, K. D. (2001). Bad is stronger than good. *Review of General Psychology, 5*, 323–370.

Bayen, U. J., Murnane, K., & Erdfelder, E. (1996). Source discrimination, item detection, and multinomial models of source monitoring. *Journal of Experimental Psychology: Learning, Memory, and Cognition, 22*, 197–215.

Bayen, U. J., Nakamura, G. V., Dupuis, S. E., & Yang, C. L. (2000). The use of schematic knowledge about sources in source monitoring. *Memory & Cognition, 28*, 480–500.

Bell, R., & Buchner, A. (2010a). Justice sensitivity and source memory for cheaters. *Journal of Research in Personality, 44*, 677–683.

Bell, R., & Buchner, A. (2010b). Valence modulates source memory for faces. *Memory & Cognition, 38*, 29–41.

Bell, R., & Buchner, A. (2011). Source memory for faces is determined by their emotional evaluation. *Emotion, 11*, 249–261.

Bell, R., Buchner, A., Erdfelder, E., Giang, T., Schain, C., & Riether, N. (2012). How specific is source memory for faces of cheaters? Evidence for categorical emotional tagging. *Journal of Experimental Psychology: Learning, Memory, and Cognition, 38*, 457–472.

Bell, R., Buchner, A., Kroneisen, M., & Giang, T. (2012). On the flexibility of social source memory: A test of the emotional incongruity hypothesis. *Journal of Experimental Psychology: Learning, Memory, and Cognition, 38*, 1512–1529.

Bell, R., Buchner, A., & Musch, J. (2010). Enhanced old-new recognition and source memory for faces of cooperators and defectors in a social-dilemma game. *Cognition, 117*, 261–275.

Bell, R., Giang, T., & Buchner, A. (2012). Partial and specific source memory for faces associated to other- and self-relevant negative contexts. *Cognition and Emotion, 36*, 1036–1055.

Bornstein, R. F. (1989). Exposure and affect: Overview and meta-analysis of research, 1968-1987. *Psychological Bulletin, 106*, 265–289.

Buchner, A., Bell, R., Mehl, B., & Musch, J. (2009). No enhanced recognition memory, but better source memory for faces of cheaters. *Evolution and Human Behavior, 30*, 212–224.

Buller, D. J. (2005). Evolutionary psychology: The emperor's new paradigm. *Trends in Cognitive Sciences, 9*, 277–283.

Carlisle, E., & Shafir, E. (2005). Questioning the cheater-detection hypothesis: New studies with the selection task. *Thinking and Reasoning, 11*, 97–122.

Carlston, D. E., & Skowronski, J. J. (1994). Savings in the relearning of trait information as evidence for spontaneous inference generation. *Journal of Personality and Social Psychology, 66*, 840–856.

Carlston, D. E., Skowronski, J. J., & Sparks, C. (1995). Savings in relearning: II. On the formation of behavior-based trait associations and inferences. *Journal of Personality and Social Psychology, 69*, 429–436.

Cartwright, J. (2008). *Evolution and human behavior: Darwinian perspectives on human nature*. Cambridge, MA: MIT Press.

Chang, L. J., & Sanfey, A. G. (2009). Unforgettable ultimatums? Expectation violations promote enhanced social memory following economic bargaining. *Frontiers in Behavioral Neuroscience, 3*, 36, 1–12.

Chiappe, D., Brown, A., Dow, B., Koontz, J., Rodriguez, M., & McCulloch, K. (2004). Cheaters are looked at longer and remembered better than cooperators in social exchange situations. *Evolutionary Psychology, 2*, 108–120.

Cook, G. I., Marsh, R. L., & Hicks, J. L. (2003). Halo and devil effects demonstrate valence-based influences on source-monitoring decisions. *Consciousness and Cognition: An International Journal, 12*, 257–278.

Cosmides, L., & Tooby, J. (1989). Evolutionary psychology and the generation of culture: II. Case study: A computational theory of social exchange. *Ethology and Sociobiology, 10*, 51–97.

Cosmides, L., & Tooby, J. (1992). Cognitive adaptations for social exchange. In J. H. Barkow, L. Cosmides, & J. Tooby (Eds.), *The adapted mind: Evolutionary psychology and the generation of culture* (pp. 163–228). New York: Oxford University Press.

Cosmides, L., & Tooby, J. (1997). The multimodular nature of human intelligence. In A. Schiebel & J. W. Schopf (Eds.), *Origin and evolution of intelligence* (pp. 71–101). Center for the Study of the Evolution and Origin of Life, UCLA.

Cosmides, L., & Tooby, J. (2005). Neurocognitive adaptations designed for social exchange. In D. M. Buss (Ed.), *The handbook of evolutionary psychology* (pp. 584–627). Hoboken, NJ: John Wiley & Sons Inc.

Darwin, C. (1871). *The descent of man, and Selection in relation to sex*. London, England: John Murray.

Dodson, C. S., Holland, P. W., & Shimamura, A. P. (1998). On the recollection of specific- and partial-source information. *Journal of Experimental Psychology: Learning, Memory, and Cognition, 24*, 1121–1136.

Dunbar, R. I. M. (2003). The social brain: Mind, language, and society in evolutionary perspective. *Annual Review of Anthropology, 32*, 163–181.

Flykt, A., Esteves, F., & Öhman, A. (2007). Skin conductance responses to masked conditioned stimuli: Phylogenetic/ontogenetic factors versus direction of threat? *Biological Psychology, 74*, 328–336.

Fodor, J. (2000). Why we are so good at catching cheaters. *Cognition, 75*, 29–32.

Garcia, J., & Koelling, R. A. (1966). Relation of cue to consequence in avoidance learning. *Psychonomic Science, 4*, 123–124.

Geary, D. C. (2007). An integrative model of human brain, cognitive, and behavioral evolution. *Acta Psychologica Sinica, 39*, 383–397.

Gould, S. J., & Lewontin, R. C. (1979). The spandrels of San Marco and the Panglossian paradigm: a critique of the adaptationist programme. *Proceedings of the Royal Society of London Series B, 205*, 581–598.

Hugdahl, K., & Johnsen, B. H. (1989). Preparedness and electrodermal fear-conditioning: Ontogenetic vs phylogenetic explanations. *Behaviour Research and Therapy, 27*, 269–278.

Johnson, M. K., Kim, J. K., & Risse, G. (1985). Do alcoholic Korsakoff's syndrome patients acquire affective reactions? *Journal of Experimental Psychology: Learning, Memory, and Cognition, 11*, 22–36.

Kensinger, E. A. (2007). Negative emotion enhances memory accuracy: Behavioral and neuroimaging evidence. *Current Directions in Psychological Science, 16*, 213–218.

Kensinger, E. A., & Corkin, S. (2003). Memory enhancement for emotional words: Are emotional words more vividly remembered than neutral words? *Memory & Cognition, 31*, 1169–1180.

Kensinger, E. A., & Schacter, D. L. (2006). Reality monitoring and memory distortion: Effects of negative, arousing content. *Memory & Cognition, 34*, 251–260.

Klauer, K. C., & Wegener, I. (1998). Unraveling social categorization in the "who said what?" paradigm. *Journal of Personality and Social Psychology, 75*, 1155–1178.

Kroneisen, M., & Bell, R. (2013). Sex, cheating, and disgust: Enhanced source memory for trait information that violates gender stereotypes. *Memory, 21*,167–181.

Kroneisen, M., & Erdfelder, E. (2011). On the plasticity of the survival processing effect. *Journal of Experimental Psychology: Learning, Memory, and Cognition, 37*, 1553–1562.

Mealey, L., Daood, C., & Krage, M. (1996). Enhanced memory for faces of cheaters. *Ethology and Sociobiology, 17*, 119–128.

Mehl, B., & Buchner, A. (2008). No enhanced memory for faces of cheaters. *Evolution and Human Behavior, 29*, 35–41.

Nairne, J. S. (2010). Adaptive memory: Evolutionary constraints on remembering. In B. H. Ross (Ed.), *The psychology of learning and motivation: Advances in research and theory (Vol 53)* (pp. 1–32). San Diego, CA: Elsevier Academic Press.

Nairne, J. S., & Pandeirada, J. N. S. (2010). Adaptive memory: Nature's criterion and the functionalist agenda. *American Journal of Psychology, 123*, 381–390.

Nairne, J. S., Pandeirada, J. N. S., & Thompson, S. R. (2008). Adaptive memory: The comparative value of survival processing. *Psychological Science, 19*, 176–180.

Oda, R. (1997). Biased face recognition in the prisoner's dilemma game. *Evolution and Human Behavior, 18*, 309–315.

Öhman, A., Fredrikson, M., Hugdahl, K., & Rimmö, P.-A. (1976). The premise of equipotentiality in human classical conditioning: Conditioned electrodermal responses to potentially phobic stimuli. *Journal of Experimental Psychology: General, 105*, 313–337.

Öhman, A., & Mineka, S. (2001). Fears, phobias, and preparedness: Toward an evolved module of fear and fear learning. *Psychological Review, 108*, 483–522.

Öhman, A., & Mineka, S. (2003). The malicious serpent: Snakes as a prototypical stimulus for an evolved module of fear. *Current Directions in Psychological Science, 12*, 5–9.

Panksepp, J., & Panksepp, J. B. (2000). Seven sins of evolutionary psychology. *Evolution and Cognition, 6*, 108–131.

Peeters, G., & Czapinski, J. (1990). Positive-negative asymmetry in evaluations: The distinction between affective and informational negativity effects. In W. Stroebe & M. Hewstone (Eds.), *European Review of Social Psychology. Vol. 1* (pp. 33–60). Chichester, UK: John Wiley & Sons.

Pratto, F., & John, O. P. (1991). Automatic vigilance: The attention-grabbing power of negative social information. *Journal of Personality and Social Psychology, 61*, 380–391.

Ranganath, C., & Rainer, G. (2003). Neural mechanisms for detecting and remembering novel events. *Nature Reviews Neuroscience, 4*, 193–202.

Rothbart, M., & Park, B. (1986). On the confirmability and disconfirmability of trait concepts. *Journal of Personality and Social Psychology, 50*, 131–142.

Rothermund, K., Voss, A., & Wentura, D. (2008). Counter-regulation in affective attentional biases: A basic mechanism that warrants flexibility in emotion and motivation. *Emotion, 8*, 34–46.

Schmitt, M., Gollwitzer, M., Maes, J., & Arbach, D. (2005). Justice sensitivity: Assessment and location in the personality space. *European Journal of Psychological Assessment, 21*, 202–211.

Singer, T., Kiebel, S. J., Winston, J. S., Dolan, R. J., & Frith, C. D. (2004). Brain responses to the acquired moral status of faces. *Neuron, 41*, 653–662.

Skowronski, J. J., & Carlston, D. E. (1989). Negativity and extremity biases in impression formation: A review of explanations. *Psychological Bulletin, 105*, 131–142.

Sperber, D., Cara, F., & Girotto, V. (1995). Relevance theory explains the selection task. *Cognition, 57*, 31–95.

Stevens, J. R., Volstorf, J., Schooler, L. J., & Rieskamp, J. (2010). Forgetting constrains the emergence of cooperative decision strategies. *Frontiers in Psychology, 1*, 235.

Suzuki, A., & Suga, S. (2010). Enhanced memory for the wolf in sheep's clothing: Facial trustworthiness modulates face-trait associative memory. *Cognition, 117*, 224–229.

Tausch, N., Kenworthy, J. B., & Hewstone, M. (2007). The confirmability and disconfirmability of trait concepts revisited: Does content matter? *Journal of Personality and Social Psychology, 92*, 542–556.

Todorov, A., & Olson, I. R. (2008). Robust learning of affective trait associations with faces when the hippocampus is damaged, but not when the amygdala and temporal pole are damaged. *Social Cognitive and Affective Neuroscience, 3*, 195–203.

Todorov, A., & Uleman, J. S. (2002). Spontaneous trait inferences are bound to actors' faces: Evidence from a false recognition paradigm. *Journal of Personality and Social Psychology, 83*, 1051–1065.

Todorov, A., & Uleman, J. S. (2004). The person reference process in spontaneous trait inferences. *Journal of Personality and Social Psychology, 87*, 482–493.

Tooby, J., & Cosmides, L. (1992). The psychological foundations of culture. In J. H. Barkow, L. Cosmides, & J. Tooby (Eds.), *The adapted mind: Evolutionary psychology and the generation of culture* (pp. 19–136). New York: Oxford University Press.

Tooby, J., & Cosmides, L. (2005). Conceptual foundations of evolutionary psychology. In D. M. Buss (Ed.), *The handbook of evolutionary psychology* (pp. 5–67). Hoboken, NJ: John Wiley & Sons Inc.

Trivers, R. L. (1971). The evolution of reciprocal altruism. *Quarterly Review of Biology, 46*, 35–57.

van Kesteren, M. T. R., Ruiter, D. J., Fernandez, G., & Henson, R. N. (2012). How schema and novelty augment memory formation. *Trends in Neurosciences*, *35*, 211–219.

Volstorf, J., Rieskamp, J., & Stevens, J. R. (2011). The good, the bad, and the rare: Memory for partners in social interactions. *PLoS ONE*, *6*, e18945.

Wason, P. C. (1968). Reasoning about a rule. *Quarterly Journal of Experimental Psychology*, *20*, 273–281.

Zajonc, R. B. (1968). Attitudinal effects of mere exposure. *Journal of Personality and Social Psychology*, *9*, 1–27.

4

Living, Dying, and Remembering: The Encoding Processes Involved in Survival Processing

DANIEL J. BURNS AND JOSHUA HART

Species evolve by accumulating adaptations to problems faced in an ever-changing environment. To the extent that those adaptations successfully solve the problems, members of a particular species are better able to survive and reproduce, so the adaptations eventually become characteristic of the species. Adaptations are often physical (e.g., skin color) but may also be cognitive (e.g., mate selection, cheater detection; e.g., Cosmides & Tooby, 1989, 1992). Although a matter of some debate, several putatively cognitive adaptations, such as self-awareness and future planning, seem to be relatively unique to human beings (e.g., Hampton, 2009; van Horik, Clayton, & Emery, 2012). The ability to remember past events in the environmental context in which they occurred—episodic memory—is another human skill that likely evolved as a consequence of evolutionary pressures (e.g., Tulving, 2002). A primary function of episodic memory, it sometimes has been argued, is to allow for mental time travel and future planning (e.g., Klein, Robertson, & Delton, 2010; Tulving, 2002). The remembered details of prior events can be used to guide planning for similar or related future events, thereby giving humans a selective advantage by allowing them to, among other benefits, avoid life-threatening situations (e.g., Klein, Robertson, & Delton, 2010).

It seems likely that an episodic memory system molded by evolution would not function arbitrarily, but rather would afford an advantage to events that contain information that might be used to forestall death in the future. Specifically, an episodic memory system that favors survival-relevant information might be particularly beneficial. Nairne, Thompson, and Pandeirada (2007; see also Nairne, Pandeirada, & Thompson, 2008) have tested this hypothesis extensively by requiring participants to imagine themselves stranded in the grasslands of a foreign land without any supplies and to rate items for their survival relevance in the imagined situation. Following the rating task and a short distractor-filled delay period, participants were given a 10-minute free recall test. Results showed that processing the items for their survival value (i.e., *survival processing*) produced retention superior to that found with numerous other processing conditions, many of which are known to produce particularly good memory. For example, the survival orienting task was superior to (1) a pleasantness rating task; (2) a nonsurvival moving task in which participants imagined moving to a foreign land and then rated items for their relevance to the move; (3) a self-reference task in which items were rated for their ability to bring to mind personal events (but see Klein, 2012); and (4) even an intentional learning task. These findings led Nairne and colleagues to suggest that natural selection may have sculpted memory structures to better retain information presented in survival-relevant contexts. They speculated that processing items for their survival value may activate survival-related modules (e.g., a predator-retention system) that enhance retention (Nairne & Pandeirada, 2008a).

Whereas some variant of Nairne and colleagues' conjecture about modules may prove to be the ultimate explanation for the memory enhancement associated with survival processing, little is known about these hypothetical modules. Moreover, other possible evolutionary explanations have been offered. For example, Klein et al. (2010; and Chapter 2 in this volume) suggested that survival processing may invoke more future planning than other processing techniques. Because episodic memory may well have evolved as a mechanism for future planning, they argued that it should be most efficient when used for its original purpose. Hence, Klein et al. proposed that it may be the planning aspect of survival processing, and not survival processing *per se*, that leads to superior retention.

Another possibility, noted by Soderstrom and McCabe (2011), is that the survival processing scenario activates a death and disgust system. Presumably this system evolved to protect humans from poisonous foods (e.g., rotting flesh) and disease (e.g., Oaten, Stevenson, & Case, 2009). Perhaps retention is enhanced when the death and disgust system is activated, and any scenario that instills thoughts of death and disgust will produce the retention advantage. In other words, any scenario involving death or disgust, regardless of its explicit survival content, might enhance retention. Of course, other, less evolutionarily based explanations are also plausible. For example, survival processing may simply increase arousal or stress levels just as some nonsurvival situations do, and the higher level of arousal or stress is responsible for the enhanced retention (but see Kang, McDermott, & Cohen, 2008; Smeets, Otgaar, Raymaekers, Peters, & Merckelbach, 2012).

In this chapter, rather than focus on ultimate explanations of the survival processing retention advantage, we concentrate on a possible proximate cause of the effect. We suggest that the survival-orienting task tends to encourage the encoding of both item-specific and relational properties of list items, whereas control conditions, including those involving nonsurvival scenarios, have tended to focus on either item-specific or relational properties. We also discuss recent research showing that simply priming individuals with mortality-salient themes enhances memory for subsequently processed material. We then discuss whether this newly discovered effect is related to survival processing, and if so, whether that insight might shed light on the origins of the survival processing advantage, as well as its mediating mechanisms. We conclude that the two phenomena may be related and that studying the effects of mortality salience on retention may help to illuminate nature's role in shaping memory.

SURVIVAL PROCESSING: PROXIMATE MECHANISMS

It is clear that survival processing is a particularly powerful mnemonic device. It is so powerful, in fact, that Nairne et al. (2008) concluded that "survival processing is one of the best—if not the best—encoding procedures yet identified in human memory research" (p. 180). Consequently, even if current evolutionary explanations are eventually discarded, an understanding of the proximate causes will still be important. Several potential explanations have been explored. For example, researchers have suggested that the survival scenario might promote more thematic processing than nonscenario control tasks, or even control tasks involving other nonsurvival scenarios. However, when nonsurvival scenarios that were designed to be equivalent in thematic processing were used, the survival processing advantage remained (Weinstein, Bugg, & Roediger, 2008). Another possibility is that the recent increase in survival-related television series and movies (e.g., "Survivor," "Lost," "Walking Dead," and "Cast Away") have made the survival scenario more familiar than the control scenarios. To test this hypothesis, Kang, McDermott, and Cohen (2008) used a bank robbery scenario that was rated as equivalently novel and exciting as the survival scenario, as well as being encountered slightly more often in the media (e.g., in movies such as Inside Man, The Town, and Dog Day Afternoon). Despite these controls, the survival scenario still produced higher recall and recognition scores than the robbery scenario.

One interesting possibility that has received some support is that processing words for their survival value promotes the processing of both the item-specific (i.e., individual characteristics of the items) and relational (i.e., relations between list items) properties of items, whereas control tasks have tended to promote only item-specific or only relational processing. Each type of processing enhances long-term retention (Einstein & Hunt, 1980; Hunt & Einstein, 1981; Hunt & McDaniel, 1993), albeit in different ways. Relational processing presumably facilitates retention by providing an organized plan for efficient retrieval of the items, whereas item-specific processing presumably facilitates discrimination of individual items on the list from other items, as well as providing alternative retrieval routes for individual items (e.g., Burns et al., 2006; Burns & Gold, 1999; Hunt & McDaniel, 1993). Einstein and Hunt (1980; Hunt & Einstein, 1981) have demonstrated repeatedly that the combination of item-specific and relational processing is particularly beneficial to recall.

Nairne and Pandeirada (2008b) tested whether the survival processing advantage found in recall resulted from the survival task inducing both item-specific and relational processing. They compared survival processing to a pleasantness rating task known to induce item-specific processing. In addition, they used a categorically related list of words, ensuring that all participants would encode relational information. This procedure of requiring pleasantness rating of categorically related items is the technique of choice for recruiting both item-specific and relational processing, and it results in recall performance superior to that of conditions promoting only item-specific or only relational processing (e.g., Burns, 1993; Burns & Schoff, 1998; Einstein & Hunt, 1981; Hunt & Einstein, 1981; Klein, Loftus, Kihlstrom, & Aseron, 1989). In fact, as Nairne and Pandeirada noted, pleasantness rating of categorically related words is the "gold standard for enhancing free recall" (p. 377). Still, the results revealed a recall advantage for the survival processing group, suggesting that survival processing produces recall performance even above that produced by the combined processing of item-specific and relational information.

However, Burns, Burns, and Hwang (2011) noted that the survival processing advantage obtained by Nairne and Pandeirada (2008b) was somewhat smaller than typically observed. Burns et al. also argued that the amount of relational processing in the survival and pleasantness rating groups may not have been fully equated. Nairne and Pandeirada used list items selected from four categories, all of which were quite relevant to survival (fruits, vegetables, four-legged animals, and human dwellings), which may have resulted in more relational (or more congruous) processing for the survival group than for the pleasantness group. In four different experiments, Burns et al. compared survival processing to encoding conditions presumed to encourage only item-specific processing, only relational processing, or both types of processing. An important feature of the Burns et al. study was that it used several different indices of item-specific and relational processing to verify that the survival task enhanced the processing of both types of information. Because these measures of item-specific and relational processing are central to the remainder of this chapter, we will describe several of them in detail next before describing the results of the Burns et al. study.

Assessing Item-Specific and Relational Processing

Perhaps the most preferred measures of relational processing are category clustering indices, which measure the extent to which items from the same categories are written in adjacent output positions during free recall. Many of the clustering measures, including the popular adjusted ratio of clustering (ARC) score (Roenker, Thompson, & Brown, 1971), take into account the amount of clustering one would expect by chance if recall output were completely random. Category clustering measures are believed to be a good index of the amount of relational processing performed when categorically related lists are used (see Burns, 2006). One shortcoming, however, is that they only work with lists for which there is a known categorical structure.

Two additional measures of relational processing can be used for unrelated as well as related lists as long as a repeated testing procedure is used in which recall is attempted more than once. These measures are subjective organization, which will not be discussed here, and item losses. An item loss occurs when a previously recalled item is forgotten on a later recall test. Burns (1993) showed that conditions that increase relational processing result in fewer item losses. Presumably relational processing protects against losses by providing a stable retrieval plan that participants use across repeated tests (see also Hunt & McDaniel, 1993; McDaniel, Moore, & Whiteman, 1998). Because they use the same relational retrieval cues to retrieve the items on each test, the same items tend to be recalled, producing few losses. The item loss measure has been used with considerable success (e.g., Engelkamp & Seiler, 2003; Klein, Loftus, & Schell, 1994; Mulligan, 2000, 2001, 2002).

Differences in item-specific processing are generally assessed with an item recognition test. Hunt and Einstein (1981) showed that conditions differing only in the amount of relational processing performed produced no recognition differences, whereas conditions differing only in the amount of item-specific processing showed large recognition differences. It is known, however, that relational processing can, under some circumstances, improve recognition, so it is helpful to use other measures as well (see Burns, 2006). One measure that can be used under conditions of repeated testing is item gains, a measure of the number of new items recalled on a later test that were not recalled earlier. Burns (1993) showed that manipulations resulting in an increase in item-specific processing tended to increase item gains, whereas relational processing manipulations did not (see also, for example, Engelkamp & Seiler, 2003; Klein et al., 1994; Mulligan, 2001, 2001, 2002). Presumably, the numerous item-specific retrieval cues typically are not exhausted by the end of the first recall period and continue to facilitate the recall of new items on later tests.

Finally, Burns and Schoff (1998) argued that recording the number of items recalled during each minute of a long recall period and then plotting the resulting cumulative-recall curves can be useful for detecting both item-specific and relational processing differences. They showed that groups encoding primarily relational processing produced high levels of recall during the first few minutes of the recall period but reached asymptotic levels quite quickly. Groups performing primarily item-specific processing approached asymptote much more slowly. Finally, groups performing both types of processing produced high initial levels of recall but also tended to recall new items throughout the recall period.

Typical cumulative-recall curves are shown in Figure 4.1, which is adapted from Burns and Schoff (1998). The top panel compares a group performing relational processing with one performing item-specific processing. The curve produced following relational processing is initially steeper but reaches asymptote more quickly than the group performing item-specific processing. The middle panel shows that a group performing both item-specific and relational processing produced initial recall levels similar to a group performing only relational processing. However, after the first minute or two, the former group began to recall more words per minute than the latter group. Conversely, when a group performing both types of processing is compared to one performing only item-specific processing (bottom panel), the groups diverged in the early portion of the recall period but then produced parallel curves throughout the remainder of the period. Hence, close scrutiny of the cumulative-recall curves can provide evidence about the type of processing (see also Burns, Jenkins, & Dean, 2007; Burns, Martens, Sweeney, Bertoni, & Lividini, 2006).

Item-Specific and Relational Indices Applied to Survival Processing

As noted earlier, Burns et al. (2011) tested whether survival processing would produce recall superior to conditions that encode both item-specific and relational processing. They also used several measures of item-specific and relational processing in their experiments to determine whether survival processing enhances item-specific or relational processing relative to control tasks. In the first two experiments, a list of categorically related items was used and in the remaining two experiments the list consisted of seemingly unrelated items from *ad hoc* categories (e.g., things that are green, things made of wood). Previous research has shown that categorically related list items inherently foster relational processing, whereas unrelated lists foster item-specific processing (Hunt & Einstein, 1981).

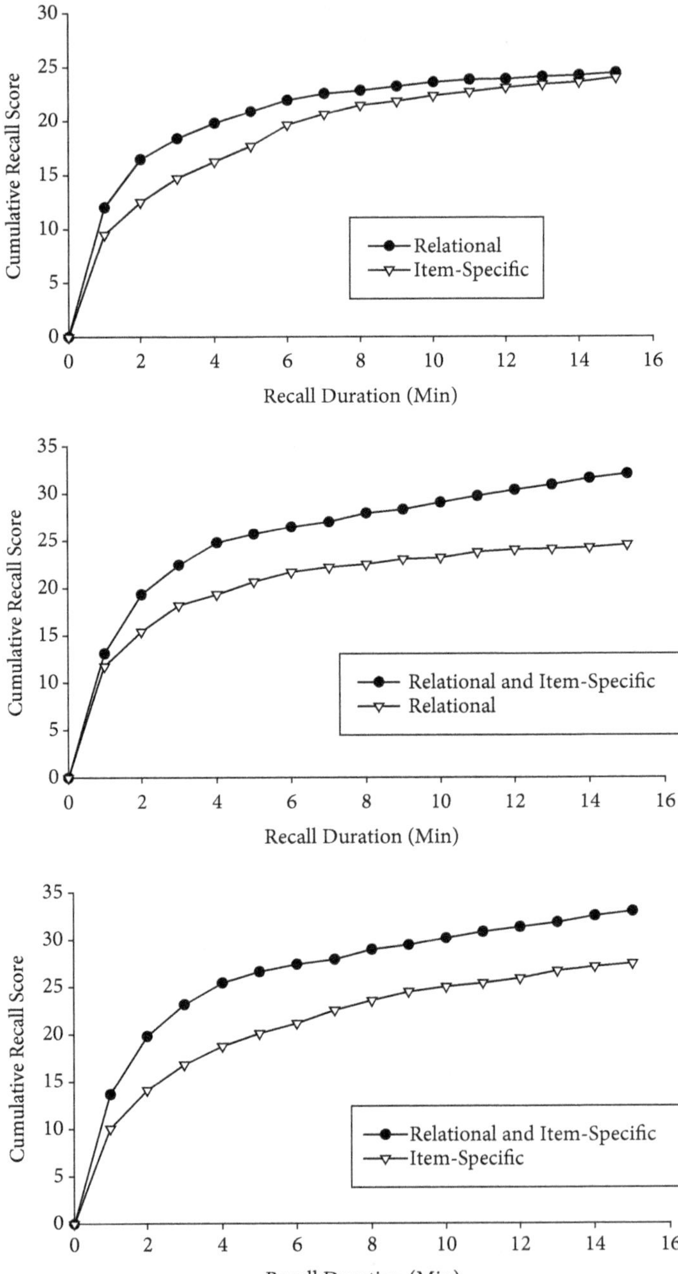

Figure 4.1 Mean cumulative-recall scores for various conditions tested in Burns and Schoff's (1998) experiments. Please see text for a full description.

In the first two experiments using the related list of words, survival processing was contrasted with a condition presumed to perform only relational processing (category sorting) and a condition presumably performing both types of processing (pleasantness rating). The results showed a recall advantage for survival processing over category sorting, but not over pleasantness rating. The various measures of item-specific processing used across the two experiments supported the authors' contention that both the pleasantness rating and survival processing conditions performed more item-specific processing than the category sorting group. First, final recognition scores were higher for the survival processing and pleasantness rating conditions than for the category sorting condition. Additionally, a minute-by-minute cumulative-recall analysis was conducted for one of the first two experiments. The curves, which are reproduced in the top panel of Figure 4.2, show that the survival

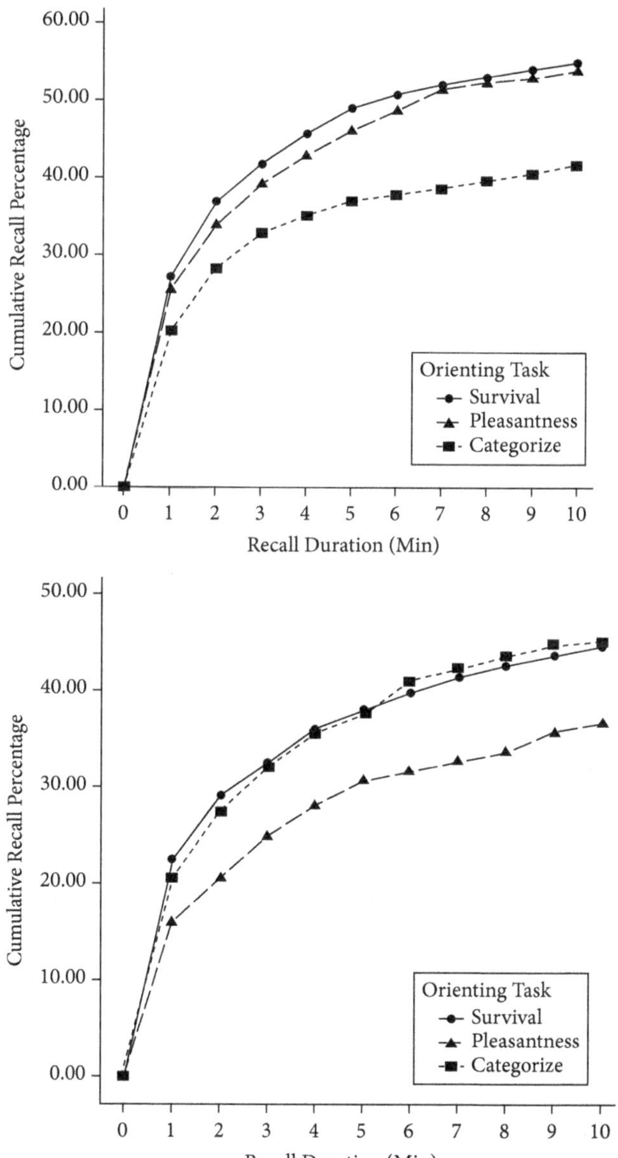

Figure 4.2 Mean cumulative-recall scores for the conditions tested in Burns, Burns, and Hwang's (2011) experiments. Top panel = Experiment 2, bottom panel = Experiment 4.

and pleasantness groups produced nearly identical cumulative recall throughout the recall period, and both groups produced curves that tended to diverge from the category sorting group more and more as the recall period continued. These results are similar to the pattern shown in the middle panel of Figure 4.1 and suggest that both the survival processing and pleasantness rating groups performed more item-specific processing than the category sorting group. Moreover, because a related list of words was used, ARC scores were calculated to assess relational processing in Experiment 2 (Experiment 1 employed a mixed-list design, so ARC scores could not be calculated). The ARC scores showed that all three groups performed high levels of clustering, suggesting that all three performed a great deal of relational processing. Additionally, the survival group actually clustered the items significantly more than the category sorting group, with neither of these groups differing from the pleasantness group.

In Experiments 3 and 4, in which a list of words from *ad hoc* categories was used, Burns et al. (2011) compared survival processing to an encoding condition promoting only item-specific processing (pleasantness rating) and to an encoding condition promoting both item-specific and relational processing (category sorting). The results showed the typical survival processing effect, with the survival task producing higher recall than the pleasantness rating task. However, the category sorting task resulted in recall scores as high as survival processing, suggesting that survival processing is not superior to conditions promoting both item-specific and relational processing. The recognition results showed that the category-sorting task produced only slightly lower performance than the other two groups in Experiment 3, with no difference between the three conditions found in Experiment 4. The recognition results suggested that the use of items from *ad hoc* categories minimized, if not eliminated, item-specific processing differences between conditions. The cumulative-recall curves, shown in the bottom panel of Figure 4.2, also suggested that item-specific processing was similar among conditions, as well as that the pleasantness rating group performed less relational processing than the other two groups, with the bulk of the recall difference occurring during the first two minutes of the recall period.

Burns et al.'s (2011) experiments provide evidence that the survival task enhances processing of both item-specific and relational information. They argued that many of the control conditions tested in previous studies resulted in the processing of only one type of information, affording survival processing a retention advantage. They proposed a two-process account of the survival processing effect, suggesting that the survival scenario recruits both item-specific and relational processing, whereas control conditions have usually recruited only one type of processing or the other. When control conditions foster both types of processing, the survival advantage is eliminated.

One problem with this explanation is that whereas Burns et al. (2011) found that pleasantness rating of a list of categorically related items produced recall levels equivalent to the survival task, Otgaar and Smeets (2010) used categorically related list items and found a recall advantage for survival processing over pleasantness rating (for similar results in recognition, see Howe & Derbish, 2010, who used semantically related lists). Otgaar and Smeets' results present a challenge for the two-process account. However, we believe that the discrepancy in results may be due to one or more of the several procedural differences between the two studies. These include the fact that Otgaar and Smeets blocked items by category, whereas Burns et al. intermixed items, and that Otgaar and Smeets used much larger category sizes (10 items from each of six categories) than Burns et al. (four items from 12 categories). As Burns, Hart, Griffith, and Burns (2013) note, it seems likely that these differences in presentation procedures caused greater relational processing by subjects in Otgaar and Smeets' study. For example, increases in category size increase relational processing regardless of orienting task (Engelkamp, Biegelmann, & McDaniel, 1998). It is possible, therefore, that in Otgaar and Smeets' experiment the extensive relational information available in the encoding context was used to a greater extent during survival processing than during pleasantness rating. In other words, when abundant relational information is available, the survival task may induce more relational processing than pleasantness rating. This conclusion is consistent with Otgaar and Smeets' conclusion that the survival task resulted in more gist processing in their experiments (and with Howe and Derbish's contention that survival processing enhanced associative processing).

Another problem facing Burns et al.'s (2011) two-process account concerns its explanation for survival processing effects with control groups involving other scenarios (e.g., moving, burglary, or vacation scenarios). The two-process account seems particularly useful for explaining the results of experiments that used non-scenario control conditions (e.g., pleasantness rating, imagery rating, generation), because research suggests these conditions enhance only item-specific processing (e.g., Burns, Curti, & Lavin, 1993; Einstein & Hunt, 1980; Hodge & Otani, 1996). It is far less obvious why the survival processing scenario would produce more item-specific or more relational processing than control conditions involving other scenarios. Why, for example, would one expect survival processing to produce more relational processing than other scenarios that were created to be equivalent in terms of schematic structure? It is also difficult to imagine why the survival scenario would promote more item-specific processing than other scenarios. Yet, according to the two-process account, the survival scenario would be expected to promote either more relational processing or more item-specific processing than other scenarios.

A study by Kroneisen and Erdfelder (2011) provides some illumination by suggesting that the survival scenario may promote encoding of more item-specific retrieval cues than nonsurvival scenarios because it asks participants to consider the relevance of items to their finding food and water *and* avoiding predators. They found that when the scenario was altered such that finding water was the only task, survival was not superior to a moving scenario. They argued that the simpler task reduced the number of item-specific retrieval cues encoded, thereby decreasing the distinctiveness of the list items. However, Kroneisen and Erdfelder did not directly assess the amount of item-specific processing performed by subjects in the various conditions of their experiments.

Burns et al. (2013) used several indices of relational and item-specific processing to directly test whether the survival scenario induces more item-specific processing than the moving scenario. They used the identical survival and moving scenarios used by Nairne et al. (2007) and others, as well as nearly identical procedures. However, they also recorded the number of items recalled during each minute of the 10-minute recall period, which allowed them to assess cumulative-recall

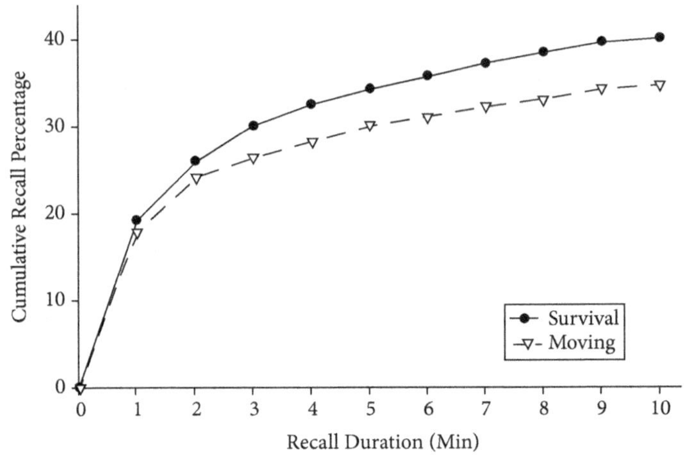

Figure 4.3 Mean cumulative-recall scores for the conditions tested in Burns et al.'s (in press) study.

performance. Additionally, they gave a recognition test following free recall. The recall results revealed the typical survival processing recall advantage. Survival processing also produced significantly higher recognition scores, suggesting that it promoted more item-specific processing than the moving scenario. Moreover, as seen in Figure 4.3, there was very little difference in the cumulative-recall curves until about the 3-minute mark, when the survival group began to recall slightly more items per minute than the moving group. Hence, the recognition and cumulative-recall scores suggest that the survival task promoted more item-specific processing than the moving task, but not more relational processing.

These findings suggest that a two-process account provides a viable explanation for the survival processing advantage relative to other scenarios; it promotes greater item-specific processing. Burns et al. replicated this pattern of results in a second experiment that manipulated the type of orienting task (survival or moving) within subjects, thus demonstrating that the two-process account also can explain the numerous survival processing effects obtained in mixed-list designs (e.g., Nairne et al., 2007).

The Self-Relevance of Survival Processing

Why would the survival scenario result in the encoding of both item-specific and relational information, whereas many control conditions only promote one type of processing? Item-specific and relational processing differences have been hypothesized to be the cause of several other list-learning phenomena, including the generation effect (e.g., Burns, 1992), the enactment effect (Engelkamp & Seiler, 2003), and the bizarreness effect (e.g., Worthen & Loveland, 2003). However, in all of these cases, a processing tradeoff was proposed in which the experimental condition was hypothesized to increase one type of processing at the expense of the other type of processing. The survival task seems to encourage both types of processing. To our knowledge, the only other phenomenon that appears to encourage both types of processing is the self-reference effect (Rogers, Kuiper, & Kirker, 1977).

In an experiment using similar procedures to those of Burns, Burns, and Hwang (2011), Klein and Loftus (1988) compared self-referential processing to a task encouraging only item-specific processing or a task encouraging only relational processing. Half of the participants in each group received a categorically related list of words, which promoted relational processing, and the other half received unrelated words, which resulted in item-specific processing. Klein and Loftus found that the self-reference task produced retention superior to conditions processing only item-specific or only relational information, but equivalent to conditions performing both types of processing. The similarity in the pattern of results found by Burns et al. and Klein and Loftus tentatively suggests that survival processing and self-referential processing might be related. In fact, because both tasks promote item-specific and relational processing, the two-process account would predict that self-referential processing would produce recall levels equivalent to that of survival processing. Nairne et al. (2007) compared survival processing to a self-referential task and found that survival processing produced superior recall (see also Nairne et al., 2008). However, the specific self-reference task used (rating the ease with which each word brings to mind personal experiences) may not be the most powerful procedure for bringing about self-referential processing (see Klein, 2012).

Further, as Burns et al. (2011) noted, the studies comparing survival and self-referential processing either failed to obtain the typical self-reference effect or else did not include control groups to test whether the self-reference effect was present. Klein (2012) compared survival processing to the self-reference task used by Nairne and colleagues as well as a similar but more powerful task that required participants to think of a personal experience related to each word prior to rating the ease with which the experience was brought to mind. Survival processing produced higher recall than the self-reference task used by Nairne et al. (2007), but not higher than the more effective self-reference task. On this basis, Klein argued that self-relevant processing may be partly responsible for the survival processing effect.

In sum, it seems that the survival scenario promotes a considerable amount of deep-level semantic processing that is both item-specific and relational and that may be self-reflective or self-relevant. But if these proximate mechanisms mediate the survival processing advantage, how is it that thoughts of survival might have engaged these mechanisms in an evolutionary context? We now turn to this question.

MORTALITY SALIENCE AND SURVIVAL PROCESSING

Of course, there are many possible reasons why thoughts of survival came to uniquely augment both item-specific and relational processing in ancestral humans, if, in fact, that is how survival processing acquired its distinct advantage. The evidence does not yet permit confident conclusions. Here, we suggest one possibility: namely, that superior memory retention occurs with the survival processing scenario because it places individuals in a heightened state of awareness of not just themselves, but also, critically, their own mortality. These thoughts of mortality might naturally induce a cognitive state that orients one to think more deeply, because the awareness of one's own mortality is, for obvious reasons, highly concerning and tends to elicit a great deal of motivated self-reflection. It is possible that the mortality-salient state induced by the survival task enhances item-specific processing, relational processing, or both.

In this vein, Hart and Burns (2012) noted that the survival processing task is likely to induce a mortality-salient state and pointed out that manipulations that lead to mortality-salient states have produced powerful effects on a multitude of psychological processes (see Pyszczynski, Greenberg, & Solomon, 2003, for a review), many of which seem oriented toward creating or preserving meaning, including an increase in (1) cognitive exploration (especially among individuals low in personal need for structure; Vess, Routledge, Landau, & Arndt, 2009), and (2) motivation to provide structure or organization (among individuals high in personal need for structure; Landau, Greenberg, & Sullivan, 2009; Landau et al., 2004), either of which might be expected to benefit encoding and retrieval.

Hart and Burns (2012) suggested that if mortality salience is the reason survival processing causes such exceptional retention, then thoughts of death in the absence of survival processing should be sufficient to produce deeper, more complex processing that would improve recall performance. In Experiment 1, they tested whether a mortality salience manipulation would improve incidental free recall. The mortality salience group was asked to write a few sentences about dying, with the control group writing about watching television. Participants were then shown a list of words and asked to rate the pleasantness of each word. Following a short period of distractor activity, an unexpected recall test was given. The results showed that the mortality salience manipulation enhanced recall. Experiment 2 replicated this "dying to remember" (DTR) effect, using a control condition thought to be as aversive and arousing as the mortality salience condition but without instilling thoughts of death: physical paralysis. Still, the mortality condition produced higher unexpected recall than the paralysis condition. In Experiment 3, a dental pain control condition that has been used extensively in the mortality salience literature replaced the paralysis condition. Additionally, rather than rating the words for pleasantness, participants were simply asked to try to remember the words for a later test. The results showed that the mortality condition produced intentional recall performance superior to that of the dental pain condition.

Analyses of the written responses to the mortality and control questions showed that participants wrote deeper, more complex responses to the mortality questions than to the control questions, and that this complexity mediated the DTR effect, consistent with Hart and Burns' (2012) speculation that mortality salience activates an encoding set oriented toward deeper processing; one that enhances information gathering. They also suggested that mortality salience may be at least partially responsible for the survival processing retention advantage.

If the survival processing effect were totally due to mortality salience, then mortality salience should, like survival processing, enhance both item-specific and relational encoding. Burns, Hart, Kramer, and Burns (2013) explored this possibility. In Hart and Burns' experiments all participants were given an unrelated list of words and either performed a pleasantness rating task or intentionally remembered the words. These procedures are likely to encourage the encoding of primarily item-specific information (e.g., Einstein & Hunt, 1980). In Experiment 1, Burns et al. tested if the DTR effect would occur with different encoding procedures, ones that encouraged both item-specific and relational processing. Following the mortality salience or control questions (dental pain), participants either rated a list of words from *ad hoc* categories for their pleasantness value or sorted the words into categories. Pleasantness rating was expected to promote item-specific processing, whereas sorting the words into *ad hoc* categories is known to encourage the processing of both item-specific and relational information.

The results showed that category sorting produced higher recall and more category clustering than pleasantness rating, suggesting that the sorting task did evoke both item-specific and relational processing, whereas the pleasantness rating task resulted in only item-specific processing. The critical finding was that mortality salience enhanced recall equivalently for the category sorting and pleasantness rating groups, suggesting that the DTR effect is not limited to encoding tasks that encourage only item-specific processing.

In Experiment 2, Burns et al. (2013) employed several different measures of item-specific and relational processing to determine which type of processing is enhanced when one is in a mortality-salient state. Participants were first given either the mortality salience or dental pain questions and then were asked to rate the pleasantness of a list of categorically related words, a procedure known to promote both item-specific and relational processing. Two successive free recall tasks followed the encoding task. The recall results replicated Experiment 1 in that mortality salience enhanced recall under encoding conditions that fostered both item-specific and relational processing. Moreover, measures of relational processing (category clustering and item loss scores) showed no effect of mortality salience. However, mortality salience did increase the number of new items recalled (gained) on the second test, indicating more item-specific processing. The cumulative-recall curves, shown in Figure 4.4, supported these conclusions, with mortality salience producing little advantage during the opening minutes of recall, but then diverging in the later portions of recall. The results of Burns et al.'s Experiment 2 suggest that, at least under encoding conditions that result in both item-specific and relational processing, mortality salience improves recall because it fosters more item-specific processing. This points to a commonality between the DTR effect and survival processing; both effects are associated with differential processing of item-specific and relational information.

In their Experiment 3, Burns et al. (2013) explored Hart and Burns' (2012) suggestion that the DTR and survival processing effects might share some of the same underlying mechanism(s). Specifically, they tested whether retention benefits of mortality salience and survival processing are additive. They did so by first requiring participants to answer questions about dying or dental pain and then requiring them to rate a list of seemingly unrelated words either for their survival value or for their pleasantness value. After the survival or pleasantness rating task, participants were given a free recall test followed by a recognition test. If the types of processing enhanced by mortality salience and survival processing are similar, then one would expect that the combined effects of the two manipulations would be less than additive, because redundancy in processing has repeatedly been shown to produce only minor retention benefits, at best (e.g., Hunt & Einstein, 1981; McDaniel, Einstein, Dunnay, & Cobb, 1986; McDaniel, Waddill, Finstad, & Bourg, 2000). On the other hand, if the two effects are unrelated, then a more additive effect would be expected to the extent that the enhanced processing produced by the two effects is different.[1]

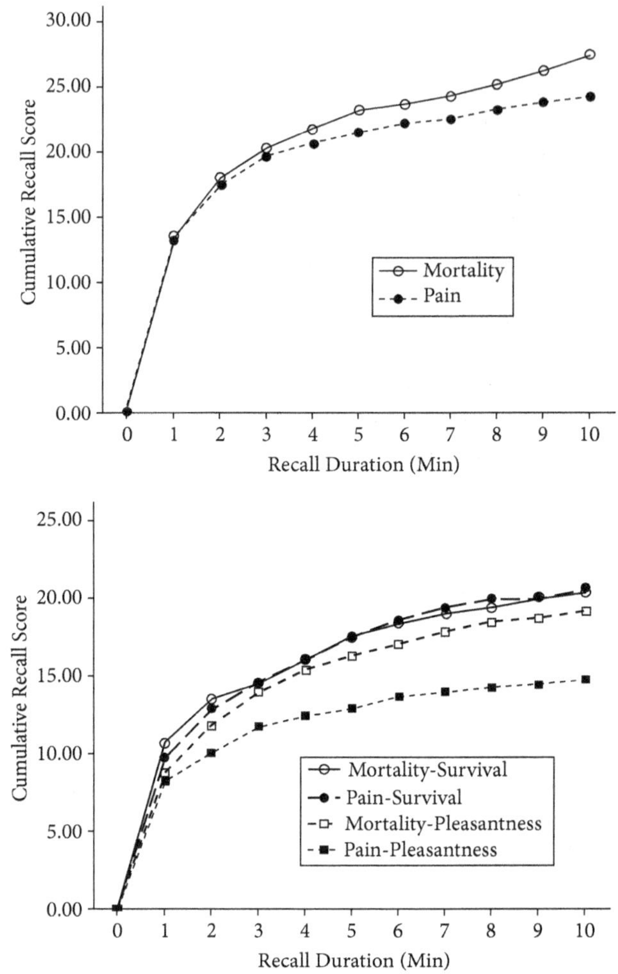

Figure 4.4 Mean cumulative-recall scores for the conditions tested in Burns et al.'s (2013) experiments. Top panel = Experiment 2, bottom panel = Experiment 3.

The experiment produced several interesting findings. First, for participants given the pleasantness rating task, the results closely replicated those of the previous experiment, showing that mortality salience enhanced recall, and that the enhanced recall was associated with greater item gains in the later portions of the cumulative recall period, not in the earlier portion. The final recognition results also showed a 4.3 percent advantage for the mortality salience group, but this difference did not reach significance. However, when participants were given the survival processing task, Burns et al. (2013) found that the DTR effect was completely eliminated, with the two groups producing nearly identical cumulative-recall curves and only a 1.48 percent difference in recognition. These results are consistent with Experiment 2, suggesting that mortality salience enhances item-specific processing during a pleasantness rating task. However, the results also show that the recall benefit produced by mortality salience is completely eliminated when the orienting task involves rating words for their survival value, a finding that implies processing redundancy. Considering that both the mortality salience manipulation and survival processing are likely to instill thoughts about dying, and that the recall benefits of morality salience are completely subsumed by the effects of survival processing, we are tempted to conclude that the two effects are overlapping, even if they are not completely redundant.

Possible Connections Between Mortality Salience and Survival Processing

The studies reviewed above suggest that although both the DTR and survival processing effects seem to be the result of differential processing of item-specific and relational information, the pattern of results may not be identical for the two effects. The survival task seems to encourage both item-specific and relational processing. When the control task encourages only item-specific processing the survival task encourages more relational processing, whereas the survival task enhances item-specific processing relative to control conditions that encode only relational information. When participants in the control condition perform both types of processing, the survival processing effect is eliminated. Interestingly, a different pattern seems to occur for the DTR effect. The DTR effect occurs regardless of whether participants perform only item-specific processing or both relational and item-specific processing, and it appears to be due primarily to a difference in item-specific processing (we have yet to test whether the DTR effect occurs when the encoding context encourages only relational processing).

It is important to realize, however, that the mortality salience manipulation is a priming manipulation that occurs prior to the encoding of the list items, whereas survival processing is performed during encoding of the list items and is performed individually for each item. These procedural differences, and not a difference in the mechanisms underlying the two effects, might be responsible for the complicated pattern of results. That the DTR effect occurs when the encoding conditions encourage both item-specific and relational processing, but not when the encoding task focuses on survival, which also encourages both item-specific and relational processing, is suggestive of at least partial redundancy. In other words, the finding that mortality salience enhances recall for two different encoding procedures that promote both types of processing (category sorting of a seemingly unrelated list and pleasantness rating of an obviously related list), but not for the survival task, is consistent with the interpretation that the two effects are related. However, it seems important to verify that mortality salience priming enhances recall for other nonsurvival scenarios, because other than the survival scenario, none of our encoding conditions have included a scenario.

To the extent that mortality salience and survival processing are related, an understanding of this relationship may inform theoretical explications about the adaptations that have led to the sculpting of human memory. For instance, one open question pertains to how the processing of information relevant to survival would have come to be associated with a memory advantage compared to the processing of information relevant to similar problems that do not have an impact on one's chances of surviving. Of course, one possibility is that this tendency appeared randomly in some humans but not others and resulted in greater reproductive success for individuals with the tendency. Alternatively, the tendency to remember information better in survival scenarios may be a byproduct of some other feature that has always naturally distinguished survival problems from other fitness-relevant problems such as mate selection.[2]

Our research suggests the possibility that one of the critical features that sets survival processing apart is its evocation of existential concerns. The survival problem is cognitively motivating, perhaps more so than any other problem, particularly because it is ultimately unsolvable (even if death can be staved off for the time being), in addition to being a surefire thwarting of all the rest of one's (unaccomplished) goals. This tendency to be highly motivated by death concerns is to be expected in a self-aware animal whose instinctual behavioral systems are inclined toward self-preservation; indeed, an animal who is both aware of and indifferent to its own mortality may be expected not to exist for very long in dangerous environments. If death thoughts naturally engage deep processing, then a direct consequence might be that survival scenarios would also engage deep processing, thereby resulting in memory improvements. If this is true, an adapted module or modules pertaining to the processing of survival-relevant information *per se* would not be required for the tendency for survival scenarios to improve memory; all that is necessary is that humans continue to be cognitively motivated by death thoughts, which could be a function of much more primitive self-preservation tendencies (e.g., fear responses) combined with an awareness of inevitable mortality (cf. terror management theory; Pyszczynski et al., 2003).

Whereas the idea that mortality salience may be partly responsible for the survival processing effect seems to warrant investigation, it is unlikely that mortality salience is the only mechanism at play. Whereas survival processing seems to induce both item-specific and relational processing, mortality salience seems primarily to increase only item-specific processing. Burns et al. (2013) suggested that the survival task may involve (at least) two factors. One factor, mortality salience, may promote item-specific processing. A second factor unrelated to mortality salience (e.g., future planning) may enhance relational processing.

There may be ways to empirically separate the effects of mortality salience from those of survival processing. For example, participants could process list items for their relevance to a dying scenario, one that is similar to the survival scenario but that instills only thoughts of dying with no hope of surviving (e.g., a terminal case of cancer). The dying scenario would be expected to induce a mortality-salient state without involving survival processing. If the mortality-salient state is what drives the survival processing effect, then this dying scenario should produce recall scores at least as large as the survival scenario.

Along these lines, Klein (in press) compared a dying scenario to the traditional survival scenario as well as a pleasantness rating control group. His dying scenario simply asked participants to imagine they were about to die and to rate the words for their relevance to the circumstances surrounding their death. Klein found that whereas the dying scenario produced recall as high as the pleasantness rating group, the survival scenario produced the highest recall. These results indicate that the recall advantage associated with survival processing is not due simply to mortality salience, leading Klein to suggest that survival and death-related processing rely on different neurocognitive resources.

There are three major reasons why we believe this conclusion is premature. First, Klein's survival scenario was more concrete, detailed, and extensive than his dying scenario. Whereas the survival scenario asked about the relevance of each word to finding food and water *and* avoiding predators in the grasslands, his dying scenario simply required rating the words for their relevance to "the circumstances surrounding your death." Kroneisen and Erdfelder (2011) showed that rating words for their survival relevance across several dimensions (i.e. food, water, predators) produces superior retention to simpler survival scenarios, so at least part of Klein's survival advantage was likely due to this confound. Second, Klein did not report relevance ratings for the survival and dying scenarios. There is ample evidence that congruity critically affects the magnitude of the survival processing advantage and therefore must be controlled (e.g., Butler et al., 2009). Indeed, we had 62 participants rate the words used by Klein for their relevance to his survival or dying scenarios and found that the words were far more relevant to the survival scenario than to the dying scenario (Burns, Hart, & Kramer, 2013). This word congruity difference may be responsible for the lower recall following death processing. Finally, we have conducted a series of experiments using dying scenarios matched for complexity and detail with the traditional survival scenario and words that were equally congruent to each scenario. Results showed that "death processing" can, under certain conditions, produce the same level of recall as survival processing.

Another possible approach would be to prime participants with survival thoughts by having them answer questions about a survival situation (e.g., asking how they would feel if they were stranded in the grasslands without supplies and it was up to them to try to survive). Preferably, the questions would instill thoughts of survival while minimizing thoughts of dying. If thoughts of survival are critical for the DTR effect, then the survival questions should produce a recall advantage for a list of words learned afterwards. We are also planning to conduct a series of experiments to test this hypothesis. Regardless of the procedures used to do so, developing a better understanding of the relationship between the two effects is likely to enhance our understanding of both.

SUMMARY

We have argued that the proximate mechanism responsible for the survival processing retention advantage may be differential processing of item-specific and relational processing. Whereas most control tasks promoted one type of processing or the other, the survival task induces both types of processing. Many nonscenario control tasks (e.g., pleasantness rating, item generation, and imagery rating) tend to promote only item-specific processing. Control tasks involving nonsurvival scenarios (i.e., moving) seem to promote less item-specific processing than the survival scenario. At present, we cannot explain why survival processing induces both types of processing, whereas control tasks, including the moving task, only induce one type of processing. The discovery that simply thinking about dying enhances retention of a list of words completely unrelated to the mortality-salience manipulation and that this effect is likely the result of differential processing of item-specific information is intriguing. Even more intriguing is the finding that the DTR effect is completely eliminated when the encoding task involves survival processing; this finding suggests a relationship between the two effects. Exploration of this relationship may be helpful in explaining both the proximate and ultimate mechanisms responsible for the memorial benefits associated with survival processing, as well as providing greater insight about the way evolutionary pressures have shaped episodic memory function.

NOTES

1. As we have previously noted, when the information available in the encoding context overlaps with that required by the orienting task, recall is not improved (e.g., Hunt & Einstein, 1981). Hunt and Einstein also showed that for repeated presentations, completing the same orienting task twice produced no retention benefit, whereas using an item-specific orienting task for one presentation and a relational task for the other presentation markedly improved recall. However, simply altering the type of orienting task without altering the underlying type of processing performed does not improve retention (see Maskarinec & Thompson, 1976 Shaughnessy, 1976). The ineffectiveness of redundant processing has also been shown in research arenas other than those concerned with item-specific and relational processing. For example, Paivio and Csapo showed that performing a verbal orienting task on words was ineffective compared to performing a verbal task on pictures or performing an imaginal task on words.
2. To date, despite several attempts, there is no published evidence that a retention advantage occurs for nonsurvival, fitness-relevant scenarios (e.g., mate selection scenarios).

REFERENCES

Burns, D. J. (1992). The consequences of generation. *Journal of Memory and Language, 3*, 615–633.

Burns, D. J. (1993). Item gains and losses during hypermnesic recall: Implications for the item-specific—relational information distinction.*Journal of Experimental Psychology: Learning, Memory, and Cognition, 19*, 163–173.

Burns, D. J. (2006). Assessing distinctiveness: Measures of item-specific and relational processing. In R. R. Hunt & J. B. Worthen (Eds.), *Distinctiveness and memory* (pp. 109–130). New York: Oxford University Press.

Burns, D. J., Burns, S. A., & Hwang, A. J. (2011). Adaptive memory: Determining the proximate mechanisms responsible for the memorial advantages of survival processing. *Journal of Experimental Psychology: Learning, Memory, and Cognition, 37*, 206–218.

Burns, D. J., Curti, E. T., & Lavin, J. C. (1993). The effects of generation on item and order retention in immediate and delayed recall. *Memory & Cognition, 21*, 846–852.

Burns, D. J., & Gold, D. E. (1999). An analysis of item gains and losses in retroactive interference. *Journal of Experimental Psychology: Learning, Memory, and Cognition, 25*, 978–985.

Burns, D. J., Hart, J., Griffith, S. E., & Burns, A. D. (2013). Adaptive memory: The survival scenario enhances item-specific processing relative to the moving scenario. *Memory 21*, 695–706.

Burns, D. J., Hart, J., & Kramer, M. E. (2013). Dying scenarios improve recall as much as survival scenarios. *Memory*, DOI:10.1080/09658211.2013.796973

Burns, D. J., Hart, J., Kramer, M. E. & Burns, A. D (2013). Dying to remember, remembering to survive: Mortality salience and survival processing. *Memory*, DOI:10.1080/09658211.2013.788660.

Burns, D. J., Jenkins, C. L., & Dean, E. E. (2007). Falsely recalled items are rich in item-specific information. *Memory & Cognition, 35*, 1630–1640.

Burns, D. J., Martens, N.J., Bertoni, A. A., Sweeney, E. J., & Lividini, M. D. (2006). An item gains and losses analysis of false memories suggests critical items receive more item-specific processing than list items. *Journal of Experimental Psychology: Learning, Memory, and Cognition, 32*, 277–289.

Burns, D. J., & Schoff, K. M. (1998). Slow and steady often ties the race: The effects of item-specific and relational processing on cumulative recall. *Journal of Experimental Psychology: Learning, Memory, and Cognition, 24*, 1041–1051.

Butler, A. C., Kang, S. H. K., & Roediger, H. L. (2009). Congruity effects between materials and processing tasks in the survival processing paradigm. *Journal of Experimental Psychology: Learning, Memory and Cognition, 35*, 1477–1486.

Cosmides, L., & Tooby, J. (1989). Evolutionary psychology and the generation of culture, Part 11. Case study: A computational theory of social exchange. *Ethology and Sociobiology 10*, 51–97.

Cosmides, L., & Tooby, J. (1992). Cognitive adaptations for social exchange. In J. Barkow, L. Cosmides, & J. Tooby (Eds.), *The adapted mind: Evolutionary psychology and the generation of culture* (pp. 163–228). New York: Oxford University Press.

Einstein, G. O., & Hunt, R. R. (1980). Levels of processing and organization: Additive effects of individual item and relational processing. *Journal of Experimental Psychology: Human Learning and Memory, 6*, 588–598.

Engelkamp, J., & Seiler, K. H. (2003) Gains and losses in action memory. *Quarterly Journal of Experimental Psychology, 56*, 829–848.

Engelkamp, J., Biegelmann, U., & McDaniel, M. A. (1998). Relational and item-specific information: trade-off and redundancy. *Memory, 6*, 307–333.

Hampton, R. R. (2009). Multiple demonstrations of metacognition in nonhumans: Converging evidence or multiple mechanisms? *Comparative Cognition & Behavior Reviews, 4*, 17–28.

Hart, J., & Burns, D. J. (2012). Nothing concentrates the mind: Thoughts of death improve recall. *Psychonomic Bulletin & Review, 19*, 264–269.

Hodge, M. H., & Otani, H. (1996). Beyond category sorting and pleasantness rating: Inducing relational and item-specific processing. *Memory & Cognition, 24*, 110–115.

Howe, M. L., & Derbish, M. H. (2010). On the susceptibility of adaptive memory to false memory illusions. *Cognition, 115*, 252–267.

Hunt, R. R., & Einstein, G. O. (1981). Relational and item-specific information in memory. *Journal of Verbal Learning and Verbal Behavior, 20*, 497–514.

Hunt, R. R., & McDaniel, M. A. (1993). The enigma of organization and distinctiveness. *Journal of Memory and Language, 32*, 421–445.

Kang, S. H. K., McDermott, K. B., & Cohen, S. M. (2008). The mnemonic advantage of processing fitness-relevant information. *Memory & Cognition, 36*, 1151–1156.

Klein, S. B. (in press). The effects of thoughts of survival and thoughts of death on recall in the adaptive memory paradigm. *Memory.*

Klein, S. B. (2012). A role for self-referential processing in tasks requiring participants to imagine survival on the savannah. *Journal of Experimental Psychology: Learning, Memory & Cognition, 38*, 1234–1242.

Klein, S. B., & Loftus, J. (1988). The nature of self-referent encoding: The contributions of elaborative and organizational processes. *Journal of Personality and Social Psychology, 55*, 5–11.

Klein, S. B., Loftus, J., Kihlstrom, J. F., & Aseron, R. (1989). Effects of item-specific and relational information on hypermnesic recall. *Journal of Experimental Psychology: Learning, Memory, and Cognition, 15*, 1192–1197.

Klein, S. B., Loftus, J., & Schell, T. (1994). Repeated testing: A technique for assessing the roles of elaborative and organizational processing in the representation of social knowledge. *Journal of Personality and Social Psychology, 66*, 830–839.

Klein, S. B., Robertson, T. E., & Delton, A. W. (2010). Facing the future: Memory as an evolved system for planning future acts. *Memory & Cognition, 3*, 13–22.

Kroneisen, M., & Erdfelder, E. (2011). On the plasticity of the survival processing effect. *Journal of Experimental Psychology: Learning, Memory, and Cognition, 37*, 1553–1562.

Landau, M. J., Greenberg, J., & Sullivan, D. (2009). Defending a coherent autobiography: When past events appear incoherent, mortality salience prompts compensatory bolstering of the past's significance and the future's orderliness. *Personality and Social Psychology Bulletin, 35*, 1012–1020.

Landau, M. J., Johns, M., Greenberg, J., Pyszczynski, T., Martens, A., Goldenberg, J. L., & Solomon, S. (2004). A function of form: Terror management and structuring the social world. *Journal of Personality and Social Psychology, 87*, 190–210.

Maskarinec, A. S., & Thompson, C. P. (1976). The within-list distributed practice effect: Tests of the varied context and varied encoding hypotheses. *Memory & Cognition, 41*, 741–746.

McDaniel, M. A., Einstein, G. O., Dunay, P. K., & Cobb, R. E. (1986). Encoding difficulty and memory: Toward a unifying theory. *Journal of Memory and Language, 25*, 645–656.

McDaniel, M. A., Moore, B. A., & Whiteman, H. L. (1998). Dynamic changes in hypermnesia across early and late tests: A relational/item-specific account. *Journal of Experimental Psychology, 24*, 173–185.

McDaniel, M. A., Waddill, P. J., Finstad, K., & Bourg, T. (2000). The effects of text-based interest on attention and recall. *Journal of Educational Psychology, 92*, 492–502.

Mulligan, N. W. (2000). Perceptual interference at encoding enhances item-specific encoding and disrupts relational encoding: Evidence from multiple recall tests. *Memory & Cognition, 28*, 539–546.

Mulligan, N. W. (2001). Generation and hypermnesia. *Journal of Experimental Psychology: Learning, Memory, and Cognition, 27*, 436–450.

Mulligan, N. W. (2002). The emergence of item-specific encoding effects in between-subjects designs: Perceptual interference and multiple recall tests. *Psychonomic Bulletin & Review, 9*, 375–382.

Nairne, J. S., & Pandeirada, J. N. S. (2008a). Adaptive memory: Remembering with a Stone-Age brain. *Current Directions in Psychological Science, 17*, 239–243.

Nairne, J. S., & Pandeirada, J. N. S. (2008b). Adaptive memory: Is survival processing special? *Journal of Memory and Language, 59*, 377–385.

Nairne, J. S., Pandeirada, J. N. S., & Thompson, S. R. (2008). Adaptive memory: The comparative value of survival processing. *Psychological Science, 19*, 176–180.

Nairne, J. S., Thompson, S. R., & Pandeirada, J. N. S. (2007). Adaptive memory: Survival processing enhances retention. *Journal of Experimental Psychology: Learning, Memory, and Cognition, 17*, 239–243.

Oaten, M., Stevenson, R. J., & Case, T. I. (2009). Disgust as a disease-avoidance mechanism. *Psychological Bulletin, 135*, 303–321.

Otgaar, H., & Smeets, T. (2010). Adaptive memory: Survival processing increases both true and false memory in adults and children. *Journal of Experimental Psychology: Learning, Memory, and Cognition, 36*, 1010–1016.

Pyszczynski T., Solomon S., & Greenberg J. (2003). *In the wake of 9/11: The psychology of terror*. Washington DC: APA.

Renkema, L. J., Stapel, D. A., Maringer, M., & van Yperen, N. W. (2008). Terror management and stereotyping: Why do people stereotype when mortality is salient? *Personality And Social Psychology Bulletin, 34*, 553–564.

Roenker, D. L., Thompson, C. R., & Brown, S. C. (1971). A comparison of measures for the estimation of clustering in free recall. *Psychological Bulletin, 76*, 45–48.

Rogers, T. B., Kuiper, N. A., & Kirker, W. S. (1977). Self-reference and the encoding of personal information. *Journal of Personality and Social Psychology, 35*, 677–688.

Soderstrom, N. C. & McCabe, D. P. (2011). Are survival processing memory advantages based on ancestral priorities? *Psychonomic Bulletin and Review, 18*, 564–569.

Smeets, T., Otgaar, H., Raymaekers, L., Peters, M. V., & Merckelbach, H. (2012). Survival processing in times of stress. *Psychonomic Bulletin & Review, 19*, 113–118.

Tulving, E. (2002). Episodic memory: From mind to brain. *Annual Review of Psychology, 53*, 1–25.

van Horik, J. O., Clayton, N. S., & Emery, N. J. (2012). Convergent evolution of cognition in corvids, apes and other animals. In J. Vonk & T. K. Shackelford (Eds.), *The Oxford handbook of comparative evolutionary psychology* (pp. 80–101). New York: Oxford University Press.

Vess, M., Routledge, C., Landau, M. J., & Arndt, J. (2009). The dynamics of death and meaning: The effects of death-relevant cognitions and personal need for structure on perceptions of meaning in life. *Journal of Personality and Social Psychology, 97*, 728–744.

Weinstein, Y., Bugg, J. M., & Roediger, H. L. (2008). Can the survival recall advantage be explained by basic memory processes? *Memory & Cognition, 36*, 913–919.

Worthen, J. B., & Loveland, J. M. (2003). Disruptive effects of bizarreness in free and cued recall for self-performed and other-performed acts: The costs of item-specific processing. In S. P. Shohov (Ed.), *Advances in psychology research, 24* (pp. 3–17). Hauppauge, NY: Nova Science Publishers.

SECTION 2

Adaptive Memory: A Special Case for Survival Information and Processing?

5

Adaptive Memory: Survival Processing, Ancestral Relevance, and the Role of Elaboration

MARK L. HOWE AND MARY H. DERBISH

Recent research has shown that when lists of words are processed for their relevance to a survival scenario (e.g., stranded in the grasslands of a foreign land) true memory performance is higher than when the same words are processed for a number of other, well-known meaningful or deep processing scenarios (e.g., moving to a new house, rating for pleasantness, self-reference) (for a review, see Nairne, 2010).This phenomenon, known as the *adaptive memory effect,* is said to showcase the human memory system's proclivity toward more efficient processing of information for its survival relevance, something that may permit better adaptation to our environment. Further support for the adaptive significance of this type of processing can be found in studies that contrast the use of nonevolutionarily relevant survival scenarios (e.g., lost in a strange city at night) with evolutionarily relevant scenarios (e.g., stranded in the grasslands). What these studies have shown is that the ancestrally relevant scenarios continue to produce higher rates of true memories than those survival scenarios that have little evolutionary significance (see Nairne, 2010, for a review).

More recently, there has been a growing body of evidence concerning alternative interpretations of this survival processing advantage, ones that do not invoke an evolutionary explanation. For example, Tse and Altarriba (2010) found no survival processing advantage when an implicit memory task was employed rather than an explicit memory task. This is inconsistent with an strict evolutionary interpretation because

> a flexible and adaptive memory system should not depend on retrieval intentionality, because many advantages that a memory system confers to an organism are more primitive in nature and less dependent on higher order processing.

> An organism can be benefitted by the retrieval of previous survival-relevant episodes, independently of whether retrieval is explicit or not. (Tse & Altarriba, 2010, p. 1119)

Not only should an adaptive memory system operate at an implicit as well as an explicit level but so too should it be tuned to the source of information stored in memory. That is, the location of fitness-relevant stimuli should be a priority in an adaptive memory system. As Bröder, Krüger, and Schütte (2011, p. 897) argue, "remembering an item is only helpful if and only if I also remember the context in which I encountered it. It is important to know where the waterhole is and when deer usually graze near the forest." However, adaptive memory advantages do not accrue to either implicit measures of memory or memory for source information (Bröder et al., 2011; Tse & Altarriba, 2010).

Other research has found that the adaptive memory effect may simply reflect the heightened operation of more global memory processes. Although these processes, which affect all of memory, were sculpted through evolutionary pressures, they are not specialized for the storage and retrieval of fitness-relevant information. For example, Burns, Burns, and Hwang (2011) examined the role of well-known variables that affect memory performance, namely item-specific and relational processing. They found that when both relational (i.e., elaborative) and item-specific (i.e., distinctive) processing occurred in the same memory task, performance levels mirrored that found in the survival-processing paradigm. Similarly, Kroneisen and Erdfelder (2011) examined the impact of elaboration (relational processing) and distinctiveness (item-specific processing) and also found that the adaptive memory effect could be accounted for by a combination of these two well-known variables that affect memory processes more generally. Thus, adaptive memory effects are not only absent in implicit and source memory tasks, but when they do appear in explicit memory tasks, their origins may have more to do with more global, well-known memory processes (elaborative and distinctive processing) than to an evolved module for processing fitness-relevant information.

Finally, what about the argument that the adaptive memory advantage is content-specific? That is, is this advantage specific to ancestrally relevant scenarios or stimuli? Recent evidence suggests that this too may not be true. For example, Soderstrom and McCabe (2011) examined the ancestral relevance claim by having participants rate words for their relevance to scenarios where the environment was either ancestrally relevant (i.e., grasslands) or irrelevant (i.e., city) as well as whether threats arose either from ancestrally relevant beings (i.e., from predators) or ancestrally irrelevant beings (i.e., from zombies). Compared to a control task (i.e., rating words for pleasantness), memory did not differ as a function of location (grasslands or city) but was better for ancestrally irrelevant threats (from zombies) than for ancestrally relevant threats (from predators). Thus, ancestral relevance may not play a role in the adaptive memory effect either.

Although elaboration and ancestral relevance are important to the research presented in this chapter, there is another critical variable that was originally discussed by Howe and Derbish (2010; also see Otgaar & Smeets, 2010). Specifically, the majority of adaptive memory studies have focused on the amount of information correctly remembered. However, before we can conclude that survival processing leads to improved memory processes, it is important to explore what effect survival processing has on memory errors. If adaptive memory effects enhance accuracy, then one would expect to see not only improved true memory but also reduced rates of memory errors (e.g., false memory). However, in the first report of the adaptive memory effect, Nairne, Thompson, and Pandeirada (2007) found not only enhanced true memory but, counterintuitively, enhanced memory errors in the survival processing paradigm.

Howe and Derbish (2010) extended these findings in a series of experiments that exploited the Deese/Roediger-McDermott (DRM) paradigm (Deese, 1959; Roediger & McDermott, 1995). Here, participants were presented with lists of words (e.g., *bed, rest, awake, blanket, doze, slumber*)that are associates of a critical unpresented word (e.g., *sleep*). Although the word *sleep* never appears on the list, when tested, participants often falsely remember *sleep* as having been presented. Howe and Derbish (2010) hypothesized that the context in which the material was processed should have an effect on memory performance. Indeed, they demonstrated that true and false memories increased when information was processed for its survival relevance regardless of the nature of the material being rated (neutral, negative, and survival-related words). Moreover, these effects were robust in both incidental and intentional memory paradigms. Finally, they found that survival-related material (e.g., lists with critical lures such as *injury, death, fight*) produced higher rates of true and false memories independent of the processing effect. Thus, both *survival processing* and *survival materials* contribute independently to increases in true and false memory.

Howe and Derbish (2010) focused on the effects of survival materials on memory performance. Although Nairne et al. (2007) have suggested that a stimulus' survival relevance is context-dependent, we would argue that terms such as *injury, death,* and *fight* may be survival-relevant independent of context (also see Wilson, Darling, & Sykes, 2011). Howe and Derbish (2010) further suggested that the increase in both true and false memory performance for words related to survival was not a product of unique memory processes evolutionarily designed for survival, but rather because the survival lists are associated with fewer possible themes (also see, Arndt & Reder, 2003), they are more likely to activate the critical lure as well as other related words on the list. Indeed, across four experiments, they found that increases in true and false memories were inversely related to the number of competing themes. Importantly, Howe and Derbish (2010) consistently reported a processing effect independent of their materials effect. That is, regardless of the strength of the materials effect, survival processing resulted in more true and false memories than other deep or meaningful encoding strategies (e.g., moving to a new house, pleasantness).

In the current chapter, we take up the question of whether this robust processing effect can be explained, like the materials effect was, without reference to an adaptive memory mechanism designed specifically for processing information for its survival relevance. In particular, we wondered whether this processing effect had more to do with the unusualness of the survival processing scenarios than their adaptive significance, an unusualness that might require more extensive and elaborate processing of information than that needed for scenarios that are more familiar to participants (e.g., going on a holiday or moving house). This interpretation is consistent with Kroneisen and Erdfelder's (2011) finding that elaborative processing was a likely candidate for survival processing effects.

Consider one alternative but plausible explanation for an increase in both true and false memory production as a function of processing task (e.g., survival vs. pleasantness), namely that such encoding requires increased relational processing (elaboration). We know that when participants are asked to use imagery in the service of remembering word lists, recollection rates improve (Paivio, 2007). When participants are encouraged to process information by using related information (i.e., elaborating, inflating, or imagining) it can prime networks of interrelated concepts already stored in memory, causing them to become activated (Kroneisen & Erdfelder, 2011). This, in turn, can and often does lead to people falsely report that these unpresented but activated items were actually presented (Collins & Loftus, 1975; Karpicke, McCabe, & Roediger, 2008; Thomas, Hannula, & Loftus, 2007).

We hypothesize that the seemingly unique effect of survival processing on true and false memory production may be due to higher levels of elaboration needed to complete the survival rating task. To see why, consider that although most college students are familiar with scenarios such as moving house, going on holiday, and even feeling lost in a city (the typical deep processing scenarios that are used as controls and are compared to survival processing), few, if any, students have been stranded in the grasslands of a foreign country or on a desert island. This fundamental difference in scenario familiarity could lead participants to spontaneously engage in higher levels of elaboration and imagination in the less familiar grasslands or desert island scenarios than in the more familiar moving house or going on holiday conditions. Indeed, it might be likely that participants have to resort to considerable imaginative processing in order to accomplish these rating tasks, as when, for example, they are asked to determine the importance of various concepts to survival on a desert island. In lieu of having experienced survival on a desert island, participants, to successfully rate these items, may have to rely on information from their memories of literary classics such as *Treasure Island* and *Robinson Crusoe* or previously popular TV series (and their reruns) such as "Gilligan's Island."

OVERVIEW OF EXPERIMENTS 1 TO 4

To test whether or not elaborative processing can account for the standard adaptive memory processing effect, we present a series of four experiments. In these experiments, we use DRM lists in which the items on each list are related to an unpresented critical lure. In Experiment 1, we replicated and extended the processing effect found in Howe and Derbish (2010) using a fully crossed type of processing (pleasantness, survival) and type of material (neutral, negative, survival) design with a different survival scenario than they used (stranded on a desert island instead of in the grasslands of a foreign country).In the next three experiments, we focused on articulating the memory mechanisms that underlie these robust survival processing effects. In Experiment 2, we tested the hypothesis that the survival rating task inherently encourages greater levels of elaboration than other scenarios. In Experiment 3, we replicated and extended the findings from Experiment 2 using a within-participant design. As it turned out, regardless of whether we used a between- or within-participant design, when participants were encouraged to use greater levels of elaboration, the effect of survival processing on memory performance disappeared. Finally, Experiment 4 was designed to provide additional tests of our elaboration hypothesis and to determine whether processing of ancestrally irrelevant survival scenarios produces the adaptive memory effect.

EXPERIMENT 1

Method

Participants

Twenty undergraduates took part in this experiment (5 male, 15 female).

Materials, Design, and Procedure

The experiment was a 3(list type: neutral, negative, survival) × 2(rating task: pleasantness, survival) design in which list type was a within-participant factor and rating task was a between-participants factor. Twelve lists of eleven items each were selected from the word association norms of Nelson, McEvoy, and Schreiber (1998). Four DRM-type lists were constructed from neutral critical lures (associates of *fruit*, *window*, *music*, and *bread*), four lists were constructed from critical lures that were determined to be emotionally negative (associates of *sad*, *cry*, *lie*, and *bad*), and the final four lists were constructed from critical lures that were determined to be related to survival (associates of *fight*, *sick*, *death*, and *pain*) (see Appendix A). The survival and negative lures chosen for this experiment were taken directly from Howe and Derbish (2010), in which analyses of participants' survival ratings confirmed that these lists were appropriately labeled as either survival terms or negative terms. Only the ten words with the greatest backward associative strength (BAS) to the unpresented critical lures were presented at study (the eleventh word from each list was used as an unpresented related associate in the subsequent recognition test).

Critical lures for the negative ($M = 5.65$) and survival ($M = 5.63$) lists were equated for arousal using the ANEW norms (Bradley & Lang, 1999). To rule out BAS as a confounding variable across all list types, a one-way analysis of variance (ANOVA) on the mean BAS between the list items and their lures revealed no significant differences across the three list types (neutral, negative, survival), $F(2, 9) = .72$. To eliminate any other possible item selection confounds we calculated, for each list type, the average semantic density using the Nelson et al. (1998) norms, word length and frequency using the e-lexicon project from Balota et al. (2007), and familiarity using the Clark and Paivio (2004) norms. Separate one-way ANOVAs were conducted for each word characteristic with the factor list type (neutral, negative, or survival). Semantic density ($F(2, 9) = 1.65, p =.25$), word frequency ($F(2, 9) = 1.38, p = .30$), word length ($F(2, 9) = .89, p =.44$), and familiarity ($F(2, 9) = 1.33, p = .31$) yielded no significant differences. Overall, then, all three list types were equated on level of BAS, semantic density, word frequency, word length, and familiarity, and the two valenced list types (negative and survival) were equated for arousal.

The rating task consisted of 120 items (the top 10 from each of the 12 DRM-like lists) compiled into a single list, presented eight words to a page, in random order. To the right of each word was a standard 7-point scale on which participants could rate the item for pleasantness or for its survival value. Specifically, participants were instructed that they would be rating words for a particular quality. Upon arrival, participants were randomly assigned to either the pleasantness rating condition or the survival rating condition. Participants in the survival condition were given the following instructions:

> "We would like you to imagine that you are stranded on a desert island without any food or supplies and you will have to survive on this island for some amount of time alone. We are going to show you a list of words and we would like you to rate them for how relevant each word would be for your survival on this island."

The survival scale ranged from 1 (least relevant) to 7 (most relevant).

Participants in the pleasantness condition were given the following instructions:

> "We are going to show you a list of words and we would like you to rate each word for how pleasant you find that word."

The pleasantness scale ranged from 1 (least pleasant) to 7 (most pleasant).

After completing the rating task, participants received a recognition test consisting of a sheet of paper containing 96 items (arranged randomly) and asked to indicate whether an item had appeared on the list (marking "yes") or had not (marking "no"). The list comprised the 12 critical lures, 4 words of varying associative strength that were studied from each list (48 items), the unpresented, related associate (11th item) from each list (12 items), and 24 unrelated (neutral and negatively valenced) items.

Results and Discussion

The proportion of words correctly and falsely recognized was calculated and analyzed separately using a 3(list type: neutral, negative, survival) × 2(rating task: pleasantness, survival) ANOVA. The results for correct recognition are summarized in Figure 5.1. Although correct recognition tended to be higher for survival processing than pleasantness ratings, there were no significant main effects or interactions

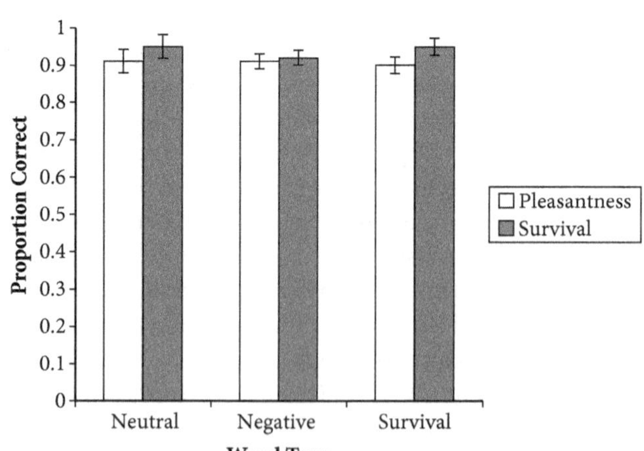

Figure 5.1 Proportion of words correctly recognized as a function of rating task in Experiment 1.

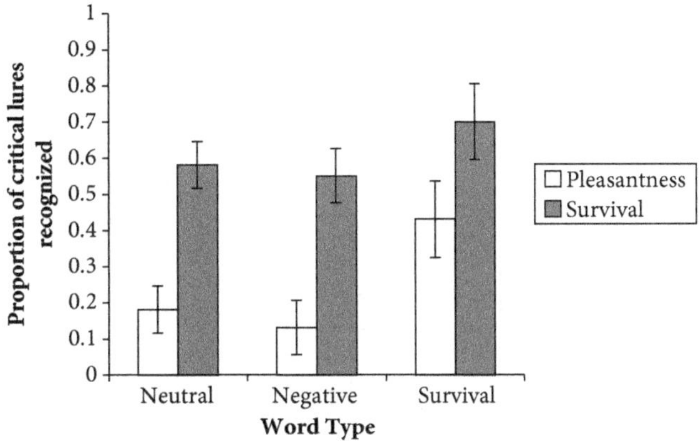

Figure 5.2 Proportion of critical lures falsely recognized as a function of rating task in Experiment 1.

(largest F = 1.38, p = .26), something that might not be unexpected given the near-ceiling performance of our participants. The proportion of critical lures falsely recognized is summarized in Figure 5.2. The ANOVA revealed a significant main effect of list type, $F(2, 36) = 6.61$, $p<.01$, $\eta_p^2 = .268$. *Posthoc* comparisons revealed that survival lists produced significantly more false memories than neutral ($p<.01$) and negative lists ($p<.01$). There was no significant difference in false memory production between neutral and negative lists ($p = .59$). There was a significant main effect of rating task on false memory production such that participants who rated the words for survival were more likely to falsely recognize the critical lures than those who rated them for pleasantness, $F(1, 18) = 16.85$, $p<.001$, $\eta_p^2 = .484$. As can be seen in Figure 5.2, the interaction between list type and rating task was not significant, $F(2, 36) = .80$, $p = .46$.

Because the ceiling effect for the correct recognition data would mask any effect the rating task might have on participants' overall response bias, the interpretation of the false recognition data becomes difficult. If participants in the survival rating task were biased toward making more "yes" responses on the recognition test, then we would expect to see a greater proportion of noncritical intrusions as well as critical lures. To determine whether the survival rating task created a response bias, the proportion of noncritical lures was calculated for the pleasantness (M = .13) and survival (M = .09) rating tasks. A one-way ANOVA with the factor rating task (pleasantness or survival) yielded no main effect, $F(1, 19) = .50$, $p = .49$.[1]

Finally, to answer our key questions—whether adaptive memories were more accurate than other memories—we computed composite accuracy scores using both true and false (critical lure) recognition measures as used in Howe and Derbish (2010):

$$\text{Accuracy} = \frac{(\text{true recognition})}{(\text{true recognition} + \text{false recognition})}$$

in which the higher the score, the greater the accuracy. The results of a 3(list type: neutral, negative, survival) × 2(rating task: pleasantness, survival) ANOVA

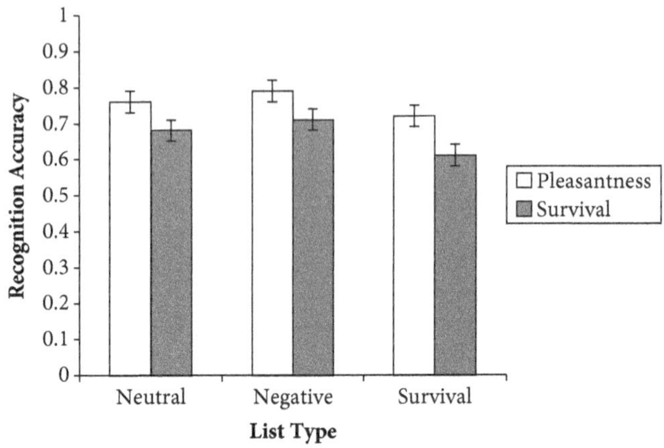

Figure 5.3 Overall accuracy as a function of rating task in Experiment 1.

revealed a main effect for list, $F(2, 36) = 6.18$, $p<.01$, $\eta_p^2 = .256$, where *post-hoc* tests ($p<.05$) showed that recognition accuracy was lowest for survival lists ($M = .75$), with significantly higher accuracy rates for both neutral ($M = .83$) and negative lists ($M = .84$), and these latter two lists did not differ. There was also a main effect for rating task, $F(1, 18) = 17.83$, $p = .001$, $\eta_p^2 = .498$, in which items rated for their survival value ($M = .73$) were recognized less accurately than items rated for pleasantness ($M = .88$). As can be seen in Figure 5.3, the interaction was not significant. Overall, there were more false memories and lower accuracy for information related to survival. More importantly, regardless of the type of material being recollected (neutral, negative, or survival), higher rates of false recognition and lower levels of accuracy were observed when that information was processed for its relevance to survival than pleasantness.

The results of Experiment 1 are consistent with previous data reported by Howe and Derbish (2010) in which no ceiling effects were present in true recognition rates and when a grasslands scenario was used instead of the desert island scenario used here. Indeed, the current data provide further support for the hypothesis that processing information for its survival relevance can alter memory performance. Moreover, these results show that the effects of item processing (pleasantness, survival) can be dissociated from the nature of the item (neutral, negative, survival). Indeed, type of processing and type of material did not interact. What this means is that survival processing and survival materials independently give rise to more false memories and therefore lower net accuracy.

EXPERIMENT 2

Having replicated the dissociation between materials and processing effects originally obtained by Howe and Derbish (2010) in our first experiment, Experiment 2 was designed to determine whether the adaptive memory processing effect was due to increased elaboration in memory. To examine this, we wondered whether the adaptive memory effect could be attenuated or even eliminated when participants were encouraged to engage in greater elaboration of nonsurvival scenarios (e.g., moving, going on holiday). If it is true that the desert island (or grasslands) condition inherently encourages more elaboration than the nonsurvival scenarios, then explicitly encouraging greater elaboration in all conditions should minimize differences in spontaneous elaboration that we hypothesize occur between survival and nonsurvival rating tasks. If the adaptive memory effect is attenuated or disappears in the absence of different rates of spontaneous elaboration, then that suggests that elaboration plays an important role when remembering information processed under survival conditions, perhaps because survival scenarios are less familiar.

In addition, we examined whether familiarity or ancestral relevance was important to the adaptive memory effect (also see Soderstrom & McCabe, 2011). To do this, we added two survival conditions that could not be considered relevant to our evolutionary past: stranded on a foreign planet and stranded under the sea. These nonevolutionarily relevant survival scenarios should be equally unfamiliar, thus requiring similar amounts of elaboration as the traditional survival scenarios (grassland and desert island scenarios), and differ from them only in their ancestral relevance. Our intuition was that the moving and holiday scenarios, being familiar to the average person, would require little elaborative processing when completing the rating task. We expected that the grasslands, desert island, foreign planet, and undersea scenarios, being less familiar, should require more elaboration. Additionally, as Nairne et al. (2007) have suggested, the grasslands and desert island scenarios represent traditional ancestral environments (Tooby & Cosmides, 2005). Thus, to rule out any confounds between adaptive relevance, familiarity, and elaboration, we added the foreign planet and undersea scenarios. Both scenarios are unfamiliar and would therefore require significant elaboration. However, they are clearly not evolutionarily encoded in our memory as ancestrally relevant survival scenarios. If adaptive significance is important, then being stranded on a foreign planet or under the sea should not produce the same level of adaptive memory effects as the grasslands and desert island scenarios. Alternatively, if familiarity is important, then all of the less familiar, survival-related scenarios should produce the adaptive memory effect.

Method

Participants

A new sample of 60 undergraduates participated in this experiment (29 male, 31 female).

Materials, Design, and Procedure

We used 12 of the lists reported in Howe and Derbish (2010), four that were neutral (*sweet, money, mountain, school*), four that were negative (*sad, fat, cry, anger*), and four that were survival-related (*war, hurt, sick, death*) (see Appendix B). Only the ten words with the greatest BAS were presented at study (the 11th word from each list was used as an unpresented associate in the recognition test). The negative (M = 5.95) and survival (M = 5.44) lists were equated for arousal (Bradley & Lang, 1999). Additionally, we ruled out BAS as a confounding variable across all three listtypes. As before, average BAS was calculated for each list and submitted to a one-way ANOVA, which revealed no significant differences across the three list types ($F(2, 9) = 0.50$, $p = .95$). To eliminate any other possible item selection confounds we calculated the average semantic density using the Nelson et al. (1998) norms, frequency and word length using the e-lexicon project from Balota et al. (2007), and familiarity, imageability, concreteness, and number of attributes from the Clark and Paivio (2004) norms. Separate one-way ANOVAs were conducted for each word characteristic. There were no significant differences among the list types for any of the factors (see Table 5.1 for details). Thus, we conclude that all three list-types were equated on BAS, semantic density, word length, word frequency, familiarity, concreteness, imageability, and number of attributes. The two emotionally charged list types (negative, survival) were equated for valance and arousal.

The lists were presented as part of a single 120-item rating task (the top 10 items from each of the 12 lists) in random order and displayed eight words to a page. To the right of each word was a standard 7-point scale on which participants could rate the item for its relevance value. Specifically, participants were instructed that they would be rating words for their relevance to a particular scenario. Upon arrival participants were assigned to one of six unique scenarios that were described as part of the rating task instructions. The moving, holiday, and desert island scenarios are taken directly from Nairne et al.'s (2007) studies and the grasslands scenario from Howe and Derbish (2010). The foreign planet and undersea scenarios serve as untested survival scenarios that we constructed for this study. The addition of these

Table 5.1. CONTRASTS OF LIST CHARACTERISTICS IN EXPERIMENTS 2 AND 4

Characteristic	**Source**	**List Type**			
		Neutral	**Negative**	**Survival**	**Contrast**
Arousal	ANEW Norms (Bradley & Lang, 1999)	N/A	5.95	5.44	$t(3) = 1.87$ $p = .16$
BAS	Nelson et al., 1998	.29	.30	.30	$F(2, 9) = 0.50$ $p = .95$
Semantic Density	Nelson et al., 1998	.06	.06	.09	$F(2,9) = .48$ $p = .63$
Frequency (LogFreq)	Balota et al., 2007	9.34	9.35	8.70	$F(2, 9) = .64$ $p = .55$
Word Length (Num Letters)	Balota et al., 2007	5.62	5.32	5.58	$F(2, 9) = .10$ $p = .91$
Familiarity	Clark & Paivio, 2004	6.00	5.95	5.93	$F(2, 9) = .04$ $p = .97$
Imageability	Clark & Paivio, 2004	5.13	4.64	5.05	$F(2, 9) = 3.14$ $p = .09$
Concreteness	Clark & Paivio, 2004	5.15	3.85	4.52	$F(2, 9) = 4.15$ $p = .06$
Number of Attributes	Clark & Paivio, 2004	4.25	3.72	4.14	$F(2, 9) = 1.03$ $p = .40$

new scenarios will allow us to discriminate between memory processes designed for adaptive survival information, as Nairne has suggested, or the result of greater elaboration during processing.

Each scenario was then framed with both positive and negative valances to control for the possible confound of valence direction. Upon arrival, participants were randomly assigned to one of the 12 experimental conditions. The instructions for each condition were as follows:

- *Moving positive.* In this task we would like you to imagine that you are planning to move to your dream home in a foreign country. Over the next couple weeks, you'll need to locate and purchase your new home and transport your belongings. We are going to show you a list of words and we would like you to rate how relevant each of these words would be for you in this scenario. Some of the words will be more relevant than others, and some may be easier to rate than others. Remember, there are no correct or incorrect answers—it's completely up to you.
- *Moving negative.* In this task we would like you to imagine that you have recently been through a difficult divorce and must now move out of your home. Over the next couple weeks, you'll need to locate and purchase your new home and transport your belongings. We are going to show you a list of words and we would like you to rate how relevant each of these words would be for you in this scenario. Some of the words will be more relevant than others, and some may be easier to rate than others. Remember, there are no correct or incorrect answers—it's completely up to you.
- *Holiday positive.* In this task we would like you to imagine that you are enjoying an extended holiday at a fancy resort with all your basic needs taken care of. Over the next couple weeks, you will want to find different activities to pass the time and maximize your enjoyment. We are going to show you a list of words and we would like you to rate how relevant each of these words would be for you in this scenario. Some of the words will be more relevant than others, and some may be easier to rate than others. Remember, there are no correct or incorrect answers—it's completely up to you.
- *Holiday negative.* In this task we would like you to imagine that you have had a terrible year. You've dropped out of school and are working in a job you hate. You've decided to use the last of your savings to go on holiday to forget your troubles. You'll be staying in a fancy resort where all your basic needs will be taken care of. Over the next couple weeks, you will want to find different activities to pass the time and maximize your enjoyment. We are going to show you a list of words and we would like you to rate how relevant each of these words would be for you in this scenario. Some of the words will be more relevant than others, and some may be easier to rate than others. Remember, there are no correct or incorrect answers—it's completely up to you.
- *Grasslands positive.* In this task we would like you to imagine that you have been awarded the opportunity to explore the grasslands of a foreign country. You will be all alone with nature the entire time. Obviously, you

can't carry enough supplies for the entire trip, so over the next couple weeks you'll need to find steady supplies of food and water and protect yourself from predators. We are going to show you a list of words and we would like you to rate how relevant each of these words would be for you in this scenario. Some of the words will be more relevant than others, and some may be easier to rate than others. Remember, there are no correct or incorrect answers—it's completely up to you.

- *Grasslands negative.* In this task we would like you to imagine that you have been stranded in the grasslands of a foreign country. You are all alone and have no supplies or basic survival materials, so over the next couple weeks you'll need to find steady supplies of food and water and protect yourself from predators. We are going to show you a list of words and we would like you to rate how relevant each of these words would be for you in this scenario. Some of the words will be more relevant than others, and some may be easier to rate than others. Remember, there are no correct or incorrect answers—it's completely up to you.
- *Desert Island positive.* In this task we would like you to imagine that you have been awarded the opportunity to take a cruise to a desert island. You will be all alone on the beach the entire time. After dropping you off on the island, the boat will leave so over the next couple weeks you'll need to find steady supplies of food and water and protect yourself from predators. We are going to show you a list of words and we would like you to rate how relevant each of these words would be for you in this scenario. Some of the words will be more relevant than others, and some may be easier to rate than others. Remember, there are no correct or incorrect answers—it's completely up to you.
- *Desert Island negative.* In this task we would like you to imagine that your cruise ship has been wrecked and you are stranded on a desert island. You are alone and have no supplies or basic survival materials, so over the next couple weeks you'll need to find steady supplies of food and water and protect yourself from predators. We are going to show you a list of words and we would like you to rate how relevant each of these words would be for you in this scenario. Some of the words will be more relevant than others, and some may be easier to rate than others. Remember, there are no correct or incorrect answers—it's completely up to you.
- *Foreign Planet positive.* In this task we would like you to imagine that you have been awarded the unique opportunity to travel to a foreign planet. The planet can sustain life, but you'll be all alone on the trip. Obviously you won't be able to carry enough supplies for the entire trip, so over the next couple weeks you'll need to find steady supplies of food and water and protect yourself from predators. We are going to show you a list of words and we would like you to rate how relevant each of these words would be for you in this scenario. Some of the words will be more relevant than others, and some may be easier to rate than others. Remember, there are no correct or incorrect answers—it's completely up to you.
- *Foreign Planet negative.* In this task we would like you to imagine that the ozone layer on Earth has become so thin that it's no longer safe for humans.

You and some other people have been sent to live on a foreign planet, but when landing the space shuttle crashes and you are stranded on the planet alone. Luckily the planet can sustain life, but you have no supplies or basic survival materials. So over the next couple weeks you'll need to find steady supplies of food and water and protect yourself from predators. We are going to show you a list of words and we would like you to rate how relevant each of these words would be for you in this scenario. Some of the words will be more relevant than others, and some may be easier to rate than others. Remember, there are no correct or incorrect answers—it's completely up to you.

- *Under Sea positive.* In this task we would like you to imagine that explorers have finally discovered the Lost City of Atlantis at the bottom of the sea. You have been awarded the opportunity to explore the city all on your own. Obviously you won't be able to carry enough supplies for the entire trip, so over the next couple weeks you'll need to find steady supplies of food and protect yourself from predators. We are going to show you a list of words and we would like you to rate how relevant each of these words would be for you in this scenario. Some of the words will be more relevant than others, and some may be easier to rate than others. Remember, there are no correct or incorrect answers—it's completely up to you.
- *Under Sea negative.* In this task we would like you to imagine that the ozone layer on Earth has become so thin that it's no longer safe for humans. You and a few other people have been sent to start a new civilization under the ocean in the recently discovered Lost City of Atlantis. Unfortunately, along the way your submarine crashes and you are stranded in the underwater city. You weren't able to carry enough supplies for the entire trip, so over the next couple weeks you'll need to find steady supplies of food and protect yourself from predators. We are going to show you a list of words and we would like you to rate how relevant each of these words would be for you in this scenario. Some of the words will be more relevant than others, and some may be easier to rate than others. Remember, there are no correct or incorrect answers—it's completely up to you.

The relevance scales in all conditions ranged from 1 (least relevant) to 7 (most relevant). To obtain an empirical measure of prior familiarity with the scenarios and to double-check that our intuition about the average familiarity of each scenario was correct, we asked participants to rate their familiarity with the scenario they would be imagining. After receiving the instructions for the rating task, participants were asked to rate on a scale from 1 to 7 how familiar they felt they were with the location or concept associated with the condition to which they were assigned. The question was the same for both positive and negative valances for each scenario. The wording of the questions for the six scenarios were as follows:

- *Moving.* Before you begin the Rating Task we would like to ask you about your familiarity with moving house. You don't actually need first-hand experience, familiarity can be knowledge gained from television, film, or study. On a scale from 1 to 7 (1 being completely unfamiliar and 7 being extremely familiar) how familiar would you say you are with this concept?

- *Holiday.* Before you begin the Rating Task we would like to ask you about your familiarity with going on holiday. You don't actually need first-hand experience, familiarity can be knowledge gained from television, film, or study. On a scale from 1 to 7 (1 being completely unfamiliar and 7 being extremely familiar) how familiar would you say you are with this concept?
- *Grasslands.* Before you begin the Rating Task we would like to ask you about your familiarity with living in the grasslands of a foreign country. You don't actually need first-hand experience, familiarity can be knowledge gained from television, film, or study. On a scale from 1 to 7 (1 being completely unfamiliar and 7 being extremely familiar) how familiar would you say you are with this concept?
- *Desert Island.* Before you begin the Rating Task we would like to ask you about your familiarity with living on a desert island. You don't actually need first-hand experience, familiarity can be knowledge gained from television, film, or study. On a scale from 1 to 7 (1 being completely unfamiliar and 7 being extremely familiar) how familiar would you say you are with this concept?
- *Foreign Planet.* Before you begin the Rating Task we would like to ask you about your familiarity with space and other planets. You don't actually need first-hand experience, familiarity can be knowledge gained from television, film, or study. On a scale from 1 to 7 (1 being completely unfamiliar and 7 being extremely familiar) how familiar would you say you are with this concept?
- *Under Sea.* Before you begin the Rating Task we would like to ask you about your familiarity with deep-sea life and exploration. You don't actually need first-hand experience, familiarity can be knowledge gained from television, film, or study. On a scale from 1 to 7 (1 being completely unfamiliar and 7 being extremely familiar) how familiar would you say you are with this concept?

After completing the rating task, all participants received the same recognition test consisting of 96 items (arranged randomly) and asked to indicate whether an item had appeared on the list (marking "yes") or had not (marking "no"). The list comprised the 12 critical lures, 4 words of varying associative strength that were studied from each list (48 items), the unpresented, related associate (11th item) from each list (12 items), and 24 unrelated (6 neutral, 6 negative, 6 positive, and 6 survival related) items.

Because we are making predictions based on the amount of elaboration required by the various scenarios, we added a measure to check that what was actually imagined was consistent with our expectations. Thus, after the recognition test participants were asked to describe, in as much detail as possible, the scenario they had imagined during the rating task. Participants' responses were recorded so that idea units could be counted and scored during later analyses.

Results and Discussion

Scenario Familiarity

The familiarity ratings were averaged for each scenario and submitted to a one-way ANOVA. The test yielded a significant main effect of scenario on familiarity rating,

$F(5, 90) = 18.31$, $\eta_p^2 = .50$, $p = .001$, such that the moving ($M = 5.9$) and holiday ($M = 6.6$) scenarios were rated as more familiar than the grasslands ($M = 4.3$), desert island ($M = 4.6$), foreign planet ($M = 5.0$), and undersea ($M = 4.3$) scenarios (largest $p = .05$). There were no significant differences among the latter four scenarios (smallest $p = .19$). These data confirm our initial predictions that the moving and holiday scenarios are more commonplace than the survival-related scenarios. The four survival scenarios were rated similarly familiar, which leads us to predict that they will produce true and false memories at similar rates.

Scenario Elaboration

The descriptions of the scenarios provided by the participants at the end of the experiment were scored according to the number of idea units (or clauses) used in the description (e.g., Mayer, 1980). The number of idea units were summed for each participant and then averaged across scenario. These data were then submitted to a one-way ANOVA that yielded no significant differences in idea units across the six scenarios, $F(5, 90) = 1.34$, $p = .24$, $\eta_p^2 = .04$. Although, numerically, the moving ($M = 7.63$) and holiday ($M = 8.18$) scenarios produced fewer idea units than the grasslands ($M = 9.44$), desert island ($M = 9.58$), foreign planet ($M = 9.06$), and undersea ($M = 9.75$) scenarios, the contrasts were not significant (smallest $p = .20$). These data, again, confirm the initial hypothesis that our instruction manipulation increased the amount of elaboration used by the participants in the nonsurvival scenarios, decreasing the difference across the six scenarios.

True and False Recognition

The proportion of words correctly and falsely recognized was calculated and analyzed separately using a 3(material type: neutral, negative, survival) × 6(rating task scenario: moving, holiday, grassland, desert island, foreign planet, undersea) × 2(valance: positive, negative) repeated-measures ANOVAs. We found no significant effects of valance in any of the results, so for the remainder of the chapter, we collapsed across this variable. The results for correct recognition performance as a function of rating task are summarized in Figure 5.4. As can be seen, there was no significant main effect of rating task scenario ($p > .10$), suggesting that correct recognition did not differ based on the type of encoding scenario. However, because the correct recognition performance data are approaching ceiling, these data should be interpreted with caution (but see Footnote 1). On the other hand, the standard materials effect was replicated for correct recognition, $F(2, 186) = 26.43$, $p < .001$, $\eta_p^2 = .03$, such that survival materials ($M = .94$) were better remembered than negative material ($M = .91$; $p = .003$) but not better remembered than neutral ($M = .93$; $p > .50$).

The proportion of critical lures falsely recognized by rating task scenario is summarized in Figure 5.5. As with the correct recognition data, the ANOVA revealed no significant main effect of rating task scenario ($F < 1.0$). Unlike in past research, there was no difference in false memory production for the survival-related rating task scenarios when compared to the moving and holiday scenarios. Additionally, the ANOVA revealed a significant main effect of material type, $F(2, 180) = 27.10$, $p < .001$, $\eta_p^2 = .23$, in the expected direction: survival material ($M = .60$) produced

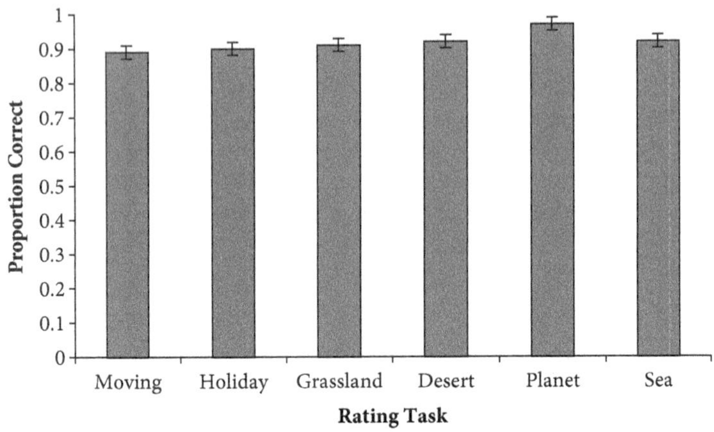

Figure 5.4 Proportion of words correctly recognized as a function of rating task in Experiment 2.

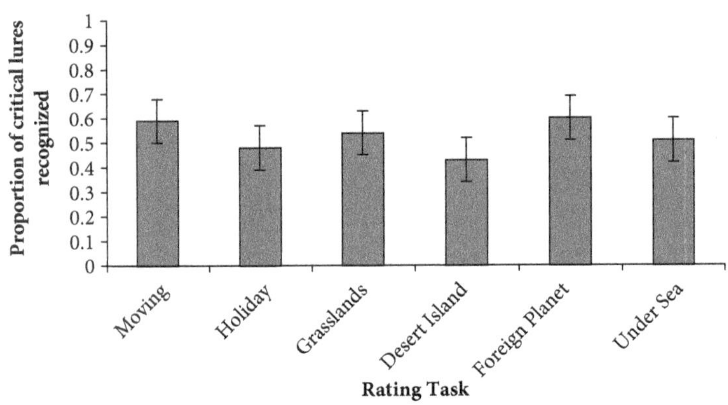

Figure 5.5 Proportion of critical lures falsely recognized as a function of rating task in Experiment 2.

significantly higher rates of false recognition than neutral materials ($M = .39$; $p < .001$) and numerically higher rates than negative materials ($M = .57$; $p > .10$).

Accuracy

Finally, to determine whether the amount of elaboration encouraged affected the overall accuracy of the various scenarios, we computed composite accuracy scores. The results of a 3(material type: neutral, negative, survival) × 6(rating task scenario: moving, holiday, grassland, desert island, foreign planet, undersea) ANOVA are summarized in Figure 5.6. The ANOVA confirmed that there was no significant main effect of rating task scenario on accuracy $F(5, 90) = 1.27$, $p = .23$, $\eta_p^2 = .07$, suggesting that increasing the amount of elaboration encouraged by the instructions removed the effect of the rating task on accuracy. As expected, there was still a significant main effect of material type, $F(2, 180) = 19.93$, $p < .001$, $\eta_p^2 = .18$, such that survival ($M = .30$) and negative ($M = .28$) materials produced significantly lower

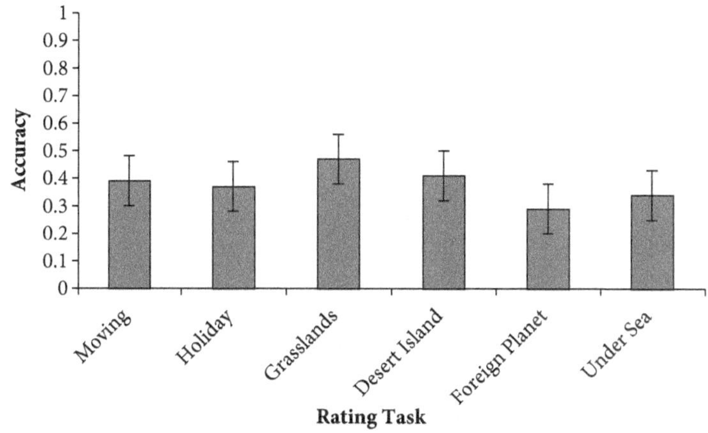

Figure 5.6 Recognition accuracy as a function of rating task in Experiment 2.

accuracy scores (largest p <.001) than neutral (M = .51) materials. The interaction was not significant.

Experiment 2 contributes several important findings to the literature. First, it replicates Howe and Derbish's (2010) materials effect with yet another combination of lists and processing conditions. Overall, survival-related materials produce greater rates of false memory production that leads to a decrease in memory accuracy. More importantly, the addition of new instructions that encouraged greater elaboration across all conditions appears to have eliminated the processing task effect reported in previous research. In previous research, as well as in Experiment 1, when participants were given no explicit elaborative instructions, the ancestrally relevant survival rating tasks generally produced higher rates of both true and false memories (e.g., Howe & Derbish, 2010; Nairne et al., 2007). However, when participants were encouraged to "fill in the details" of the scenarios for themselves, all conditions produced true and false memories at similar rates regardless of ancestral relevance.

Interestingly, the number of idea units produced during elaboration was correlated with familiarity, $r(N = 96) = .463$, p<.001, two-tailed. More importantly, when familiarity was entered as a proxy variable for idea units in an analysis of covariance, all differences in recognition rates as a function of scenario type were eliminated. That is, the covariate familiarity was significant, $F(1, 14) = 14.50$, p<.01, $\eta_p^2 = .51$, and the effects of type of scenario being rated disappeared, $F(2, 14) < 1$, $p = .733$. Together, these results suggest that the adaptive memory processing effect may not be due to unique memory processes designed specifically for ancestrally significant, survival-relevant information, but rather may be a byproduct of the novelty, and hence need for more elaborative processing, of the scenario itself to comply with the processing demands of the rating task.

EXPERIMENT 3

The results of the previous experiment are clear: When explicit elaboration instructions are given at the beginning of the rating task, the relative increase in true and false memory production normally observed for participants in the survival (e.g., grasslands, desert island) versus nonsurvival (e.g., moving, holiday) conditions disappears. When participants were encouraged to elaborate, all conditions, regardless of their survival relevance or familiarity, exhibited similar rates of true and false memory production. However, because some of the survival conditions were tested for the first time in this study, we thought it would be important to show that these results are robust. Therefore, Experiment 3 was designed to replicate and extend the results of Experiment 2 by retesting three of the six scenarios from Experiment 2 (holiday, grasslands, undersea) using a within-participant design. We split the rating task into three sections, each with different scenario instructions, so that each participant rated one third of the words in the grasslands scenario, one third of the words in the holiday scenario, and one third of the words in the undersea scenario. We anticipated that Experiment 3 should replicate the results of Experiment 2 and confirm that when participants are given the elaborative instructions, the usual adaptive memory effects disappear.

Method

Participants

A new sample of 25 undergraduates took part in this experiment (16 male, 9 female).

Materials, Design, and Procedure

The experiment was a 3(rating task scenario: holiday, grasslands, undersea) × 3(material type: neutral, negative, survival) × 3(order) design where rating task and material type were manipulated within each participant and order was manipulated between participants. The factor of order was included as a counterbalancing condition to control for any order effects in the presentation of scenario types across participants.(There were no theoretical hypotheses concerning the effects of this variable on performance.)

Nine of the twelve 10-item lists from Experiment 2 (3 neutral: *school, mountain, sweet*; 3 negative: *cry, sad, bad*; 3 survival: *death, hurt, sick*) were used to construct three separate 30-item rating tasks (see Appendix C).Each 30-item section included one list of each type. The words in each rating task were presented in random order but held constant across participants and conditions. Upon arrival, participants were assigned to one of three order conditions that determined the order that they would receive their scenarios. Across participants, one third received their scenarios as holiday, grasslands, undersea, another third received their scenarios as grasslands, undersea, holiday, and the final third received their scenarios as undersea, holiday, grasslands. Because the order of the words and sections was held constant, this meant that across participants each word was rated in all three conditions.

The rating task procedure was identical to that in Experiment 2.To the right of each word was a standard 7-point scale on which participants could rate the item for its relevance value. Before each individual rating task participants were given a brief description of their scenario and asked to imagine themselves in that situation. The instructions for the three scenarios were:

- *Holiday.* In this task we would like you to imagine that you have had a terrible year. You've dropped out of school and are working in a job you hate. You've decided to use the last of your savings to go on holiday to forget your troubles. You'll be staying in a fancy resort where all your basic needs will be taken care of. Over the next couple weeks, you will want to find different activities to pass the time and maximize your enjoyment. We are going to show you a list of words and we would like you to rate how relevant each of these words would be for you in this scenario. Some of the words will be more relevant than others, and some may be easier to rate than others. Remember, there are no correct or incorrect answers—it's completely up to you.

- *Grasslands*. In this task we would like you to imagine that you have been stranded in the grasslands of a foreign country. You are all alone and have no supplies or basic survival materials, so over the next couple weeks you'll need to find steady supplies of food and water and protect yourself from predators. We are going to show you a list of words and we would like you to rate how relevant each of these words would be for you in this scenario. Some of the words will be more relevant than others, and some may be easier to rate than others. Remember, there are no correct or incorrect answers—it's completely up to you.
- *Under Sea*. In this task we would like you to imagine that the ozone layer on Earth has become so thin that it's no longer safe for humans. You and a few other people have been sent to start a new civilization under the ocean in the recently discovered Lost City of Atlantis. Unfortunately, along the way your submarine crashes and you are stranded in the underwater city. You weren't able to carry enough supplies for the entire trip, so over the next couple weeks you'll need to find steady supplies of food and protect yourself from predators. We are going to show you a list of words and we would like you to rate how relevant each of these words would be for you in this scenario. Some of the words will be more relevant than others, and some may be easier to rate than others. Remember, there are no correct or incorrect answers—it's completely up to you.

Before beginning the first rating task section, participants were asked to rate how familiar they felt they were with the concept described in each scenario. We used the same familiarity instructions and scale as in Experiment 2, except that now participants rated all three scenarios. After rating their familiarity, participants began the first rating task section. After completing the first 30-item section they were given a second set of instructions with a new scenario for them to imagine and the second set of 30 items. This procedure was repeated until participants had completed all three rating task sections. Finally, participants were asked to complete a cumulative 96-item recognition test in the same format as Experiment 2.

Results and Discussion

Scenario Familiarity

As in Experiment 2, the average familiarity ratings were calculated for each of the three scenarios. The data were then submitted to a one-way ANOVA, which confirmed that, as in Experiment 2, the holiday (M = 5.91) scenario was rated more familiar on average than the grasslands ($M = 3.61$) and undersea ($M = 3.56$) scenarios, $F(2, 44) = 38.95, p = .001, \eta_p^2 = .64$. These data replicate the results of Experiment 2 and confirm our initial intuition about the familiarity and novelty of the scenarios.

True and False Recognition

The proportion of words correctly recognized was calculated and analyzed using a 3(material type: neutral, negative, survival) × 3(rating task scenario: holiday, grassland, undersea) ANOVA. Correct recognition performance was near ceiling and showed no significant main effects or interactions (smallest $p = .06$). There was a marginally significant effect of material type, $F(2, 44) = 2.93, p = .06, \eta_p^2 = .20$, such that negative ($M = .97$) and survival ($M = .96$) materials produced better true memory performance than neutral ($M = .93$) materials (largest $p = .05$). All other effects and interactions were not significant (smallest $p > .10$). Although these results confirm the data reported in Experiment 2, they are still difficult to interpret because of the overall high levels of performance (but see Footnote 1).

As with the correct recognition data, the proportion of critical lures falsely recognized was calculated and submitted to a 3(material type) × 3(rating task scenario) ANOVA. The results are summarized in Figure 5.7. As can be seen, and as confirmed by the ANOVA, there was no main effect of rating task scenario on false memory production ($F < 1.0$). These results are consistent with the findings reported in Experiment 2: When given explicit elaborative instructions before the rating task, participants perform the same in all conditions regardless of survival or ancestral relevance. Interestingly, there was a main effect of material type, $F(2, 44) = 4.42, p = .02, \eta_p^2 = .17$, such that survival material ($M = .67$) produced significantly higher rates of false memories than neutral ($M = .45; p < .01$) and negative ($M = .48; p = .02$) materials. These results are consistent with Howe and Derbish's (2010) findings using rather different conditions.

Accuracy

Finally, we computed composite accuracy scores. The results of the 3(material type) × 3(rating task scenario) ANOVA are summarized in Figure 5.8. The ANOVA

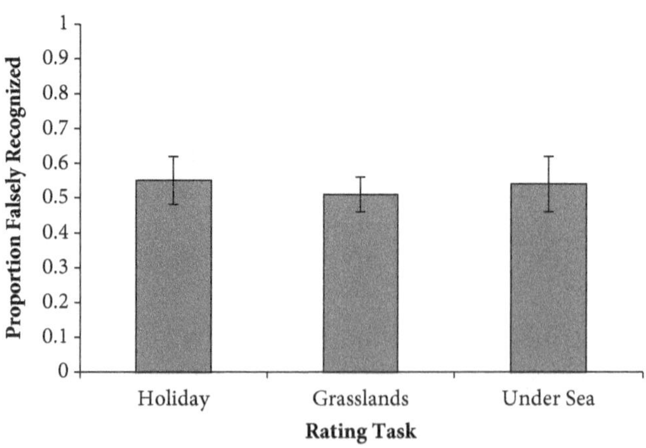

Figure 5.7 Proportion of critical lures falsely recognized as a function of rating task in Experiment 3.

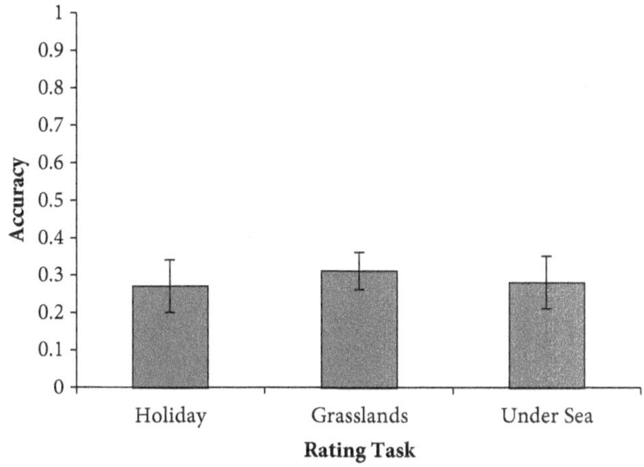

Figure 5.8 Recognition accuracy as a function of rating task in Experiment 3.

confirmed that there was no significant main effect of rating task scenario on accuracy, $F(2, 22) = 1.23$, $p = .31$, $\eta_p^2 = .10$. These results resemble the accuracy results from Experiment 2, even though the current experiment used a within-participant design. What this suggests is that explicit elaboration instructions do, in fact, have a significant impact on overall memory accuracy regardless of whether survival and nonsurvival scenarios are processed within or across participants.

One of the drawbacks to using a within-participants design is the possibility of order effects contaminating the results. The ANOVAs for both correct recognition and false recognition yielded significant order by rating task scenario interactions, $F(4, 44) = 2.96$, $p = .03$ and $F(4, 44) = 4.18$, $p = .01$, respectively. These results are summarized in Table 5.2. As can be seen, whichever scenario was rated first typically exhibited lower true and higher false memory performance. This effect is consistent across the three scenarios and does not suggest any differences in overall true or false memory production for any one scenario. Most likely participants underestimate the difficulty of the task to begin with and improve with time and practice.

In summary, the results of Experiment 3 confirm what we found in Experiment 2 despite using a within-participant design. Using a variety of scenarios we have found that the adaptive memory effect can be eliminated by increasing the amount of elaboration produced at the beginning of the rating task. However, it is always difficult to compare results across studies for a variety of reasons, one of which concerns any inconsistency in control conditions and methodology. Although our results suggest

Table 5.2. Order Effects in Experiment 3

	Holiday 1st		**Grasslands 1st**		**Under Sea 1st**	
Scenario	**True**	**False**	**True**	**False**	**True**	**False**
Holiday	.94	.61	.95	.63	.95	.41
Grasslands	.98	.38	.95	.78	.96	.37
Under Sea	.97	.43	.96	.52	.91	.67

that when given elaborative instructions participants in the nonsurvival (moving and holiday) conditions perform more like participants in the survival conditions, we cannot conclude that our results represent true causality without establishing that in the absence of elaborative instructions our conditions would produce typical adaptive memory effects. To establish the validity of these effects, we conducted a fourth experiment.

EXPERIMENT 4

Thus far, the foreign planet and undersea conditions have never been tested without the increased elaboration instructions. So before we can determine whether the non-evolutionarily relevant survival conditions produce similar memory effects to the evolutionarily relevant conditions because of their novelty, or whether they produce similar effects to the nonsurvival conditions because they are not evolutionarily relevant, we need to establish a basic level of memory performance without any additional elaboration instructions. To do this, Experiment 4 was designed to establish a baseline for true and false memory rates for a nonsurvival (moving house), an ancestrally relevant survival (grasslands), and a survival-related but evolutionarily irrelevant (foreign planet) set of conditions. We expected that if our hypothesis is correct, when participants are given standard rating task instructions we will see the typical processing effect: That is, the moving house scenario will produce significantly lower true and false memories, but higher accuracy, than the survival conditions. Moreover, if the adaptive memory effect involves only ancestrally relevant survival scenarios, then only the grasslands, not the foreign planet, condition should exhibit enhanced true and false memory rates.

Method

Participants

A new sample of 18 undergraduates participated in this study (8 male, 10 female).

Materials, Design, and Procedure

The same 12 lists were used as in Experiment 2 and were presented as a single 120-item rating task. To the right of each word was a standard 7-point scale on which participants were asked to rate each item for its relevance to a particular scenario. Upon arrival participants were assigned to one of three scenarios, moving, grasslands, or foreign planet. Depending on their assigned condition, participants were given one of the following sets of instructions modeled on the standard adaptive memory instructions:

- *Moving.* We would like you to imagine that you have to move house and there is no one around to help you so you will have to move your things from one place to the other alone. We are going to show you a list of words and we would like you to rate them on a scale from 1–7 for how relevant the meaning of that word would be for you successfully moving house.

- *Grasslands*. We would like you to imagine that you are stranded in the grasslands of a foreign country. You are all alone and have no supplies or basic survival materials, so over the next couple of weeks you will need to find food and shelter and protect yourself from predators. We are going to show you a list of words and we would like you to rate them on a scale from 1–7 for how relevant the meaning of that word would be for your survival.
- *Foreign Planet*. We would like you to imagine that you are stranded on a foreign planet. You can breathe, but you are all alone and have no supplies or basic survival materials, so over the next couple of weeks you will need to find food and shelter. We are going to show you a list of words and we would like you to rate them on a scale from 1–7 for how relevant the meaning of that word would be for your survival.

After completing the rating task, participants were given the same 96-item recognition test as in Experiment 2.

Results and Discussion

The proportion of words correctly recognized was calculated and submitted to a 3(material type: neutral, negative, survival) × 3(rating task scenario: moving, grassland, foreign planet) mixed ANOVA. The results are summarized in Figure 5.9. There was no main effect of scenario ($F < 1.0$), but once again the data were nearing ceiling. Although the ANOVA failed to detect any effect of rating task on correct recognition, the standard main effect of materials remained, $F(2, 30) = 3.65, p = .04, \eta_p^2 = .20$, such that survival materials ($M = .93$) produced significantly higher rates of true memories than negative ($M = .86; p = .03$), but not neutral ($M = .92; p > .50$), materials.

As with the correct recognition data, the proportion of critical lures falsely recognized was calculated and submitted to the same 3(materials) × 3(rating task scenario) ANOVA. This time the ANOVA yielded a significant main effect of rating task, $F(2, 15) = 4.26, p = .03, \eta_p^2 = .36$.As can be seen in Figure 5.10, in the absence of elaborative instructions participants performed as expected: participants in the moving scenario produced significantly fewer false memories than participants in

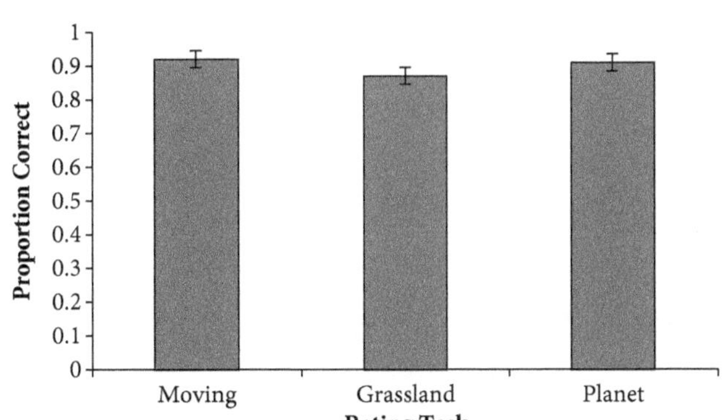

Figure 5.9 Proportion of words correctly recognized as a function of rating task in Experiment 4.

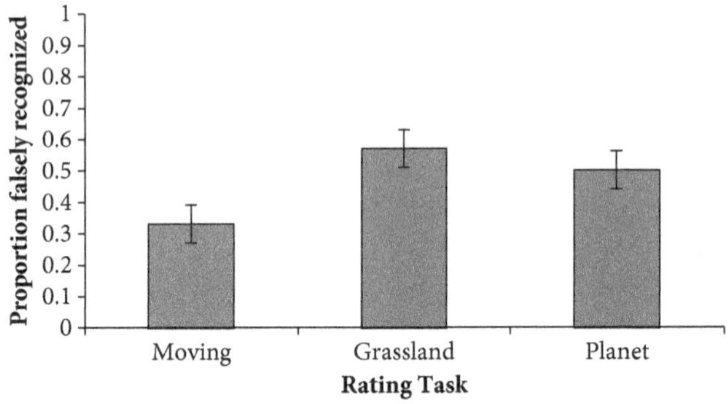

Figure 5.10 Proportion of critical lures falsely recognized as a function of rating task in Experiment 4.

the grassland ($p = .01$) and foreign planet ($p = .07$) scenarios. Unlike in Experiments 2 and 3, the results of Experiment 4 replicated Howe and Derbish's (2010) findings that the grasslands condition produced significantly more false memories than the moving scenario. More importantly, these results showed that ancestral relevance of the survival scenario is irrelevant: That is, both the foreign planet and grasslands scenarios produced equivalent true and false memory rates. Finally, there was a main effect of material, $F(2, 30) = 44.55$, $p < .001$, such that survival-related material produced the highest rate of false recognition ($M = .67$), which was significantly higher than for negative ($M = .51$) and neutral ($M = .22$) materials (all contrasts were significant; largest $F < .01$).

We calculated the overall accuracy scores and then submitted them to a 3(materials) × 3(rating task) ANOVA. There was a significant main effect of rating task scenario, $F(1, 15) = 3.87$, $p = .04$, $\eta_p^2 = .34$. As can be seen in Figure 5.11, the moving scenario was highest in overall memory accuracy, followed by the foreign planet and grasslands condition, and there was no significant difference between the latter two scenarios. Interestingly, these results are not consistent with a purely evolutionary account of the adaptive memory effect. Despite the fact that the foreign planet scenario was not evolutionarily relevant, it produced similar false memory and accuracy

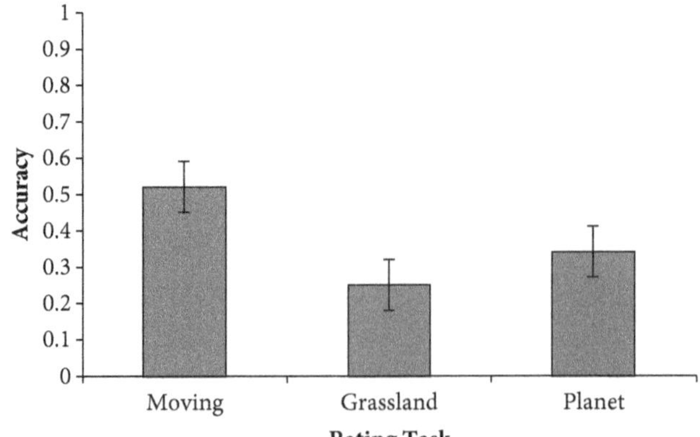

Figure 5.11 Recognition accuracy as a function of rating task in Experiment 4.

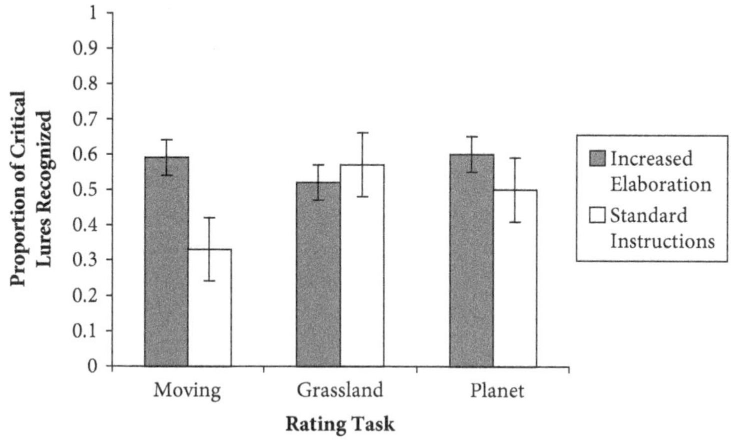

Figure 5.12 Proportion of critical lures falsely recognized as a function of rating task across Experiments 2 and 4.

rates as the evolutionarily relevant grasslands scenario. These results support the earlier claim that the results of Experiments 2 and 3 demonstrate that the adaptive memory effect observed for the grasslands condition is due primarily to the fact that it inherently encourages greater elaboration and thus greater relational processing.

COMPARING EXPERIMENTS 2 AND 4

Before we can conclude that elaboration is the cause of changes in false memory rates during survival processing, we need to compare baseline performance (with no elaboration instructions) obtained in Experiment 4 with performance in Experiment 2 when elaboration instructions were provided before the processing task. When the false recognition data from both experiments were submitted to a single ANOVA, we found that the standard moving condition (no elaboration instruction) produced significantly lower rates of false recognition than all the other conditions (largest $p<.01$). There were no other significant differences among the groups. Additionally, the only significant difference observed from manipulating elaboration instructions was in the moving scenario. All other groups performed similarly regardless of whether they were given elaborative instructions or standard instructions (Fig. 5.12). From these data we can conclude that, as predicted, only performance in the more familiar moving scenario was altered by the increased elaboration instructions. The survival (e.g., grasslands and foreign planet) conditions showed no effect due to elaboration instructions because they already required participants to spontaneously engage in significant elaboration. More importantly, these latter effects were independent of the various scenarios' evolutionary relevance.

GENERAL DISCUSSION

We began this chapter by asking several questions about the adaptive memory effect. First, we wanted to make sure that the processing effect reported in Howe and Derbish (2010) was robust. Next, because Howe and Derbish had been able to explain the survival materials effect using well-known research on the importance of themes in memory networks, we were interested in whether increased survival-related processing could be similarly explained in terms of basic memory processes without having to invoke a special adaptive memory mechanism. To that end, we suggested that the most probable basic memory variable producing these effects was elaboration during the rating task (Kroneisen & Erdfelder, 2011). Because elaboration is known to increase both true and false memory, we argued that the greater elaboration required for less familiar (e.g., grasslands and desert island) scenarios as opposed to other, more familiar (e.g., moving house, going on holiday) scenarios, would be the cause of differences in true and false memory rates. To test this, we examined if encouraging high levels of elaboration in all conditions would eliminate true and false memory performance differences regardless of scenario type. Third, we wanted to ensure that any effect due to manipulating the level of elaboration on memory performance was due to the amount of elaboration used and was not a product of individual differences. Finally, we were interested in whether the data support Nairne's (2010) original theory that human memory has developed a specific memory process for evolutionarily relevant survival information, or whether it is the unfamiliar nature of all survival scenarios, evolutionarily relevant or not, that drives the adaptive memory effect.

The present experiments provide some answers to these questions. In particular, they converge on the conclusion that the typical increase in true and false memory seen when participants are asked to rate words for their survival relevance in traditional ancestral (e.g., desert island, grasslands) scenarios could be a product of different amounts of elaboration required in these less familiar processing conditions. First, Experiment 1 replicated and extended Howe and Derbish's (2010) findings using the grasslands scenario. That is, when asked to rate the desert island scenario, participants in our experiment produced higher rates of both true (only marginally and not statistically significant) and false memories than when they were asked to rate the same words for pleasantness. This effect was independent of the type of material being processed and clearly demonstrates a robust and consistent processing effect for the desert island (survival) condition.

Next, Experiments 2 and 3 demonstrated that when the amount of elaboration is equated, differences between survival (ancestrally relevant or not) and nonsurvival conditions disappear. We hypothesized that because the survival rating conditions are more unfamiliar, regardless of their ancestral relevance, they may require greater elaboration in order to complete the standard rating task (Martin & Pressley, 1991; Willoughby, Waller, Wood, & MacKinnon, 1993; Willoughby, Wood, & Kraftcheck, 2003). Using both a between- and a within-participant design, Experiments 2 and 3 showed that when the level of elaboration was equated (using number of idea units as the relevant index), survival processing conditions no longer exhibited greater true or false memory production compared to other nonsurvival conditions. By controlling for the amount of elaboration produced in each condition, we were able to eliminate the standard adaptive memory effect, strongly suggesting that their effect was not due to an increased efficiency for processing fitness-related information. Rather, consistent with other recent research (e.g., Burns et al., 2011; Kroneisen & Erdfelder, 2011), survival conditions may be distinct from nonsurvival conditions in the amount of spontaneous elaboration normally required to accomplish such incidental memory tasks.

Experiment 4 confirmed that the results of Experiments 2 and 3 were not simply artifacts of the lists or scenarios we had chosen or some flaw in our methodology. When we removed the explicit elaboration instructions but kept all other procedures the same, the standard adaptive memory effect re-emerged. We hypothesized that if the survival scenarios were inherently producing greater elaboration due to their relative novelty, then encouraging greater elaboration in all conditions would alter performance in the more familiar scenarios but not in the survival scenarios, something that in turn should eliminate the effect of spontaneous elaboration on memory performance. This hypothesis was confirmed when we compared the results of Experiments 2 and 4 and showed that only the familiar moving scenario was affected by the elaborative instructions. That is, performance increased in the nonsurvival condition but remained unchanged in all survival conditions, regardless of ancestral relevance.

The results of these four experiments support our contention that not only do the elaboration instructions eliminate the standard adaptive memory effect, they do so by increasing the production of false memories in the more familiar moving condition. Because participants were forced to draw on more varied and complex prior knowledge, their true memory performance may be enhanced, but this comes at the cost of more memory illusions. These data support the hypothesis that the survival conditions inherently require an extensive amount of spontaneous elaboration (relational processing) and thus there was little or no benefit when an explicit elaboration instruction was added. However, in the conditions that were the most familiar, encouraging participants to use more elaboration caused them to perform more like the people in the survival conditions, conditions that already demand greater elaborative effort.

This brings us to the final and perhaps most interesting question in this chapter: Is the evolutionary explanation for the adaptive memory effect consistent with these results? Tooby and Cosmides (2005) suggest that any evolutionarily adapted memory process would be specific for information processed in a way that would have been important or relevant for our ancestors. Our data, as well as that collected by a number of other researchers (e.g., Burns et al., 2011; Kroneisen & Erdfelder, 2011; Wilson et al., 2011), suggest that the adaptive memory effect is not due to a unique memory advantage for evolutionarily adaptive survival conditions but rather is a result of the unique unfamiliarity of the usually tested survival scenarios compared to the nonsurvival control scenarios, something that leads to increases in elaborative processing. Our ancestrally irrelevant survival conditions were statistically matched to the more traditional survival conditions for familiarity and valance; they differed only in their evolutionary relevance. When compared to these otherwise matched survival conditions, the traditional survival scenarios no longer produced memory performance differences. It was only the more familiar moving scenario that showed the predicted differences in memory performance.

Although the data reported in this chapter do not support the theory that human memory has specifically adapted to uniquely process survival-related information, it is still possible that these effects are evolutionarily beneficial. The most dangerous situations are those we are unfamiliar with, so a memory process that promotes increased elaboration and activation of related information in unfamiliar scenarios could be adaptive indeed (also see Howe, 2011a, 2011b; Schacter, Guerin, & St. Jacques, 2011).

This increased elaboration might lead to a state of hypervigilance and thus more cautious behavior (for a summary of hypervigilance in stressful situations, see Howe, Toth, & Cicchetti, 2006). In this case the increase in false memories is not itself adaptive but rather is a byproduct of something that is adaptive, namely the integration of prior knowledge from diverse sources through elaboration. This integration of information may be important in solving problems, including problems arising from unfamiliar situations that are life-threatening. Indeed, there exists evidence that false memories, like true memories, can and do prime solutions to insight-based problems (see Howe, Garner, Charlesworth, & Knott, 2011; Howe, Garner, Dewhurst, & Ball, 2010).

Few would disagree with the proposition that our memory system has been sculpted by the same adaptive forces that have helped forge other fitness-relevant (survival-related) characteristics. Although there is no doubt that memory has an adaptive function, it is difficult to conclude with any degree of certainty what factors influenced the evolution of our memory systems (e.g., see Howe, 2011b; Paivio, 2007). Of course, the more information we gain about the adaptive utility of our basic memory processes, the closer we may get to understanding the adaptive function of our memory system(s). The adaptive memory literature is relatively new and thus is still in its early stages. This chapter is one of a handful of papers to show empirical evidence pertaining to the proximate mechanism responsible for the adaptive memory effect, and although there are still more experiments to be done before we can say for certain what causes these effects, the results of the current set of experiments give us a very good starting point for further study.

ACKNOWLEDGMENT

The research reported in this chapter was supported by a grant to MLH from the Economic and Social Research Council of Great Britain (RES-062-23-0452).

NOTE

1. Note that using d′ did not change the pattern of findings. Moreover, the absence of response bias held across all four of our experiments in this chapter. Although we acknowledge that near-ceiling effects for true recognition can affect the calculation of net accuracy scores, we opted to report them because the effects observed in our experiments parallel those obtained in studies in which ceiling effects were not observed (e.g., Howe & Derbish, 2010; Otgaar & Smeets, 2010). Indeed, ongoing research in our lab has shown that the effects reported in this chapter are replicable using both recall and recognition tests and occurred in the absence of any ceiling effects.

REFERENCES

Arndt, J., & Reder, L. M. (2003). The effect of distinctive visual information on false recognition. *Journal of Memory and Language, 48*, 1–15.

Balota, D. A., Yap, M. J., Cortese, M. J., Hutchison, K. A., Kessler, B., Loftis, B., Neely, J. H., Nelson, D. L., Simpson, G. B., & Treiman, R. (2007). The English Lexicon Project. *Behavior Research Methods, 39*, 445–459.

Bradley, M. M., & Lang, P. J. (1999). *Affective norms for English words (ANEW):Instruction manual and affective ratings.* Technical Report C-1, The Centerfor Research in Psychophysiology, University of Florida, Gainesville, FL.

Bröder, A., Krüger, N., & Schütte, S. (2011). The survival processing memory effect should generalize to source memory, but it doesn't. *Psychology, 2*, 896–901.

Burns, D. J., Burns, S. A., & Hwang, A. J. (2011). Adaptive memory: Determining the proximate mechanisms responsible for the memorial advantages of survival processing. *Journal of Experimental Psychology: Learning, Memory, and Cognition, 37*, 206–218.

Clark, J. M., & Paivio, A. (2004).Extensions of the Paivio, Yuille, and Madigan (1968) norms. *Behavior Research Methods, Instruments, & Computers, 36*, 371–383.

Collins, A. M., & Loftus, E. F. (1975). Spreading-activation theory of semantic memory. *Psychological Review,82*, 407–428.

Deese, J. (1959). On the prediction of occurrence of particular verbal intrusions on immediate recall. *Journal of Experimental Psychology, 58*, 17–22.

Howe, M. L. (2011a). The adaptive nature of memory and its illusions. *Current Directions in Psychological Science, 20*, 312–315.

Howe, M. L. (2011b). *The nature of early memory: An adaptive theory of the genesis and development of memory.* New York: Oxford University Press.

Howe, M. L., &Derbish, M. H. (2010). On the susceptibility of adaptive memory to false memory illusions. *Cognition, 115*, 252–267.

Howe, M. L., Garner, S. R., Charlesworth, M., & Knott, L. M. (2011). A brighter side to memory illusions: False memories prime children's and adults' insight-based problem solving. *Journal of Experimental Child Psychology, 108*, 383–393.

Howe, M. L., Garner, S. R., Dewhurst, S. A., & Ball, L. J. (2010). Can false memories prime problem solutions? *Cognition, 117*, 176–181.

Howe, M. L., Toth, S. L., & Cicchetti, D. (2006). Memory and developmental psychopathology. In D. Cicchetti & D. Cohen (Eds.), *Developmental psychopathology* (2nd Ed.). *Vol. 2: Developmental neuroscience* (pp. 629–655). New York: Wiley.

Karpicke, J. D., McCabe, D. P., & Roediger, H. L., III (2008). False memories are not surprising: The subjective experience of an associative memory illusion. *Journal of Memory and Language, 58*, 1065–1079.

Kroneisen, M., & Erdfelder, E. (2011). On the plasticity of the survival processing effect. *Journal of Experimental Psychology: Learning, Memory, and Cognition, 37*, 1553–1562.

Martin, V. L, & Pressley, M. (1991). Elaborative interrogation effects depend on the nature of the question. *Journal of Educational Psychology, 83*, 113–119.

Mayer, R. E. (1980). Elaboration techniques that increase the meaningfulness of technical text: An experimental test of the learning strategy hypothesis. *Journal of Educational Psychology, 72*(6), 770–784.

Nairne, J. S. (2010). Adaptive memory: Evolutionary constraints on remembering. In B. H. Ross (Ed.), *The psychology of learning and motivation: Advances in research and theory* (Vol. 53, pp. 1–32). San Diego, CA: Elsevier.

Nairne, J. S., Thompson, S. R., & Pandeirada, J. N. S. (2007). Adaptive memory: Survival processing enhances retention. *Journal of Experimental Psychology: Learning, Memory, and Cognition, 33*, 263–273.

Nelson, D. L., McEvoy, C. L., & Schreiber, T. A. (1998).The University of South Florida word association, rhyme, and word fragment norms. Available online at: http://www.usf.edu/FreeAssociation/.

Otgaar, H., & Smeets, T. (2010). Adaptive memory: Survival processing increases both true and false memory in adults and children. *Journal of Experimental Psychology: Learning, Memory, and Cognition*, *36*, 1010–1016.

Paivio, A. (2007). *Mind and its evolution: A dual-coding theoretical approach*.Mahwah, NJ: Erlbaum.

Roediger, H. L., III, & McDermott, K. B. (1995). Creating false memories: remembering words not presented in lists. *Journal of Experimental Psychology: Learning, Memory, & Cognition*, *21*, 803–814.

Schacter, D. L.Guerin, S. A., & St. Jacques, P. L. (2011). Memory distortion: An adaptive perspective. *Trends in Cognitive Sciences*, *15*, 467–474.

Soderstrom, N. C., & McCabe, D. P. (2011). Are survival processing memory advantages based on ancestral priorities? *Psychonomic Bulletin & Review*, *18*, 564–569.

Thomas, A. K., Hannula, D. E., & Loftus, E. F. (2007). How self-relevant imagination affects memory for behavior. *Applied Cognitive Psychology*, *21*, 69–86.

Tooby, J., & Cosmides, L. (2005). Conceptual foundations of evolutionary psychology. In D. M. Buss (Ed.), *The handbook of evolutionary psychology* (pp. 5–67). Hoboken, NJ: Wiley.

Tse, C.-S., & Altarriba, J. (2010). Does survival processing enhance implicit memory? *Memory & Cognition*, *38*, 1110–1121.

Willoughby, T., Waller, T.G., Wood, E., & MacKinnon, G.E. (1993). The effect of prior knowledge on an immediate and delayed associative learning task following elaborative interrogation. *Contemporary Educational Psychology*, *18*, 36–46.

Willoughby, T., Wood, E., & Kraftcheck, E. R. (2003).When can a lack of structure facilitate strategic processing of information? *British Journal of Educational Psychology*, *73*, 59–69.

Wilson, S., Darling, S., & Sykes, J. (2011). Adaptive memory: Fitness relevant stimuli show a memory advantage in a game of pelmanism. *Psychonomic Bulletin &Review*, *18*, 781–786.

APPENDIX A: LISTS USED IN EXPERIMENT 1

Neutral Lists			
BREAD	**WINDOW**	**MUSIC**	**FRUIT**
baker	blinds	band	berry
bun	curtain	concert	cherry
butter	door	harmony	citrus
crust	drapes	jazz	kiwi
dough	glass	orchestra	pear
garlic	ledge	radio	plum
loaf	pane	rhythm	produce
stale	shatter	stereo	raspberry
toast	shutter	symphony	strawberry
wheat	sill	tune	vegetables
Negative Lists			
SAD	**LIE**	**CRY**	**BAD**
blues	bluff	baby	attitude
depressed	deceive	emotion	awful
despair	deception	handkerchief	good
frown	deny	laugh	mischief
grief	dishonest	onion	nasty
happiness	perjury	sensitive	sin
happy	rumor	sob	terrible
lonely	truth	tears	unpleasant
misery	untrue	upset	villain
remorse	untruthful	weep	worse
Survival Lists			
FIGHT	**PAIN**	**SICK**	**DEATH**
arguing	ache	disease	burial
battle	aspirin	fever	casket
conflict	cramp	flu	cemetery
debate	headache	healthy	funeral
disagreement	hurt	hospital	grave
feud	injury	medicine	life
fists	ouch	nausea	murder
riot	sore	throw up	suicide
struggled	torture	virus	tragedy
war	ulcer	well	widow

APPENDIX B: LISTS USED IN EXPERIMENTS 2 AND 4

Neutral Lists			
SWEET	**MONEY**	**MOUNTAIN**	**SCHOOL**
honey	job	climber	campus
bitter	bill	hill	university
sugar	income	climb	semester
sour	salary	peak	education
candy	debt	hike	principal
tart	earn	top	educate
tangy	pay	valley	backpack
cute	greed	cliff	book
dessert	tax	hiker	college
nice	cent	summit	learning
Negative Lists			
SAD	**FAT**	**CRY**	**ANGER**
blues	blubber	baby	rage
depressed	thin	emotion	hostility
despair	calories	handkerchief	fury
frown	skinny	laugh	enrage
grief	slim	onion	frustration
happiness	weight	sensitive	temper
happy	diet	sob	revenge
lonely	thigh	tears	mad
misery	slender	upset	pissed
remorse	wide	weep	mean
Survival Lists			
WAR	**HURT**	**SICK**	**DEATH**
revolution	harm	disease	burial
veteran	pain	fever	casket
peace	injury	flu	cemetery
nuclear	ouch	healthy	funeral
missile	offend	hospital	grave
soldier	sore	medicine	life
tank	wound	nausea	murder
cannon	bruise	throw up	suicide
bomb	pinch	virus	tragedy
destruction	punish	well	widow

APPENDIX C: LISTS USED IN EXPERIMENT 3

Neutral Lists		
SWEET	**MOUNTAIN**	**SCHOOL**
honey	climber	campus
bitter	hill	university
sugar	climb	semester
sour	peak	education
candy	hike	principal
tart	top	educate
tangy	valley	backpack
cute	cliff	book
dessert	hiker	college
nice	summit	learning
Negative Lists		
SAD	**BAD**	**CRY**
Blues	attitude	baby
depressed	good	emotion
despair	awful	handkerchief
frown	nasty	laugh
grief	mischief	onion
happiness	sin	sensitive
happy	terrible	sob
lonely	unpleasant	tears
misery	villain	upset
remorse	worse	weep
Survival Lists		
HURT	**SICK**	**DEATH**
harm	disease	burial
pain	fever	casket
injury	flu	cemetery
ouch	healthy	funeral
offend	hospital	grave
sore	medicine	life
wound	nausea	murder
bruise	throw up	suicide
pinch	virus	tragedy
punish	well	widow

6

On the Domain-Specificity of Survival Processing Advantages in Memory

NICHOLAS C. SODERSTROM AND ANNE M. CLEARY

The past few decades have seen a change in terms of how the study of human memory is approached. Whereas it was once common for memory to be viewed as a domain-general, unitary construct, it is now much more widely assumed that memory is multifaceted, and that it likely consists of multiple interacting components that evolved for their adaptive purpose or utility (e.g., see Klein, Cosmides, Tooby, & Chance, 2002, for a review). It is now generally accepted that cognitive processes, including memory processes, are domain-specific in ways that are reflective of selection pressures from the human ancestral past. However, there is still some question about the *degree* of domain-specificity of these processes.

An adaptive memory process for which this is readily apparent is the survival processing effect (e.g., Nairne & Pandeirada, 2008a, 2008b, 2010; Nairne, Pandeirada, & Thompson, 2008; Nairne, Thompson, & Pandeirada, 2007), which is the finding that words that are processed during encoding in terms of their usefulness in a survival scenario (e.g., being stranded on the grasslands without basic survival necessities) are better remembered than words processed according to every other memory-encoding strategy known to be effective in supporting excellent retention (e.g., imagery, self-reference, rating pleasantness). On the one hand, the fact that such survival processing leads to better memory than other known mnemonic techniques suggests some degree of domain-specificity of human memory processes: Memory is best in those situations that are most similar to the situations under which memory processes likely evolved. On the other hand, the degree of domain-specificity is unclear. Is the process so specific that it actually requires one to encode items within the context of ancestral environments (e.g., grasslands)? Or is the process a more general adaptive process related to survival more broadly? This chapter reviews recent research that helps to address this question.

THE SURVIVAL PROCESSING EFFECT

Espousing a functionalist approach, Nairne, Thompson, and Pandeirada (2007) started with the hypothesis that nature—through the workings of evolution by natural selection—tuned our memory systems to solve fitness-relevant problems, such as finding food, avoiding harm, and identifying mates. Consequently, they argued, memory should be sensitive to processing information in terms of its survival utility. Thus, Nairne et al. produced *a priori* predictions based on an evolutionary hypothesis. They predicted that memory would be better in a condition meant to simulate the presumed survival processing demands of human ancestors than in comparison conditions in which mnemonic techniques already known to enhance memory were used.

The general survival processing paradigm was as follows: Participants were first assigned to one of three experimental conditions: (1) *Survival*—in which participants were instructed to imagine themselves stranded on the grasslands of a foreign land without basic survival necessities, (2) *Moving*—in which participants were to imagine themselves planning a move to a new home in a foreign land, and (3) *Pleasantness*—in which participants were instructed to simply rate the pleasantness of items. Thus, the conditions differed in terms of the instructions orienting participants to the task.

In each condition, participants were then shown single words, presented individually for a few seconds each, and were asked to rate the relevance/pleasantness of each word according to their scenario. These ratings were made on a 5-point scale, where 1 = *totally irrelevant/unpleasant* and 5 = *totally relevant/pleasant*. After this rating phase, participants engaged in a brief distracter task before taking a surprise (i.e., incidental) memory test in which the rated words were to be remembered. Memory retention was then compared across the three conditions, with the prediction being that processing items in terms of their survival utility (i.e., in the context of the survival scenario) should benefit memory more than the other, non–fitness-relevant conditions. Indeed, as shown in Figure 6.1, Nairne et al. (2007) were the first to show that survival processing did, in fact, lead to superior memory retention, a result that

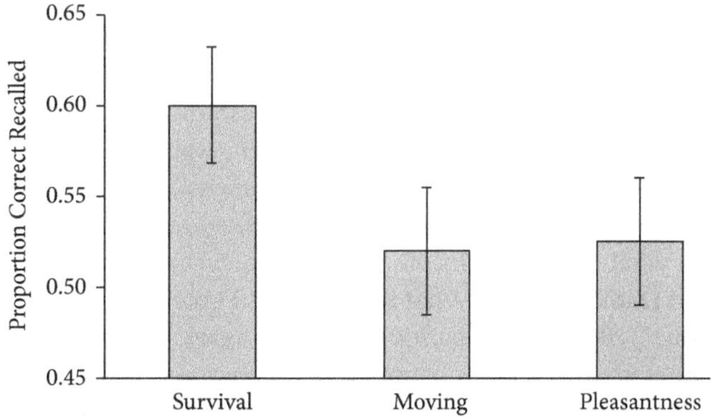

Figure 6.1 Mean proportion correct recall by rating condition (data adapted from Nairne, Thompson, & Pandeirada, 2007).

held across both between- and within-subjects designs, and for both recall and recognition memory tests. Furthermore, explanations based on congruency effects—that items given higher ratings are better remembered than those given lower ratings (see Craik & Tulving, 1975)—and study times were ruled out because average relevance ratings and response times did not differ across conditions in ways consistent with these possible explanations. Thus, this survival processing advantage was viewed as being consistent with the idea that nature sculpted our memory systems to be tuned to fitness-relevant information.

Since the first survival processing advantage in memory was reported by Nairne et al. (2007), this effect has been replicated many times by Nairne's research group (Nairne & Pandeirada, 2008, 2010, 2011; Nairne, Pandeirada, Gregory, & Van Arsdall, 2009; Nairne, Pandeirada, & Thompson, 2008; Nairne, Van Arsdall, Pandeirada, & Blunt, 2012), as well as by other laboratories (e.g., Kang, McDermott, & Cohen, 2008; Kostic, McFarlan, & Cleary, 2012; Otgaar, Smeets, & Van Bergen, 2010; Soderstrom & McCabe, 2011; Weinstein, Bugg, & Roediger, 2008), using a variety of experimental designs, control conditions, and materials. For example, both Otgaar et al. (2010) and Nairne et al. (2012) showed that the survival processing advantage generalizes to stimuli other than just words—in these cases, pictures and locations, respectively. Moreover, Nairne et al. (2008) showed that survival processing leads to better memory than some of the "deepest" levels of encoding reported in the memory literature—including imagery (Paivio, 1969), self-reference (Roger, Kuiper, & Kirker, 1977), and generation (Slamecka & Graf, 1978)—leading Nairne et al. (2008) to conclude that "survival processing is one of the best—if not *the* best—encoding procedures yet identified in human memory research, at least when free recall is used as the retention measure" (p. 180).

It should be noted, however, that important boundary conditions have been identified, which suggests that the survival processing advantage is not ubiquitous. As examples, the survival advantage has failed to materialize after processing faces (Savine, Scullin, & Roediger, 2011); on tests of implicit (Tse & Altarriba, 2010) and source (Bröder, Krüger, & Schütte, 2011) memory; and when certain types of encoding processes are equated between scenarios, such as item-specific and relational processing (Burns, Burns, & Hwang, 2011), and elaboration (Kroneisen & Erdfelder, 2011).

Although the general explanation for the survival processing advantage is that our memory systems have been tuned by nature to process fitness-relevant information, identifying the proximate mechanisms responsible for this advantage has garnered considerable interest. For example, Klein, Robertson, and Delton (2010) investigated the potential role of planning in the survival processing advantage, noting that this dimension is often confounded with survival (see also Klein, 2012). That is, asking what is relevant in the context of a survival scenario might activate future-oriented thinking (e.g., "What items would I take if I knew I was going to be in this situation?"). By using scenarios that teased apart survival and planning components, Klein et al. found that planning was, in fact, an important feature driving the memory advantage in the survival paradigm. As will be discussed in more detail later in this chapter, the idea that future-oriented thinking might play an important role in survival processing advantages is generally consistent with the recently proposed notion that memory's primary adaptive purpose is to simulate the future (see Schacter, Addis, Hassabis, Martin, Spreng, & Szpunar, 2012, for a review of recent research on future-oriented thinking, its relation to memory, and its potential adaptiveness).

Other investigated mechanisms of the survival processing effect include arousal (e.g., Nairne et al., 2007; Soderstrom & McCabe, 2011; Otgaar et al., 2010), stress (Smeets, Otgaar, Raymaekers, Peters, & Merckelbach, 2012), isolationism (Kostic et al., 2012), congruity (Butler, Kang, & Roediger, 2009; Nairne & Pandeirada, 2011), thoughts of one's own mortality (Hart & Burns, 2012; see also Soderstrom & McCabe, 2011), and familiarity (Nairne & Pandeirada, 2010). Regardless of the responsible proximate mechanism(s), however, survival processing clearly has mnemonic benefits, a finding that fits nicely within an evolutionary framework.

THE SURVIVAL PROCESSING ADVANTAGE: HOW DOMAIN-SPECIFIC?

One strong evolutionary hypothesis regarding the survival processing advantage is that inducing problems specifically faced by our Pleistocene ancestors should lead to greater memorial benefits than inducing more modern problems, a notion captured in Nairne and Pandeirada's (2010) *ancestral priorities framework*. Indeed, Nairne and Pandeirada state that "optimal cognitive performance should be induced by ancestrally-based problems, particularly those present in ancestral foraging environments, rather than by problems faced in modern environments" (p. 2). Such an idea is rooted in the common belief that the majority of modern humans' (*Homo sapiens sapiens*) cognitive development occurred during the Pleistocene era (approximately from 1.8 million to 10,000 years ago). Consequently, it is argued, our memory systems should be tuned to the problems faced by our ancestors during that time—problems that included finding food and avoiding predators on the grasslands of the African savannah.[1] This reasoning follows from a more general idea posited by evolutionary theory, which states that specific adaptations reflect the specific problems that originally drove them into existence—that is, that adaptations are domain-*specific* as opposed to domain-*general* (see Cosmides & Tooby, 1994; Futuyma & Moreno, 1988).

Weinstein et al. (2008) were the first to show compelling evidence in favor of this strong evolutionary account using the survival processing paradigm. In both conditions described below, participants rated the survival utility of words according to a given scenario. Critically, these scenarios were set in either an ancestral or modern context. In the ancestral condition, participants imagined themselves stranded on the grasslands of a foreign land without survival materials, with their goal being to find basic necessities and to evade predators (this was identical to the standard survival scenario introduced by Nairne et al. [2007]). In the modern condition, participants read the exact same scenario with only two changes: the word *grasslands* was replaced with the word *city*, and the word *predators* was replaced with the word *attackers*. Consistent with the ancestral priorities hypothesis, Weinstein et al. found that evading *predators* on the *grasslands* led to better memory retention than did evading *attackers* in a *city* (Fig. 6.2), concluding that "our memory has adapted very well to self-preservation in the type of setting in which we might have found ourselves until very recently (i.e., the grasslands), but has not yet evolved to function optimally in more modern contexts (i.e., the city)" (p. 918).

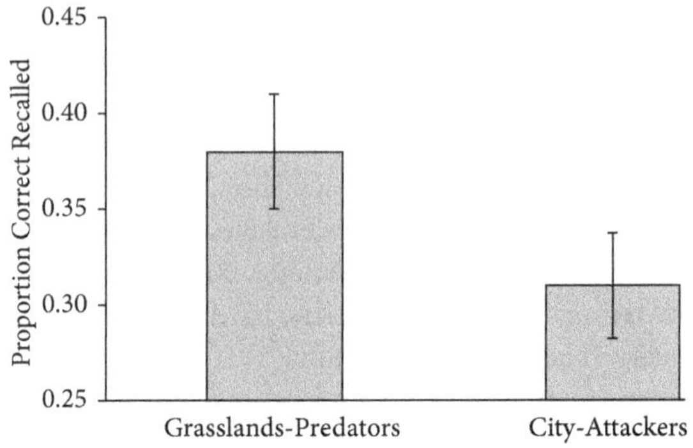

Figure 6.2 Mean proportion correct recall by rating condition (data adapted from Weinstein, Bugg, & Roediger, 2008).

Nairne and Pandeirada (2010) provided additional evidence in favor of the idea that survival processing should be sensitive to ancestral priorities. Similar to Weinstein et al. (2008), participants rated the relevance of words in either an ancestral or modern context and then took a memory test. In one experiment, for example, participants imagined having to stave off an infectious disease by finding either "relevant medicinal plants" (ancestral) or "relevant antibiotics" (modern). In yet another experiment, participants imagined being hunger-stricken, needing to either "search for and gather edible plants" (ancestral) or "search for and buy food" (modern). In each case, the ancestral condition led to better memory retention than the modern condition, leading Nairne and Pandeirada to state that "our cognitive systems must have evolved to handle survival situations engendered in natural environments because those were the situations in play during the environment of evolutionary adaptation" (p. 19).

These data reported by Weinstein et al. (2008), and Nairne and Pandeirada (2010) appear to lend credence to the notion that ancestral priorities play a role in the survival processing advantage and that our memory systems reflect domain-specific adaptations, congruent with the precise environmental problems that drove them into existence. However, recent survival processing data from our laboratories and others (e.g., Klein, 2013) cast doubt on the idea that the survival processing effect is so domain-specific as to require a scenario that describes an evolutionarily relevant context (i.e., grasslands).

In one such study, Kostic, McFarlan, and Cleary (2012) demonstrated the survival processing advantage using survival scenarios that, unlike grasslands scenarios, were not part of the human ancestral past. In their study, participants rated words according to their usefulness in various survival scenarios, comparing the grasslands scenario originally used by Nairne et al. (2007) to other scenarios in which the environment was not evolutionarily relevant. Specifically, participants rated words according to their survival utility while lost at sea, while lost in the jungle, and while lost in outer space. Although there may be some question about the evolutionary relevance of a jungle scenario, lost at sea and lost in outer space should not be evolutionarily relevant contexts; these were not contexts in which early human ancestors evolved. The various instructions for these conditions were as follows:

> "In this task we would like you to imagine that you are ***stranded in the grasslands of a foreign land/lost at sea/stranded in the jungle of a foreign land/stranded in a spaceship in space*** without any basic survival materials. Over the next few months, you'll need to find steady supplies of food and water and protect yourself from ***predators/predators/predators/danger***. We are going to show you a list of words, and we would like you to rate how relevant each of these words would be for you in this survival situation. Some of the words may be relevant and others may not—it's up to you to decide."

As shown in Figure 6.3, not only did rating items in the grasslands scenario lead to superior memory compared to the control conditions of pleasantness ratings and self-relevance ratings, but so did rating words in the lost at sea, the lost in the jungle, and the lost in outer space scenarios. Also evident from inspection of Figure 6.3 is that these scenarios (lost at sea, lost in the jungle, lost in outer space) produced equivalent levels of recall compared to the grasslands scenario, leading the authors to conclude that "the survival advantage in memory does not appear to be contingent upon the use of ancestral scenarios; it generalizes across a wide range of survival scenarios" (Kostic et al., 2012, p. 1097).

Kostic et al. (2012) hypothesized that a driving factor behind the survival processing advantage might be the implied social isolation in the survival scenarios. They noted that in all of their survival scenarios (grasslands, jungle, sea, and outer space), there is an implication that the participant is alone in these survival situations. Kostic et al. suggested that, because social isolation would likely have been extremely detrimental to survival in the human ancestral past, perhaps it is the implied social isolation that drives the survival processing effect, and not the particular environmental context used in the scenario. They suggested that the survival effect might be domain-specific insofar as it activates "isolation avoidance" mechanisms, or survival mechanisms related to perceived social isolation. They tested this idea by comparing survival scenarios in which the only difference was whether the scenario implied that the participant would be socially isolated within it or implied that the participant would be with a group of other people in it. For example, in one experiment, participants imagined being either lost at sea alone or lost at sea with a group of other people, using the scenario below, where the only difference between groups of

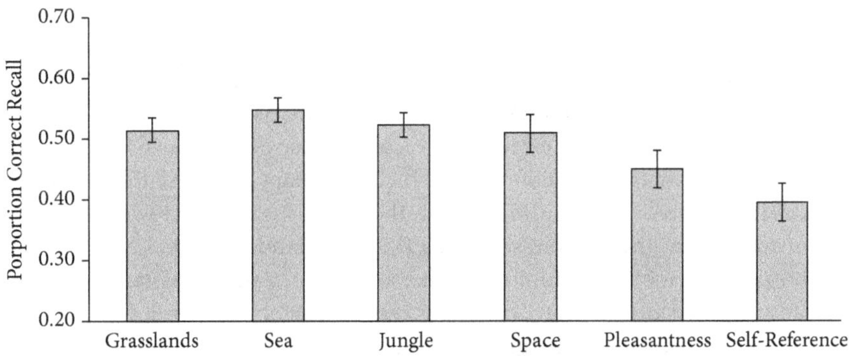

Figure 6.3 Mean proportion correct recall by rating condition (data adapted from Kostic, McFarlan, & Cleary, 2012).

participants was whether or not the scenario involved being stranded alone or with a group:

> "In this task, we would like you to imagine that you are stranded ***alone at sea/ with a group of people at sea*** on a raft, without any basic survival materials. Over the next few months, you will need to find steady supplies of food and water and protect yourself from predators. We are going to show you a list of words…"

Inconsistent with an isolation explanation, Kostic et al. found that it did not matter if the survival scenario implied social isolation or the presence of a group of people. Thus, they were able to rule out this explanation for the survival processing effect. From these results, the survival processing effect appears to be a more general survival benefit. It occurs in a wide range of survival scenarios, both ancestrally relevant (e.g., grasslands) and modern (e.g., lost in outer space), and in group situations (e.g., lost with a group of people at sea) and situations of implied social isolation (e.g., lost alone at sea). Survival processing clearly benefits memory and may be one of the most effective memory-encoding strategies currently known; however, it is not so domain-specific as to require ancestrally relevant contexts or situations of implied social isolation.

Other evidence that the survival processing advantage is not so domain-specific that it requires ancestrally relevant contexts like grasslands scenarios is provided by Soderstrom and McCabe (2011). They had participants rate the relevance of words according to a number of scenarios that varied in terms of the environmental setting and the to-be-avoided threat. Critically, these varying dimensions rendered the scenarios either ancestral or non-ancestral. Denoted by the environmental setting and threat type, the between-subject scenarios included (1) grasslands–predators, (2) grasslands–zombies, (3) city–attackers, and (4) city–zombies. Below we provide the environment and threat-type manipulations in the context of the scenarios' full text:

> "In this task we would like you to imagine that you are stranded in the ***grasslands/ city*** of a foreign land, without any basic survival materials. Over the next few months, you'll need to find steady supplies of food and water and protect yourself from ***predators/zombies/attackers***. We are going to show you a list of words…"

Note that the grasslands–predators and city–attackers scenarios were identical in wording to those used by Nairne et al. (2007) and Weinstein et al. (2008), respectively.

According to the strong domain-specificity hypothesis, the grassland–predators scenario should show the highest levels of recall because this scenario is the only one that is a specific evolutionary context in which our adaptations are thought to have evolved. However, as shown in Figure 6.4, the scenarios that tied with the highest levels of recall were the zombie scenarios (i.e., grasslands–zombies and city–zombies); these scenarios led to higher recall than the ancestral grasslands–predators scenario and the more modern city–attackers scenario. Moreover, the previously reported recall advantage of the grasslands–predators scenario over the city–attackers scenario (Weinstein et al., 2008) was not replicated as these scenarios showed equivalent recall, possibly owing to experimental design differences between studies.

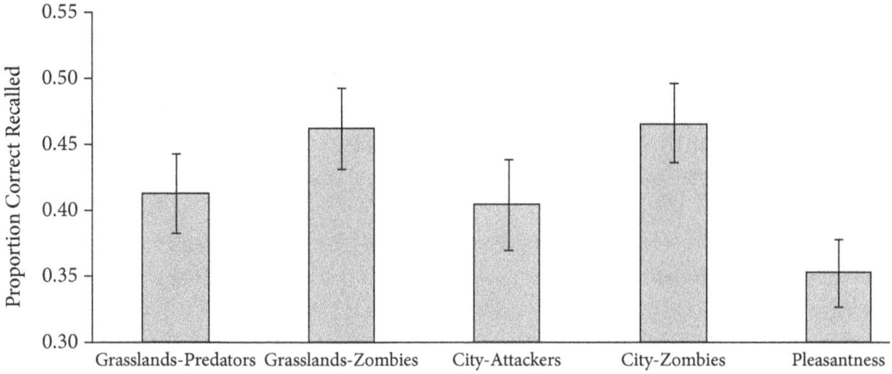

Figure 6.4 Mean proportion correct recall by rating condition (data adapted from Soderstrom & McCabe, 2011).

All scenarios, however, showed better recall than a pleasantness-rating control condition, bolstering the claim that survival processing, in general, is one of the most effective encoding techniques for benefiting memory. The strong domain-specificity hypothesis, however, was not supported.

Two aspects of Soderstrom and McCabe's (2011) study make their results particularly compelling with regard to the idea that the survival processing benefit to memory is not so domain-specific as to require an evolutionarily relevant context. First, the fact that the zombie scenarios (e.g., grasslands–zombie or city–zombie) showed *better* recall than the ancestral scenario (e.g., grasslands–predator) is the opposite of what is predicted by a strong domain-specificity hypothesis. Whereas grasslands are thought to be a context in which early humans evolved many of their current adaptations, zombies are fictitious creatures that have never played a role in any human ancestral environment; also, the city scenario, much like the outer space scenario used by Kostic et al. (2012), is a modern context. Second, the recall advantage for either zombie scenario over the others—such as when comparing the grasslands–zombies and grasslands–predators scenarios—can be attributed to one word: *zombies*! There were no other methodological or procedural differences between these conditions.

It is important to note that several explanations for these data can be ruled out. First, average ratings and their corresponding response times did not vary across conditions in ways consistent with congruency effects (see Craik & Tulving, 1975) or a study time explanation, respectively. Second, participants in each condition rated the same words, ruling out potential item effects. Third, Soderstrom and McCabe (2011) collected self-reported ratings of arousal and valence for each scenario and found that, although the zombie scenarios were perceived as more arousing and more negative than the others, these ratings could not account for the recall differences between scenarios when added as covariates (for similar data inconsistent with an emotional processing explanation of the general survival processing effect, see Kang et al., 2008; Nairne & Pandeirada, 2010). Finally, it is conceivable that the zombie scenarios led to greater distinctive processing as a result of zombies being relatively more familiar and/or bizarre. Indeed, zombies have been highly popularized in the media recently. Although possible, these explanations seem unlikely given that

(1) previous work has shown that survival processing effects persist despite modern scenarios being rated as more familiar than survival scenarios (see Nairne & Pandeirada, 2010, Exp. 4) and (2) both distinctiveness and bizarreness effects are typically found only in within-subjects designs, not in between-subjects designs like that employed by Soderstrom and McCabe (see McDaniel & Bugg, 2008, for a discussion of within- and between-subjects designs and their effect on memory performance).

Those data reported by Soderstrom and McCabe (2011) clearly challenge the specificity of Nairne and Pandeirada's (2010) ancestral priorities framework (see also Weinstein et al., 2008). This then leaves the primary question of to what extent the survival processing advantage in memory is domain-specific. Though zombies themselves were never actually encountered by our Pleistocene ancestors, perhaps the fact that they are a perceived threat to one's survival activates a more general survival processing mechanism that in turn benefits memory. For example, perhaps zombies are perceived as predators and thus activate more general predator-avoidance processes that evolved for adaptive purposes and thus also benefit memory for adaptive purposes. In fact, one could argue that zombies might even *resemble* predators from our distant past, and as a result, activate the same "predator-avoidance" or "self-preservation" systems that evolved to lead to superior memory. It is also possible that both the zombie and standard survival scenarios activate a similar "death and disgust" system, whereby thoughts of one's own mortality during encoding lead to enhanced retention of that encoded material. Indeed, recent work has shown that increasing mortality saliency during encoding is positively related to subsequent memory performance (Hart & Burns, 2012). From these perspectives, then, a degree of domain-specificity is conceivable.

One final study that fits with the idea that an ancestrally relevant environment need not be induced to confer the benefits of survival processing was conducted by Klein (2013). Instead of *adding* potentially confounding variables (e.g., zombies, space travel) that may, for example, increase the familiarity of the scenarios' content, Klein approached the issue by *subtracting* from the standard grasslands scenario. Specifically, one group of participants rated words according to their relevance when the survival context was specified (i.e., survival on the grasslands; identical in wording to Nairne et al., 2007), whereas another group rated these words according to a scenario in which no survival context was specified (i.e., no environment was mentioned):

> "In this task, I want you to imagine that you are trying to stay alive. I am going to show you (on the next page) a list of words, and I would like you to rate how relevant each of these words might be to your staying alive. Some words might be relevant and others may not—it is up to you to decide."

The critical finding was that these conditions yielded essentially identical levels of recall, suggesting that the reinstatement of an ancestrally relevant environment during the encoding is not necessary to produce the survival processing effect.

Taken together, the results of Kostic et al. (2012), Soderstrom and McCabe (2011), and Klein (2013) suggest that the survival processing advantage in memory is not the result of a strong form of domain-specificity: It does not require a specific environmental context in which humans evolved (e.g., grasslands). To the extent that survival processing advantages in memory generalize to a wide range of circumstances—not just those specifically faced by our Pleistocene ancestors as implied by the ancestral priorities framework (Nairne & Pandeirada, 2010)—survival processing might be described as "domain-general." However, we should be clear that we are not arguing for the domain-generality of memory broadly. As we argued in the introduction of this chapter, the collective evidence suggests that cognitive processes, including memory processes, are domain-specific in ways that are reflective of selection pressures from the human ancestral past. Our main argument is that there is still some question about the degree of domain-specificity of memory processes, particularly with regard to the survival processing advantage in memory, and that the evidence suggests that the advantage is due to mechanisms that are more domain-general than may have initially been thought; the advantage actually extends to many types of survival situations, even situations not encountered by early human ancestors at critical times during evolution. To illustrate our point, people are not afraid of only rattlesnakes; people are usually afraid of *all* snakes, even those species encountered for the first time and that were never part of our ancestral past.

Thus, such a domain-general adaptation may actually be advantageous because it prepares us for a wide range of *potential* situations. In helping a species survive a multitude of potential situations and environments, including novel ones, the likelihood that the species transmits its genes to the next generation may be increased. In this way, our perspective on survival processing is very much in line with Schacter and Addis' (2007) view that "information about the past is useful only to the extent that it allows us to anticipate what may happen in the future" (p. 27). Without such an adaptation, it would be difficult for us to accurately judge the benefits and perils of novel situations. Indeed, a case can be made that this type of future-orienting thinking has afforded our species the ability to spread both quickly and broadly by granting us the flexibility to survive in many types of environments, including novel ones that have never before been specifically encountered.

In a related vein, it is possible that explicit planning for the future plays a role in the survival processing advantage. As previously mentioned, Klein et al. (2010) found support for the idea that planning, or future-oriented thinking, plays a role in the survival processing advantage. Thus, because planning may frequently be confounded with survival in studies examining the survival processing effect, the effect may be due to the elicitation of planning or future-oriented thinking. If asking what is relevant in the context of a survival scenario indeed activates future-oriented thinking (e.g., "What items would I take if I knew I was going to be in this situation?"), then this may be one way in which the survival processing effect reflects an adaptive aspect of human memory (e.g., see Schacter et al., 2012, for a review of recent research on future-oriented thinking, its relation to memory, and its potential adaptiveness), while at the same time not being quite as domain-specific as requiring the instantiation of a particular ancestral context as the perceived environment. Future research should further investigate the idea that survival contexts induce imaginative, planful, future-oriented thinking to a greater degree than other contexts.

SUMMARY AND CONCLUSIONS

Exploring *why* our memory systems evolved into their current states is arguably just as important as understanding *how* these memory systems function. The recently introduced survival processing paradigm (Nairne et al., 2007) provides an empirical method for studying memory functionality, and its use has consistently demonstrated that processing information in terms of its survival utility enhances memory performance, a finding consistent with evolutionary theory. More controversial, however, is the idea that survival processing effects are optimally achieved when specific, ancestrally consistent problems are induced as opposed to problems faced in more modern contexts (Nairne & Pandeirada, 2010; Weinstein et al., 2008). Kostic et al. (2012) found no recall differences between ancestor-consistent and ancestor-inconsistent survival scenarios defined in terms of their physical environments (e.g., stranded on the grasslands vs. lost in space); Soderstrom and McCabe (2011) found that inducing a never-before-experienced survival problem by our species (i.e., evading zombies) produced a recall advantage over an ancestor-consistent survival scenario (i.e., evading predators); and Klein (2013) showed that even when the survival context (i.e., the grasslands) is stripped away from the scenario, an equivalent survival processing effect persists. Taken together, these data suggest that survival processing advantages in memory are domain-general insofar as they are not optimally suited for any one environment or situation in particular, but rather operate effectively in a wide range of scenarios. Such an adaptation might aid in our ability to extract important, survival-relevant information from novel situations.

Although the survival processing paradigm has generated a wealth of interest and intriguing data, it is important to note that establishing a *definitive* evolutionary account of survival processing advantages in memory is difficult—if not impossible—for two main reasons. First, we may never know for certain what environmental pressures sculpted our cognitive machinery, and thus we cannot make any definitive claims about the function(s) of memory. Second, just because something serves a function that we can currently observe does not necessarily mean that that function was produced by natural selection. A classic example of such an *exaptation* (Gould & Vrba, 1982) is how feathers, which are thought to have initially evolved for heat regulation, are now also used for flight and display purposes. Thus, one must always consider these indirect adaptations when considering functionality.

Finally, we also need to recognize and appreciate the scope of what it is we are investigating. The topics of evolution and memory are, by themselves, ubiquitous and often nebulous, and thus any attempt to empirically link the two is inherently difficult. Regardless, we still believe that pursuing the study of memory functionality, while challenging, is a commendable pursuit worthy of future investigation.

NOTE

1. The assumption that our ancestors spent the majority of their lives on the "grasslands" during the Pleistocene era, and earlier time periods, has been challenged. For example, analyses of the recent fossil discovery of "Ardi" (*Ardipithecus ramidus*), an early hominid found in Ethiopia, have led scientists to conclude that the *woodlands*, not the grasslands, served as the primary environment for early humans (see Gibbons, 2009).

REFERENCES

Bröder, A., Krüger, N., & Schütte, S. (2011). The survival processing memory effect should generalise to source memory, but it doesn't. *Psychology, 2*, 896–901.

Burns, D. J., Burns, S. A., & Hwang, A. J. (2011). Adaptive memory: Determining the proximate mechanisms responsible for the memorial advantages of survival processing. *Journal of Experimental Psychology: Learning, Memory, and Cognition, 37*, 206–218.

Butler, A. C., Kang, S. H. K., & Roediger, H. L., III (2009). Congruity effects between materials and processing tasks in the survival processing paradigm. *Journal of Experimental Psychology: Learning, Memory, and Cognition, 35*, 1477–1486.

Cosmides, L. & Tooby, J. (1994). Origins of domain-specificity: The evolution of functional organization. In L. Hirschfeld & S. Gelman (Eds.), *Mapping the mind: Domain-specificity in cognition and culture* (pp. 85–116). New York: Cambridge University Press.

Craik, F. I. M., & Tulving, E. (1975). Depth of processing and the retention of words in episodic memory. *Journal of Experimental Psychology: General, 104*, 268–294.

Futuyma, D. J., & Moreno, G. (1988). The evolution of ecological specialization. *Annual Review of Ecology and Systematics, 19*, 207–233.

Gibbons, A. (2009). A new kind of ancestor: *Ardipithecus* revealed. *Science, 326*, 36–40.

Gould, S. J., & Vrba, E. S. (1982). Exaptation: A missing term in the science of form. *Paleobiology, 8*, 4–15.

Hart, J., & Burns, D. J. (2012). Nothing concentrates the mind: Thoughts of death improve recall. *Psychonomic Bulletin & Review, 19*, 264–269.

Kang, S. H. K., McDermott, K. B., & Cohen, S. M. (2008). The mnemonic advantage of processing fitness-relevant information. *Memory and Cognition, 36*, 1151–1156.

Klein, S. B. (2012). A role for self-referential processing in tasks requiring participants to imagine survival on the Savannah. *Journal of Experimental Psychology: Learning, Memory, and Cognition, 38*, 1234–1242.

Klein, S. B. (2013). Does optimal recall performance in the adaptive memory paradigm require the encoding context to encourage thoughts about the environment of evolutionary adaptation? *Memory & Cogntion, 41*, 49–59.

Klein, S. B., Cosmides, L., Tooby, J., & Chance, S. (2002). Decisions and the evolution of memory: Multiple systems, multiple functions. *Psychological Review, 109*, 306–329.

Klein, S. B., Robertson, T. E., & Delton, A. W. (2010). Facing the future: Memory as an evolved system for planning future acts. *Memory & Cognition, 38*, 13–22.

Kostic, B., McFarlan, C. C., & Cleary, A. M. (2012). Extensions of the survival advantage in memory: Examining the role of ancestral context and implied social isolation. *Journal of Experimental Psychology: Learning, Memory, and Cognition, 38*, 1091–1098.

Kroneisen, M., & Erdfelder, E. (2011). On the plasticity of the survival processing effect. *Journal of Experimental Psychology: Learning, Memory, and Cognition, 37*, 1553–1562.

McDaniel, M. A., & Bugg, J. M. (2008). Instability in memory phenomena: A common puzzle and a unifying explanation. *Psychonomic Bulletin & Review, 15*, 237–255.

Nairne, J. S., & Pandeirada, J. N. S. (2008). Adaptive memory: Is survival processing special? *Journal of Memory and Language, 59*, 377–385.

Nairne, J. S., & Pandeirada, J. N. S. (2010). Adaptive memory: Ancestral priorities and the mnemonic value of survival processing. *Cognitive Psychology, 61*, 1–22.

Nairne, J. S., & Pandeirada, J. N. S. (2011). Congruity effects in the survival processing paradigm. *Journal of Experimental Psychology: Learning, Memory, and Cognition, 37*(2), 539–549.

Nairne, J. S., Pandeirada, J. N. S., Gregory, K. J., & Van Arsdall, J. E. (2009). Adaptive memory: Fitness relevance and the hunter-gatherer mind. *Psychological Science, 20*, 740–746.

Nairne, J. S., Pandeirada, J. N. S., & Thompson, S. R. (2008). Adaptive memory: The comparative value of survival processing. *Psychological Science, 19*, 176–180.

Nairne, J. S., Thompson, S. R., & Pandeirada, J. N. S. (2007). Adaptive memory: Survival processing enhances retention. *Journal of Experimental Psychology: Learning, Memory, and Cognition, 33*, 263–273.

Nairne, J. S., VanArsdall, J. E., Pandeirada, J. N. S., & Blunt, J. R. (2012). Adaptive memory: Enhanced location memory after survival processing. *Journal of Experimental Psychology: Learning, Memory, and Cognition, 38*, 495–501.

Otgaar, H., Smeets, T., & Van Bergen, S. (2010). Picturing survival memories: Enhanced memory for fitness-relevant processing occurs for verbal and visual stimuli. *Memory & Cognition, 38*, 23–28.

Paivio, A. (1969). Mental imagery in associative learning and memory. *Psychological Review, 76*, 241–263.

Rogers, T. B., Kuiper, N. A., & Kirker, W. S. (1977). Self-reference and the encoding of personal information. *Journal of Personality and Social Psychology, 35*, 677–688.

Savine, A. C., Skullin, M. K., & Roediger, H. L., III (2011). Survival processing of faces. *Memory & Cognition, 39*, 1359–1373.

Schacter, D. L., & Addis, D. R. (2007). The ghosts of past and future. *Nature, 445*, 27.

Schacter, D. L., Addis, D. R., Hassabis, D., Martin, V. C., Spreng, R., & Szpunar, K. K. (2012). The future of memory: Remembering, imagining, and the brain. *Neuron, 76*, 677–694.

Slamecka, N. J., & Graf, P. (1978). The generation effect: Delineation of a phenomenon. *Journal of Experimental Psychology: Human Learning and Memory, 4*, 592–604.

Smeets, T., Otgaar, H., Raymaekers, L., Peters, M. J. V., & Merckelbach, H. (2012). Survival processing in times of stress. *Psychonomic Bulletin & Review, 19*, 113–118.

Soderstrom, N. C., & McCabe, D. P. (2011). Are survival processing memory advantages based on ancestral priorities? *Psychonomic Bulletin & Review, 18*, 564–569.

Tse, C-S., & Altarriba, J. (2010). Does survival processing enhance implicit memory? *Memory & Cognition, 38*, 1110–1121.

Weinstein, Y., Bugg, J. M., & Roediger, H. L., III (2008). Can the survival recall advantage be explained by basic memory processes? *Memory and Cognition, 36*, 913–919.

7

Survival Processing, Attention, and Interference

JEANETTE ALTARRIBA AND STEPHANIE KAZANAS

BACKGROUND AND OVERVIEW

Survival Processing

The process of evolution dictates that adaptation is necessary for the survival of a species. For example, if climate or topology changed, so too did humans adapt to those changes in order to remain fit, mate, and reproduce to ensure the passing along of genetic information and the vitality of the species as a whole. Additionally, the brain may possess features that help to detect the presence or approach of predators (Barrett, 2005) or the availability of prospective mates (Schmitt, 2005). Does human memory have adaptive features as well? That is, could one provide evidence that memory itself functions so as to take into account the selection processes that would mitigate the above evolutionary processes? In recent times, cognitive psychologists have examined these questions more closely in an effort to uncover the ways in which human memory might be attuned to information that would complement survival, thereby enhancing the *function* of memory in terms of acquisition, storage, and retrieval (see, e.g., Nairne, 2005; Nairne & Pandeirada, 2008; Nairne, Pandeirada, & Thompson, 2008; Nairne, Thompson, & Pandeirada, 2007).

Nairne et al. (2007, 2008) investigated the benefits that accrue via the processing of information in terms of its survival value. That is, they investigated how well stimuli would be remembered if processed in a survival context versus one that did not overtly emphasize survival qualities. A parallel may be drawn between this type of work and an investigation of the levels-of-processing theory wherein the depth of processing of information (i.e., semantically vs. orthographically in the case of word stimuli) has been shown to enhance overall memory retrieval (Craik & Lockhart, 1972; Craik & Tulving, 1975). If an item is processed in the context of a survival scenario, will it be better remembered in a subsequent recall task than if the same information is processed in a nonsurvival context?

In a typical survival processing task, participants are asked to rate a series of words that are presented subsequent to instructions asking them to imagine either that they are stranded in a foreign land, in need of food, water, and safety, or that they are planning to move to a foreign land, in need of securing shelter and transporting their belongings. The words are rated as to their relevance to a given scenario. Nairne et al. (2007) included a third condition in which participants rated the pleasantness of each of the words. In each case, the same list of words was used. After a brief interpolated task, participants are then asked to recall all of the words that had been previously rated. Nairne et al. found that the proportion correct for free recall is significantly higher for the survival condition compared to the other two conditions (i.e., moving and pleasantness). Other studies have confirmed that survival-based processing tends to result in higher overall retention (see Howe & Derbish, Chapter 5 in this volume; Schwartz & Brothers, Chapter 9 in this volume). It is important to note that average ratings are typically similar across survival and moving conditions; thus, it is not the case that higher ratings in terms of survival importance co-occur with better memory recall. Similar results have also been reported using tasks that measure recognition memory (e.g., Nairne et al., 2007, Experiment 3). Likewise, response times for ratings do not appear to be any higher or lower for words in the survival condition versus those same words in the moving condition. Thus, thinking about words in terms of their relevance to survival enhanced later memory for those words indicating a potential bias in memory toward remembering fitness-relevant information in an adaptive context.

Further tests of the above phenomenon in memory recall included a comparison of memory for words under survival conditions versus conditions that asked participants to rate words in terms of the ease or difficulty in forming an image for them; consider past personal experiences and rate words on how easily each one would bring to mind a past personal experience; and generate word completions for partial words and then rate their pleasantness; and, finally, an intentional learning instruction asking participants to try to remember each word for a future memory test (Nairne et al., 2008). A free recall test was used to test retention of the word stimuli. Overall, survival processing resulted in the highest retention rates compared to all other encoding conditions. Although some researchers have claimed that these results may be a function of the congruity between the word stimuli and the nature of the processing tasks (Butler, Kang, & Roediger, 2009), Nairne and Pandeirada (2011) demonstrated that survival advantages in memory can be found across a wide variety of tasks and stimuli, and that in cases in which words are highly congruent to various processing scenarios, benefits are still produced for survival scenarios that are significantly stronger than across all other conditions (but see Tse & Altarriba, 2010).

To date, the above findings have been replicated in tests of memory recall and recognition, in particular those that use explicit testing conditions (though the learning is typically implicit). However, in at least one instance, no survival advantages were found when implicit tasks were used (Tse & Altarriba, 2010). After exposure to survival, moving, or pleasantness scenarios and subsequent rating tasks, participants engaged in a short filler task and were then asked to either complete a stem-cue with the first word that came to mind (the implicit condition) or with a word that had previously been rated (the explicit condition). Although experimental conditions yielded higher overall completion rates compared to the completion of a set of cues for nonstudied items, there were no differences across those conditions (i.e., survival, moving, and pleasantness) in overall performance. The typical findings were, in fact, demonstrated in the explicit conditions. Whereas this set of findings begins to delimit the conditions under which survival processing may influence memory enhancement and performance, it also begins to ask the question as to how generalizable these effects are across different types of tasks, a theme several of the chapters in this volume explore. Moreover, one might ask, does survival processing exert any influences over online or "real-time" cognitive processing? That is, can one demonstrate an effect of survival "priming" on a task that requires a good deal of attention, memory, and cognitive control? To test these notions, the current study employed the N-back task under survival and moving conditions to examine the ways in which survival processing influences performance in a task that serves to tax working memory in ways that demand attention to either spatial or verbal features of a stimulus. The next section discusses the above task and the ways in which it has been used in the past to examine cognitive processing and attention.

The N-back Task

The N-back task has at its roots an early investigation by Kirchner (1958), whose aim it was to measure short-term retention as a function of age in younger and older adults. The younger group ranged in age from 18 to 24 years, whereas the older group ranged from 60 to 84 years. Through the use of a display with a series of 12 lights, he was able to test participants' memories for the sequence in which the lights had been activated and turned off. For example, in the two-back task, participants were instructed to press the key in which the light had gone out two positions before the current one. There was also a no-back, a one-back, and a three-back condition. Whereas the younger group was able to perform all of the tasks with a certain degree of accuracy, the older group was able to perform the task only up to the two-back condition. Overall, the older group made fewer correct responses than the younger group in the one-back and in the two-back conditions. Whereas these results appear to indicate that the retention of material over the short term was accomplished with greater accuracy for the younger versus older participants, there was little by way of any prescreening of participants in terms of overall measures of memory span, health, etc. Thus, without additional information regarding the participants, it is difficult to draw conclusions from these data that could be generalized to other populations. However, this type of task has been modified over time and used in a variety of ways to test memory retention and in some investigations of cognitive control (see, e.g., Gray, 2001; Kirchner, 1958; Ladouceur et al., 2009; Reis & Gray, 2009; Schmeichel, Volokhov, & Demaree, 2008).

Thus, in this type of task, each trial consists of deciding whether the current item matches the one presented N items ago. This task has been modified for many research domains; however, to our knowledge, N-back performance has not been applied to survival processing, nor has the effect of survival processing on attention been investigated, overall. However, the N-back task has been used to measure the influence of anxiety on visuospatial attention (Lavric, Rippon, & Gray, 2003), to examine the interaction between emotion and higher cognition (Gray, 2001; Gray,

Braver, & Raichle, 2002), to investigate the relationship between intelligence and working memory (Shamosh et al., 2008), to measure influences of working memory processing on emotional states (Li, Li, & Luo, 2006), to examine the relationship between aging and working memory (Vaughan, Basak, Hartman, & Verhaeghen, 2008; Verhaeghen & Basak, 2005), and to measure the relationship between working memory capacity and emotion regulation (Reis & Gray, 2009; Schmeichel et al., 2008). Jonides et al. (1997, p. 471) described the two-back task, in particular, as comprising the following components: encoding, storage, rehearsal, matching (of current to stored), temporal ordering (to identify which item was two previous), inhibition (of currently irrelevant items), and response execution.

Gray (2001), for example, presented participants with both a spatial and a verbal version of the N-back task and examined the influence of varying emotional states on performance on these tasks. In the spatial N-back task, a set of 10 letters (b, c, d, f, g, h, j, k, l, m) appeared on a computer screen. A box appeared on the screen at random, in one of six positions, and the box contained a single letter that was repeated multiple times. The spatial instructions were to press the "S" key or the "D" key on a computer keyboard depending on whether the location of the box on the screen was the same as it was two times ago, while disregarding the identity of the letters in the box. The verbal instructions asked participants to press the "S" or the "D" key depending on whether the identity of the letters in the box was the same or different than it was two items ago. Prior to this task, participants saw either a neutral video, one that provided an "approach" context (i.e., a comedic video), or one that provided an "avoidance" context (i.e., a scary video). Gray was interested in investigating whether emotion would moderate a task involving controlled processing, such as the one above, and perhaps make that task more challenging to perform.

The results were quite interesting. In the verbal N-back task, participants' performance was enhanced by an approach state and impaired by an avoidance state. The negative emotional state appeared to pose a strong challenge to the ability to recollect the identities of the letters that were outlined by the box two items before. The opposite was found in the spatial N-back task—participants' performance was enhanced by an avoidance state but impaired by an approach state. Spatial processing in terms of recollecting and comparing the current location of the box versus its location two items prior was facilitated when participants had been induced into a negative state. A negative emotional state actually facilitated participants' ability to perform the N-back task when the key decision centered on the spatial locations of the box in the current and previous locations (see Sutton & Altarriba, 2011, for similar findings in a modified dot probe paradigm). These results contrast those that have reported disruption in spatial processing tasks when threat-evoked anxiety was investigated (see, e.g., Lavric et al., 2003). Though the reasons for the above dissociations are not clear, these findings indicate that emotional states can have moderating effects on tests of cognitive control, attention, and memory. Other researchers have similarly concluded that the N-back task can shed light on the nature of attentional control processes and attentional focus (Basak & Verhaeghen, 2011; Gray et al., 2002; Ladouceur et al., 2009; Verhaeghen, Cerella, & Basak, 2004) and on issues concerning spatial processing and spatial memory (Boot, Kramer, Simons, Fabiani, & Gratton, 2008).

In some ways, the inclusion by Gray (2001) of an emotional or neutral video prior to the presentation of the N-back task acted like a "prime" activating certain associations and physiological reactions that then appeared to play a role in the processing of the subsequent stimuli. In a similar vein, one might ask how a survival prime affects how we update information within working memory and manage the processes involved in navigating an N-back task. That is, by measuring the relationship between emotion and cognitive control, we believe Gray's findings have encouraged a line of related research investigating how other types of "primes" affect N-back performance, specifically when testing attentional processing. To mimic some of the procedural characteristics of Gray's work, we chose the two-back task—one that has been widely implemented by others and that has been shown to discriminate among a variety of variables, including differences in speed (but not accuracy) between expert videogame players and nongamers (Boot et al., 2008), as well as differences in the ability to suppress negative and positive emotions (Schmeichel et al., 2008).

The Current Study

In summary, research investigating the effect of survival processing on human memory performance has found an overall advantage when participants are asked to remember items for the purpose of survival, but not necessarily for other reasons, such as moving to a new city. This effect has now been well documented by Nairne and colleagues (Nairne et al., 2007; Nairne & Pandeirada, 2008; Nairne et al., 2008; Nairne, Pandeirada, Gregory, & Van Arsdall, 2009; but see Klein, Chapter 2 in this volume), though recent research has found that this effect might be restricted to explicit tasks, such as speeded item recognition (Tse & Altarriba, 2010). This limitation suggests that the survival processing advantage may be more complex than originally described. Furthermore, these findings indicate that survival processing research might benefit from an experiment that provides real-time insight into the effects of survival processing on other cognitive domains.

We sought to investigate this possibility by utilizing an N-back task used by Gray (2001), as described above. By using a survival prime and measuring spatial and verbal N-back performance, we are primarily interested in determining the relationship between survival processing and attention, such as whether a survival prime will enhance a participant's ability to attend to specific items, like the identity or location of a letter. It is clear from the research conducted by Gray and his colleagues (e.g., Gray, 2001) that the spatial version of the N-back task has important implications for survival processing research; will the survival processing advantage in memory tasks generalize to one that strictly demands cognitive control mechanisms like attention?

By directly addressing whether survival processing creates an advantage for attention, we hope not only to provide an explanation for why survival processing influences memory, but also to encourage future research utilizing the N-back task as a behavioral measure in and of itself. Prior research has focused on other aspects of the task: where the task is performed in the brain (Gray et al., 2002), whether it is a reliable measure of working memory (Conway, Kane, & Engle, 2003) and can be used to increase working memory span (Verhaeghen et al., 2004), or how aging affects a participant's ability to perform the task (Vaughan et al., 2008). To our knowledge, this review and its accompanying data are the first to use the N-back task in a way that emphasizes the basic function of the task: to discover how a given set of circumstances—such as those presented in a survival prime—affects how well a participant is able to continuously update information held within working memory.

Predictions

When one considers that spatial information (e.g., the location of a food source, shelter, predators) and the ability to consult spatial memory effectively are important to overall survival, it is expected that individuals may be faster at performing a spatial two-back task if they are primed with a survival prime compared to a moving prime. This would indeed be the case if survival or adaptive memory moderates attention much more acutely than the basic elements that are associated with moving one's home to a new city. With verbal instructions, the information provided by the prime (spatial or moving) may not be as important to processing in the two-back task, as in either case one would be attuned to the nature of the items involved (e.g., whether they are eating utensils, bedding, clothing, and the like).

GENERAL METHOD

Participants

Participants in the current experiment ($N = 120$) were University at Albany undergraduates. Each participant gave informed consent and received course credit, extra credit, or monetary compensation for his or her participation.

Procedure

The primary manipulations in this experiment included a survival/moving prime and verbal/spatial two-back test. Participants were randomly assigned to the prime and task type conditions, with an equal number of participants assigned to each condition (e.g., 30 survival–spatial, 30 survival–verbal, 30 moving–spatial, and 30 moving–verbal).

Priming

Two passages were adapted from the survival processing literature (e.g., Nairne et al., 2007) to prime participants immediately before beginning the experimental trials. Participants assigned to the survival condition read the following passage: "In this task we would like you to imagine that you are stranded in the grasslands of a foreign land, without any basic survival materials. Over the next few months, you'll need to find steady supplies of food and water and protect yourself from predators." Participants assigned to the moving condition read the following passage: "In this task we would like you to imagine that you are planning to move to a new home in a foreign land. Over the next few months, you'll need to locate and purchase a new home and transport your belongings."

Two-back Tasks

We used a two-back task in this experiment that was a modified version of Gray's (2001) programmed with E-Prime (Version 1.1) software. In a two-back task, we presented participants with a series of items and instructed them to respond to each trial by indicating on the computer keyboard whether the current trial matches (is the "same") or does not match (is "different") the trial presented two trials ago. For "same" trials, participants were instructed to press the "S" key; for "different" trials, participants were instructed to press the "D" key. We recorded accuracy and reaction times (RTs) for each trial, and participants were presented with the same set of 136 trials.

During each trial, we presented participants with an array of the following letters: B, C, F, G, H, J, K, and M (presented in a pseudo-randomized order of uppercase and lowercase). This array filled the entirety of the 15-inch monitor and served as the "background." The task included a 5-inch box containing multiple copies of any of these letters presented in one of eight locations superimposed on the array (top left, top center, top right, center left, center right, bottom left, bottom center, or bottom right). The box was presented for 500 milliseconds (ms), followed by 2,500 ms with the background alone (Fig. 7.1). The spatial location and verbal identity of the boxes moved across the screen in a pseudo-randomized order, with no repeated locations or letters.

Each letter/position was a "same" trial 17 times; letters and positions were presented as "different" trials as close to equally often as possible. Presentation of "same" and "different" trials was manipulated such that no more than 4 trials of either type were presented in a row. Thirty percent of the 136 trials required a "same" response (40 trials).

Participants were instructed to respond according to either the change in locations of the box or letters within the box. For example, participants in the spatial condition were given spatial instructions: "In this version of the experiment, you will need to

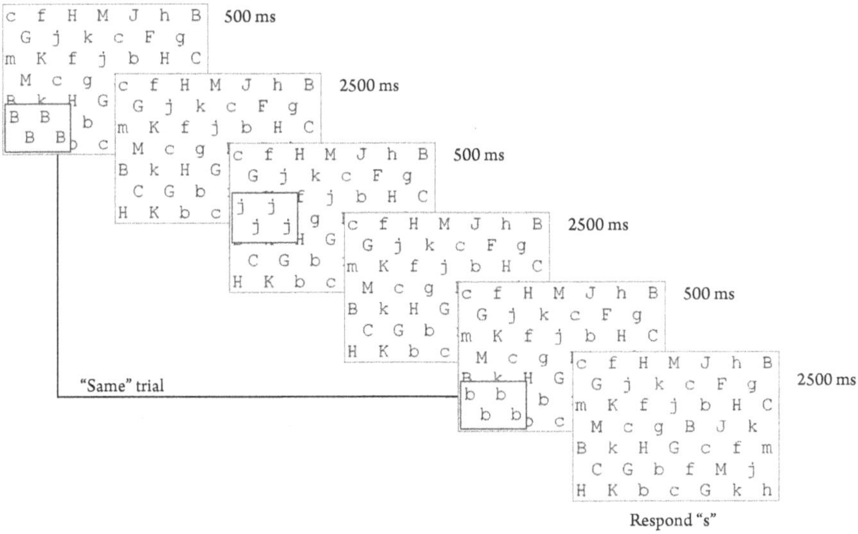

Figure 7.1 Procedure for the current experiment.

remember the location of the letters shown on the computer screen, while ignoring how the letters change from trial to trial. Try to focus on remembering the positions only." Participants in the verbal condition were given verbal instructions: "In this version of the experiment, you will need to remember the letters shown on the computer screen, while ignoring how the location of each letter changes from trial to trial. Try to focus on remembering the letters only."

Current Experiment

Each participant's data were collected separately. Participants were first presented with an informed consent form, which they signed and dated. Next, they completed two questionnaires: the Beck Depression Inventory (BDI-II; Beck, Steer, & Brown, 1996) and the "state" portion of the State-Trait Anxiety Inventory (STAI; Spielberger et al., 1983). The experimenter explained that they should answer the BDI items according to how they have felt, on average, over the past two weeks. The BDI-II is a self-report measurement intended to assess the existence of symptoms of depression. A total score of 0 to 13 is considered minimal range, 14 to 19 is mild, 20 to 28 is moderate, and 29 to 63 is severe. Participants were instructed to answer the STAI items according to how they felt at that moment. The STAI is a self-report measurement intended to assess state anxiety and trait anxiety in adults. For a college population, a score of 42 or above represents high anxiety levels (C. D. Spielberger, personal communication, September 14, 2007). Then, the participants completed two span tasks, assessing both forward and backward span. For each span task, participants began with a set of three numbers and were asked to type the numbers that they saw (in forward order or backward order, depending on the task). The experimenter encouraged each participant to increase the set of numbers as much as possible. Participants spent 5 minutes on each task, and an average forward span and backward span were computed.

After completing the span tasks, participants began the two-back task, responding according to either letters or positions. They were given 48 practice trials: 16 with the experimenter (at a slower pace) and 32 on their own (at the pace identical to that of the experimental trials). Following the practice trials, they were presented with either the survival or moving prime—which they read aloud to ensure processing of the passage—and immediately began the experimental trials. The prime passage was presented in between the practice and experimental trials to ensure that the participant had adequate experience with the task and could begin the experimental trials having just processed the meaning of the passage. In other words, we made every effort to maximize the influence of the prime passage on two-back performance. After the completion of the experimental trials, the participants were debriefed and given credit (or monetary compensation) for their participation.

RESULTS

Responses to each trial were coded according to whether the participant responded correctly or incorrectly. RTs were analyzed for correct responses, whereas a participant's percentage of incorrect responses (% errors) was analyzed separately. The first two trials were omitted from analyses because of the nature of the two-back task and the need to compare a trial with one presented two trials ago.

Trimming Procedures

Each participant's data were analyzed for a raw mean RT. Then, a range of 2.5 standard deviations above and below the mean was used to trim the participant's data. RTs outside of this range were omitted from analyses and accounted for approximately 2 percent of a participant's raw data overall. Analyses with trimmed RTs and % errors are described in the following subsections.

ANALYSES

Main Effects

Mean RTs for correct responses and % errors were analyzed for main effects of both prime (survival or moving) and instructions (spatial or verbal). See Table 7.1 for overall means. For RTs, participants in the survival condition ($M = 652$) and moving condition ($M = 648$) responded at approximately the same rate ($p = .923$). When analyzing RTs for type of instructions, participants given spatial instructions ($M = 571$) responded faster than those given verbal instructions ($M = 729$), and this difference was significant, $F(1,116) = 9.799, p <.01, \eta^2 = .076$.

In terms of % errors, participants in the survival condition (M = 34.9 percent) made no more errors than those in the moving condition (M = 32.8 percent), $p = .585$. Participants given spatial instructions ($M = 33.7$) and verbal instructions ($M = 33.9$) made approximately the same number of errors ($p = .968$).

Analyses of Variance

The overall 2 (prime) × 2 (instructions) interactions for RT data (Fig. 7.2) and % error data (Fig. 7.3) were significant, $F(1,116) = 3.610, p <.05, \eta^2 = .028$; $F(1,116) = 3.561, p <.05, \eta^2 = .030$. Means for the four conditions are included in Table 7.2.

A series of 2 (prime) × 2 (instructions) ANOVAs were conducted with the four experimental conditions to analyze differences in participants' RTs and % errors. Survival–spatial participants were compared to moving–spatial participants. For RT, participants in the survival–spatial condition ($M = 621$) responded *slower* than those in the moving–spatial condition ($M = 521$), $F(1,58) = 2.267, p <.05, \eta^2 = .038$. For % error, participants in the survival–spatial condition ($M = 38.3$) made more errors than those in the moving–spatial condition ($M = 29.2$), $F(1,58) = 2.608, p <.05, \eta^2 = .043$.

Moving–spatial participants were compared to moving–verbal participants. For RT, participants in the moving–spatial condition ($M = 521$) responded faster than

Table 7.1. Means per Condition

	Prime				Instructions			
	Survival		Moving		Spatial		Verbal	
	Mean	SD	Mean	SD	Mean	SD	Mean	SD
RT (ms)	652	284.8	648	293.6	571	261.7	729	293.4
% Errors	34.9	22.5	32.8	18.9	33.7	22.3	33.9	19.2

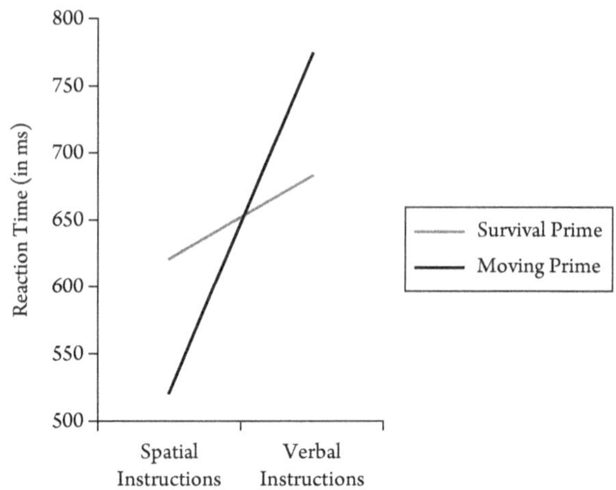

Figure 7.2 Prime type × Instruction type interaction (RTs in milliseconds).

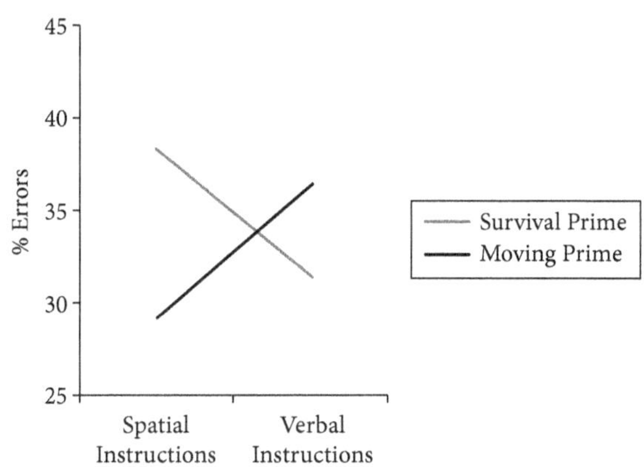

Figure 7.3 Prime type × Instruction type interaction (% errors).

Table 7.2. Two-Back Performance: Means and Error Rates for Prime × Instructions ANOVAs

	Survival Prime				Moving Prime				
	Spatial Instructions		Verbal Instructions		Spatial Instructions		Verbal Instructions		
	Mean	**SD**	**Mean**	**SD**	**Mean**	**SD**	**Mean**	**SD**	**MSE**
RT (ms)	621	304.6	683	264.9	521	203.2	774	317.4	76,255
% Errors	38.3	23.8	31.4	21.0	29.2	20.0	36.4	17.2	426

those in the moving–verbal condition ($M = 774$), and this difference was significant, $F(1,58) = 13.583$, $p < .01$, $\eta^2 = .190$. For % error, participants in the moving–spatial condition ($M = 29.2$) made fewer errors than participants in the moving–verbal condition ($M = 36.4$), $F(1,58) = 2.268$, $p < .05$, $\eta^2 = .038$. No other comparisons yielded significant effects. It is important to note that none of the speed–accuracy tradeoffs for these four conditions were significant (all *ps* >.05).

Correlations

A series of correlations were conducted between the dependent variables (RT and % error) and the four measures assessed prior to the N-back task (BDI-II, STAI, forward span, and backward span). Means for these four measures are included in Table 7.3. Significant findings from these correlations are detailed below, while the entire set of correlations is included in Table 7.4.

For survival–verbal participants, RT was significantly correlated with forward span, $r = 0.43$, $p < .05$. Specifically, greater forward span scores correlated with longer RTs. For moving–spatial participants, RT was also significantly correlated with forward span, $r = -0.50$, $p < .01$. Within this condition, as forward span increased, RTs decreased. Correlations between % errors and forward span were also significant within this condition, $r = -0.47$, $p < .05$, such that, as forward span increased, % errors decreased. Similarly, % errors was significantly correlated with backward span, $r = -0.57$, $p < .01$, and, as backward span increased, % errors decreased. For moving–verbal participants, correlations between RT and BDI-II were significant, $r = -0.55$,

Table 7.3. Participant Background Measures

	Survival–Spatial		Survival–Verbal		Moving–Spatial		Moving–Verbal		Overall	
Measure	Mean	SD	Mean	SD	Mean	SD	Mean	SD	Mean	SD
BDI-II	5.93	4.58	9.64	8.32	5.93	4.86	8.71	6.52	7.55	6.39
STAI	34.11	8.99	35.43	11.40	30.96	9.90	38.07	11.28	34.64	10.62
Forward Span	4.33	0.74	4.63	0.84	4.78	0.90	4.78	0.59	4.63	0.79
Backward Span	4.04	0.76	4.30	0.74	4.23	0.71	4.26	0.47	4.21	0.68

Table 7.4. Correlations Within Condition

	Survival–Spatial		**Survival–Verbal**		**Moving–Spatial**		**Moving–Verbal**	
Measure	**RT (ms)**	**% error**	**RT (ms)**	**% error**	**RT (ms)**	**% error**	**RT (ms)**	**% error**
BDI-II	−0.14	0.31	−0.31	0.01	0.12	−0.08	−0.55**	−0.21
STAI	−0.02	0.04	−0.32	−0.10	0.13	0.11	−0.44*	−0.13
Forward Span	0.22	−0.10	0.43*	−0.15	−0.50**	−0.47*	−0.23	0.16
Backward Span	0.01	−0.06	0.22	−0.27	−0.36	−0.57**	−0.28	−0.04

* Denotes significance at the $p < .05$ level.
** Denotes significance at the $p < .01$ level.

$p < .01$. Specifically, as scores on the BDI-II increased, RTs decreased. Additionally, within this condition, correlations between RT and STAI scores were significant, $r = -0.44$, $p < .05$. Within this condition, as STAI scores increased, RTs decreased.

DISCUSSION

The current work used the N-back task to investigate the potential influence of survival processing on aspects of attention and cognitive control at both a spatial and verbal level. Participants were asked to report whether or not the identity and location of a target matched that of the item presented two steps back. This task has been used multiple times in the past as a measure of the ability to engage memory and attentional processes, as participants are asked to draw comparisons between present and past events (see, e.g., Banich et al., 2009; Conway et al., 2005; Jarrold & Towse, 2006; Smith & Jonides, 1997). Prior to performing this task, participants were primed with a short statement alluding to having to survive in a foreign land or having to move from one location to another and seek shelter. A main finding of interest was the interaction between these primes and the spatial versus verbal dimension of the two-back task. For the spatial condition, reaction times were actually slowed (and less accurate) for those in the survival group versus those in the moving group. Whereas survival processing has been shown to facilitate performance in memory recall and recognition tasks (Nairne et al., 2007, 2008), in the current attentional task, *interference* results from a survival prime (but see Nairne, VanArsdall, Pandeirada, & Blunt, 2012, for an instance of better location memory—measured with a surprise retention test—for pictorial stimuli when paired with a survival scenario; their use of a surprise retention test suggests increased attention for the stimuli most relevant to survival). However, the verbal two-back task did not appear to be moderated at all by the presence of a survival or moving prime. Perhaps in a different scenario, such as the bank robbery scenario used in various other studies, it might be possible to find different effects (see, e.g., Butler et al., 2009; Kang, McDermott, & Cohen, 2008).

The fact that spatial tasks seem to be moderated by this variable has a parallel in the emotion literature, as Li et al. (2006) reported that negative emotion significantly influenced event-related potentials as a function of spatial tasks more so than for verbal tasks. Li et al. used a modified N-back task and reported that for both spatial and verbal tasks, the anterior late positive component and the posterior P300 were affected by negative emotion. These effects were greater in spatial tasks than in verbal ones, and the parietal P300 was reduced by negative emotion in spatial tasks but not in verbal ones. They conducted a current density analysis as well that revealed high current density in the frontal-parietal cortex, but only in spatial tasks. Their results suggested differential effects of negative emotion on spatial and verbal working memory.

Similar to the work of Li et al. (2006), the current research also suggests that perhaps survival processing exerts differing effects on verbal versus spatial processing within the context of a two-back task. Could it be the case that there is, in some sense, a "negative" cast associated with the notion that one's survival is at stake? Could this negativity at once slow responses in an attention-type task, leading to longer latencies and more errors, in spatial conditions? However, negativity alone likely does not underlie the interference effects found in the current study. Recall that Gray (2001) found that N-back performance tended to be facilitated in "avoidance" situations. Consider, though, that being concerned about one's survival, while at once perhaps anxiety-provoking, is perhaps not easily categorized as an "avoidance" situation. Rather, an individual would want to approach this kind of situation effectively, efficiently, and in a manner that would preserve survival above all else. Therefore, one possible explanation for the current findings is the notion that reading about one's need to engage in survival-related behaviors heightens anxiety and awareness to the point at which careful processing of spatial features of a task is engaged in, given the importance of paying attention to spatial cues and the location of target objects in an environment. Given that the nature of the two-back task as enacted in the current study is by no means a natural or typical task, the participant takes a longer amount of time to complete the task due to its novelty and difficulty compared to conditions wherein a moving scenario is present. Thus, both the more arousing, negative connotations of being unsuccessful at completing a task under a survival scenario and the need to pay close attention to the task at hand contribute to the lengthening of RTs in the spatial condition. Various other studies have confirmed the interfering effects of negative mood states or anxious states on processing in the N-back task (see, e.g., Banich, 2009, for a review). One possible issue to consider is the idea that perhaps an additional nonspatial condition other than the "moving" scenario might have served as an appropriate neutral priming condition. Researchers may want to focus on alternative conditions that may be further removed from the notion of spatial processing and representation to provide for additional comparisons to the current spatial condition. An alternative and potentially useful comparison could be made with a condition that does not contain a prime and begins immediately with the N-back task.

Although the moving condition in the current study was meant as a type of control condition, as has been used previously (Nairne et al., 2007, 2008), curiously, in the current study, the moving scenario appears to have sped up responses in the spatial condition and significantly lengthened responses in the verbal condition. Conversely, one might summarize these findings by noting that with verbal instructions, survival primes actually speed up processing compared to moving primes. It may be the case that survival primes, again, heighten awareness and perhaps even create a bit of arousal that prepares one to perform a task with greater attention; in the verbal case, given that spatial locations are not the main focus of attention, that arousal tends to facilitate processing, as compared to a moving prime, such that performance on the task is slightly facilitated for participants. In summary, it appears to be the combination of the arousal or perhaps anxiety prompted by survival primes, at once heightening awareness, that exerts an interfering influence in the spatial domain but not in the verbal domain. This effect may not be a typical prediction given extant data on both stress and survival mechanisms as boosting memory performance (see e.g., Smeets et al., 2012); however, note again that the current investigation focuses on an attentional, two-back task and not on a typical memory recall task *per se*.

What, if any, are the influences exerted by survival and moving primes on measures of memory span and emotional state? Correlations examined within the current data indicated that as forward span increased for our participants, so did RT in the survival–verbal condition. Typically, measures of working memory such as forward span may be predictive of a heightened ability to combine maintenance and processing demands on tasks in order to facilitate behavior (Jarrold & Towse, 2006). In the current case, it appears as though having a higher forward span was correlated with *slower* response times when primed by a survival prime in a verbal two-back task. Again, an aspect of the survival prime—an influence that co-occurs with being exposed to the survival passage—acts to slow reaction times in a two-back task, causing *interference* more so in individuals with higher forward memory spans. For the moving condition, however, there appears to be no correlation between memory span (forward or backward) in a verbal two-back task. Other reports in the literature have also failed to find very strong correlations between N-back performance and measures of working memory (e.g., Kane, Conway, Miura, & Colflesh, 2007).

In conclusion, the current work indicates that perhaps survival processing works to moderate attentional control in ways that are not the same as in typical tests of memory recall and recognition. If indeed some level of arousal, anxiety, or perhaps stress (Smeets et al., 2012) is associated with processing survival scenarios, then it is possible that their effects on processing will be inhibitory versus facilitatory. In the current work, it appears as though processing under a survival prime condition worked to slow responses in a spatial two-back task, as compared to a prime related to moving, and that this slowdown was also accompanied by a higher error rate. We hope that the current work will stimulate other works in the area of survival and adaptive processing using tasks that examine a greater range of human performance in the areas of attention and perception. Moreover, more work is needed examining the potential effects of survival processing on individual differences in working memory (Engle, 2010), emotional states, and performance overall, using a variety of measures and tasks to create a composite of features associated with survival processing in general.

ACKNOWLEDGMENTS

We would like to thank Elizabeth Foskett and Chantae Miller for their assistance with data collection. Additionally, portions of this work were presented at the 53rd Annual Meeting of the Psychonomic Society, Minneapolis, MN.

REFERENCES

Banich, M. T., Mackiewicz, K. L., Depue, B. E., Whitmer, A., Miller, G. A., & Heller, W. (2009). Cognitive control mechanisms, emotion & memory: A neural perspective with implications for psychopathology. *Neuroscience & Biobehavioral Reviews, 33*, 613–630.

Barrett, H. C. (2005). Adaptations to predators and prey. In D. Buss (Ed.), *The handbook of evolutionary psychology* (pp. 200–223). Hoboken, NJ: Wiley.

Basak, C., & Verhaeghen, P. (2011). Three layers of working memory: Focus-switch costs and retrieval dynamics as revealed by the N-count task. *Journal of Cognitive Psychology, 23*, 204–219.

Beck, A. T., Steer, R. A., & Brown, G. K. (1996). *BDI-II: Beck Depression Inventory manual* (2nd ed.). Boston: Harcourt, Brace.

Boot, W. R., Kramer, A. F., Simons, D. J., Fabiani, M., & Gratton, G. (2008). The effects of video game playing on attention, memory, and executive control. *Acta Psychologica, 129*, 387–398.

Butler, A. C., Kang, S. H. K., & Roediger, H. L. (2009). Congruity effects between materials and processing tasks in the survival processing paradigm. *Journal of Experimental Psychology: Learning, Memory, and Cognition*, *35*, 1477–1486.

Conway, A. R. A., Kane, M. J., Bunting, M. F., Hambrick, D. Z., Wilhelm, O., & Engle, R. W. (2005). Working memory span tasks: A methodological review and user's guide. *Psychonomic Bulletin & Review*, *12*, 769–786.

Conway, A. R. A., Kane, M. J., & Engle, R. W. (2003). Working memory capacity and its relation to general intelligence. *Trends in Cognitive Sciences*, *7*, 547–552.

Craik, F. I. M., & Lockhart, R. S. (1972). Levels of processing: A framework for memory research. *Journal of Verbal Learning and Verbal Behavior*, *11*, 671–684.

Craik, F. I. M., & Tulving, E. (1975). Depth of processing and the retention of words in episodic memory. *Journal of Experimental Psychology: General*, *104*, 268–294.

Engle, R. W. (2010). Role of working-memory capacity in cognitive control. *Current Anthropology*, *51*, S17–S26.

Gray, J. R. (2001). Emotional modulation of cognitive control: Approach-withdrawal states double-dissociate spatial from verbal two-back task performance. *Journal of Experimental Psychology: General*, *130*, 436–452.

Gray, J. R., Braver, T. S., & Raichle, M. E. (2002). Integration of emotion and cognition in the lateral prefrontal cortex. *Proceedings of the National Academy of Sciences USA*, *99*, 4115–4120.

Jarrold, C., & Towse, J. N. (2006). Individual differences in working memory. *Neuroscience*, *139*, 39–50.

Jonides, J., Schumacher, E. H., Smith, E. E., Lauber, E. J., Awh, E., Minoshima, S., & Koeppe, R. A. (1997). Verbal working memory load affects regional brain activation as measured by PET. *Journal of Cognitive Neuroscience*, *9*, 462–475.

Kane, M. J., Conway, A. R. A., Miura, T., & Colflesh, G. J. H. (2007). Working memory, attention control, and the N-back task: A question of construct validity. *Journal of Experimental Psychology: Learning, Memory, and Cognition*, *33*, 615–622.

Kang, S. H. K., McDermott, K. B., & Cohen, S. M. (2008). The mnemonic advantage of processing fitness relevant information. *Memory & Cognition*, *36*, 1151–1156.

Kirchner, W. K. (1958). Age differences in short-term retention of rapidly changing information. *Journal of Experimental Psychology*, *55*, 352–358.

Ladouceur, C. D., Silk, J. S., Dahl, R. E., Ostapenko, L., Kronhaus, D. M., & Phillips, M. L. (2009). Fearful faces influence attentional control processes in anxious youth and adults. *Emotion*, *9*, 855–874.

Lavric, A., Rippon, G., & Gray, J. R. (2003). Threat-evoked anxiety disrupts spatial working memory performance: An attentional account. *Cognitive Therapy and Research*, *27*, 489–504.

Li, X., Li, X., & Luo, Y.-J. (2006). Differential influences of negative emotion on spatial and verbal working memory: Evidence from event-related potential and source content density analysis. *Cognitive Neuroscience and Neuropsychology*, *17*, 1555–1559.

Nairne, J. S. (2005). The functionalist agenda in memory research. In A. F. Healy (Ed.), *Experimental cognitive psychology and its applications: Festschrift in honor of Lyle Bourne, Walter Kintsch, and Thomas Landauer* (pp. 115–126). Washington, DC: American Psychological Association.

Nairne, J. S., & Pandeirada, J. N. S. (2008). Adaptive memory: Remembering with a stone-age brain. *Current Directions in Psychological Science*, *17*, 239–243.

Nairne, J. S., & Pandeirada, J. N. S. (2011). Congruity effects in the survival processing paradigm. *Journal of Experimental Psychology: Learning, Memory, and Cognition*, *37*, 539–549.

Nairne, J. S., Pandeirada, J. N. S., Gregory, K. J., & Van Arsdall, J. E. (2009). Adaptive memory: Fitness relevance and the hunter-gatherer mind. *Psychological Science, 20*, 740–746.

Nairne, J. S., Pandeirada, J. N. S., & Thompson, S. R. (2008). Adaptive memory: The comparative value of survival processing. *Psychological Science, 19*, 176–180.

Nairne, J. S., Thompson, S. R., & Pandeirada, J. N. S. (2007). Adaptive memory: Survival processing enhances retention. *Journal of Experimental Psychology: Learning, Memory, and Cognition, 33*, 263–273.

Nairne, J. S., VanArsdall, J. E., Pandeirada, J. N. S., & Blunt, J. R. (2012). Adaptive memory: Enhanced location memory after survival processing. *Journal of Experimental Psychology: Learning, Memory, and Cognition, 38*, 495–501.

Reis, D. L., & Gray, J. R. (2009). Affect and action control. In E. Morsella, J. A. Bargh, & P. Gollwitzer (Eds.), *Oxford handbook of human action* (pp. 277–297). New York: Oxford University Press.

Schmeichel, B. J., Volokhov, R. N., & Demaree, H. A. (2008). Working memory capacity and the self-regulation of emotional expression and experience. *Journal of Personality and Social Psychology, 95*, 1526–1540.

Schmitt, D. P. (2005). Fundamentals of human mating strategies. In D. Buss (Ed.), *The handbook of evolutionary psychology* (pp. 258–291). Hoboken, NJ: Wiley.

Shamosh, N. A., DeYoung, C. G., Green, A. E., Reis, D. L., Johnson, M. R., Conway, A. R. A,.... Gray, J. R. (2008). Individual differences in delay discounting. *Psychological Science, 19*, 904–911.

Smeets, T., Otgaar, H., Raymaekers, L., Peters, M. J. V., & Merckelbach, H. (2012). Survival processing in times of stress. *Psychonomic Bulletin & Review, 19*, 113–118.

Smith, E. E., & Jonides, J. (1997). Working memory: A view from neuroimaging. *Cognitive Psychology, 33*, 5–42.

Spielberger, C. D., Gorsuch, R. L., Lushene, R., Vagg, P. R., & Jacobs, G. A. (1983). *Manual for the State-Trait Anxiety Inventory*. Palo Alto, CA: Consulting Psychology Press.

Sutton, T. M., & Altarriba, J. (2011). The automatic activation and perception of emotion in word processing: Evidence from a modified dot probe paradigm. *Journal of Cognitive Psychology, 23*, 736–747.

Tse, C.-S., & Altarriba, J. (2010). Does survival processing enhance implicit memory? *Memory & Cognition, 38*, 1110–1121.

Vaughan, L., Basak, C., Hartman, M., & Verhaeghen, P. (2008). Aging and working memory inside and outside the focus of attention: Dissociations of availability and accessibility. *Aging, Neuropsychology, and Cognition, 15*, 703–724.

Verhaeghen, P., & Basak, C. (2005). Ageing and switching of the focus of attention in working memory: Results from a modified N-back task. *Quarterly Journal of Experimental Psychology, 58A*, 134–154.

Verhaeghen, P., Cerella, J., & Basak, C. (2004). A working memory workout: How to expand the focus of serial attention from one to four items in 10 hours or less. *Journal of Experimental Psychology: Learning, Memory, and Cognition, 30*, 1322–1337.

8

Cognitive and Social Factors in the Study of Survival Memory

MICHAEL P. TOGLIA, AARON D. LEEDY, ADAM M. WILDE, CATHERINE M. BAKER, AND EILEEN M. DACEY

With the advent of the cognitive revolution in the mid-1950s (Mandler, 2007; Neisser, 1967) psychology's focus began to shift, especially with the appearance of seminal studies such as those by George Miller on short-term memory (Miller, 1956). The field expanded rapidly and was soon defined by Neisser's 1967 book on cognitive psychology. This opened the floodgates for cognitive researchers to study the characteristics and functions of sensory memory, short-term memory, and long-term memory. For example, it has been argued that the capacity limit of short-term memory, seven plus or minus two, is adaptive (Miller, 1956). Not only do these limits influence focusing on that which is most important to survival by concentrating on a smaller set of information, but they also increase one's ability to sense patterns and causation (Nairne, Thompson, & Pandeirada, 2007). Similar considerations should drive what information is transferred to long-term memory, such that more cognitive effort can then be dedicated to memories of greater importance.

Before the recent interest in an adaptive human memory launched by Nairne and colleagues' 2007 paper, Anderson and Schooler (2000) suggested that there is a cost associated with memory. This is seen both in the time dedicated to memory encoding and retrieval and also the metabolic expenditure (calories needed for physical exertion elsewhere) while doing so. The gain received by being able to correctly recall is offset by these costs. They argued that the cost of having memory void of error may outweigh any potential benefits. Whereas there are no limits to the capacity of long-term memory, problems arise both with the accuracy of encoding and with the ability to retrieve information from long-term memory (Brown & Craik, 2000). However, distortions and even forgetting, although usually seen as negative, may at times actually be advantageous (Howe & Derbish, 2010; Schacter, Guerin, & St. Jacques, 2011). Is it beneficial to be able to recall any past experience, if one must first sift through all unnecessary past experiences to access it? With the potentially high costs associated with memory, survival-relevant material should be especially readily available for retrieval, but also as error-free as possible.

Current theorists have suggested that most aspects of human memory evolved to facilitate survival (Anderson & Schooler, 2000; Burns, Burns, & Hwang, 2011; Nairne et al., 2007). What may be considered a "flaw" or "limitation" of memory may in fact be a developed mechanism designed to help individuals optimize their time and resources. The human mind's ability to remember, recall inaccurately, and forget may be features of human cognition that evolved for perpetuation of the human species. Thus human memory, frailties and all, may be adaptive in its makeup in more ways than initially thought.

From an adaptive point of view, instead of one general memory mechanism, the human brain is more likely composed of multiple specific mechanisms that evolved under particular environmental stressors to aid in differing survival situations (Soderstrom & Cleary, Chapter 6 in this volume). As evolution has shaped contemporary brain mechanisms to be primarily domain-specific, we selectively attend to and create memories encoded from past events. Presumably we may do this most reliably in a survival context (Nairne & Pandeirada, 2008). Thus, it is believed there are domain-specific modules, such as semantic memory, perception, and language, that are capable of addressing specific tasks as sculpted by nature. Although many of these mechanisms have evolved for functions related to issues confronting early man, their functionality is not limited to the specific domain for which it evolved (Nairne et al., 2007). These modules facilitate other situations faced by early humans, as well as contemporary circumstances. Present-day human beings are able to use these domain-specific mechanisms for current memorial or cognitive situations even though they evolved for another function. In this regard, Nairne et al. (2007) note that there are proximate mechanisms that underlie our abilities to read and write, yet these could not have evolved for those purposes.

The early human brain evolved under the constraints of the environment and therefore is best suited to function in an environment mimicking the conditions under which it developed. The human brain may have evolved to remember information most likely to facilitate the survival of the individual based on relevance to ancestral situational factors. It is believed that early humans came to best remember information relevant to their survival, such as locations of food and water, proper tool-building skills, and how to best handle predators; in short, they were probably best equipped to manage survival scenarios (Nairne et al., 2007). As with other mechanisms explained by the law of natural selection (i.e., bipedal movement and opposable thumbs), advanced memory systems capable of processing, storing, and retrieving information evolved over time, with the "fittest" of the species procreating (Darwin, 1859). Based on evolutionary theory, memory should show specificity to information that is related to human survival, whatever form survival may take.

Regarding the bias of memory with respect to survival, there has been recent interest in studying the effect of survival processing in relation to memory, beginning with the work of Nairne et al. (2007). Their procedure required participants to rate words for their relevance (on a 1-to-5 scale) to one of three scenarios, and then the participants free recalled as many of the words as possible. Encoding of the words was incidental because participants only thought they were rating words based on their relevance to any given scenario. The subsequent free recall test was thus a surprise to the participants. The key aspect of their study was that one of the scenarios was about surviving in the grasslands of a distant foreign land (Nairne et al., 2007). This scenario was compared with a control scenario about moving into a new home and a condition that required the rating of words for pleasantness, which repeatedly has been shown to improve retention (Hunt & Einstein, 1981; Nairne et al., 2007). As an aside, the Hunt and Einstein findings are particular relevant to the current chapter, as pleasantness ratings boosted free recall for related (categorized lists) compared to unrelated word lists. Returning to Nairne et al. (2007), the results of their experiment showed that participants' memory of the presented words was best in the survival condition. Furthermore, retention was superior in the survival condition even when compared with the encoding technique of self-reference, which is widely regarded as one of the best methods to enhance memory (Symons & Johnson, 1997). Moreover, survival processing still had a significantly higher retention rate compared to encoding via pleasantness, imagery, generation, and intentional learning (Nairne, Pandeirada, & Thompson, 2008).

In Nairne et al.'s (2007) original grasslands scenario experiment, they found significantly greater correct recall in the survival condition; however, participants also produced more intrusions (false recall) after engaging in survival processing than the pleasantness condition (although false memories in the survival processing group did not differ from those in the moving scenario condition). This pattern of results was largely the basis for our designing of Experiment 1. Even though this difference was not statistically significant (Nairne et al., 2007), the notion that survival processing might produce increases in memory distortion was intriguing in light of the claim that processing for survival relevance improves memory. It is not uncommon for semantically oriented tasks that support high levels of true memory to also support high levels of false memory. This pattern in which increases in veridical memory are accompanied by increases in false memory was first labeled by Toglia, Neuschatz, and Goodwin (1999) as the "more is less" effect. Their study employed the Deese–Roediger–McDermott (DRM) paradigm, a methodology that consistently produces robust levels of illusory memory (Deese, 1959; Roediger & McDermott, 1995). Thus, rather than using unrelated word lists, employed by Nairne in his early work, in Experiment 1, we had participants study thematically related lists of words, frequently referred to as DRM lists. With this approach we could study survival processing and memorial accuracy when false memory is factored in.

In the DRM paradigm, participants receive an auditory presentation of a list of words that are semantically similar to a critical nonpresented word such as "sleep," in which *bed, rest, awake, tired, dream, wake, snooze, pillow, doze, slumber, snore, nap, peace, yawn,* and *drowsy* are the related words. When participants are asked to recall and/or recognize the list of words, they often "remember" hearing the nonpresented critical item. Therefore, the reconstructive nature of memory leads participants to "remember" a word that summarizes the entire list of words (Roediger & McDermott, 1995). Toglia et al. (1999) varied levels of processing in a DRM experiment, with their results showing that not only did semantic processing of DRM word lists lead to greater veridical memory, but it also lead to greater false memory (and hence the "more is less" effect noted above). Given the similar pattern observed by Nairne et al. (2007), we extended the examination of the survival processing advantage to the DRM paradigm in Experiment 1. We are not the first to investigate this idea (e.g., Howe & Derbish, 2010; Otgaar & Smeets, 2010). Howe and Derbish conducted a series of four experiments and found that participants showed greater levels of false memory for survival-relevant DRM lists than neutral or negatively valenced lists. Survival processing also increased subjects' vulnerability to falsely remembering list items. The survival processing versus pleasantness encoding experiments targeted recognition memory and revealed that orientation to survival, compared to pleasantness, increases true recognition as well as false recognition.

Similar results were reported by Otgaar and Smeets (2010), who tested free recall in three experiments involving both adults and children. In each study they found no survival recall advantage when performance was evaluated with net accuracy scores (computed by dividing correct recall by correct plus false recall). Survival processing failed to be beneficial, as it boosted true recall while concomitantly increasing false memory. Hence, the "more is less" pattern was observed by both Howe and Derbish (2010) and Otgaar and Smeets (2010). Our Experiment 2 also crossed the DRM paradigm with the survival processing paradigm, while additionally being designed to examine memory under moderate levels of stress (see Brainerd, Holliday, Reyna, Yang, & Toglia, 2010; Pesta, Murphy, & Sanders, 2001, for DRM studies using arousal-inducing lists), as most DRM studies do not involve arousal-inducing lists, nor are they conducted under conditions that produce stress independent of characteristics of the lists. One such property is valance (negative, neutral, positive), which has been examined in a number of papers (e.g., Brainerd et al. 2010; Corson & Verrier, 2007; Howe, Candel, Otgaar, Malone, & Wimmer 2010). Brainerd et al. were careful not to confound valence and arousal, which they point out have not at times been separated, as seen in Budson et al. (2006); thus, valence should not be considered a proxy for arousal. To avoid these potential list attribute problems, in our second experiment we employed lists that were neutral in their composition. This permitted us to study environmental and/or situational arousal-stress in its own right.

We suspect that the human brain has evolved mechanisms to facilitate memory for certain information in high arousal conditions with parallels to survival situations. Multiple factors, including arousal, can affect performance on cognitive tasks in addition to those discussed (e.g., encoding context, as provided by the various scenarios used in this literature). Often, while being observed by others, individuals will perform better on certain tasks than when they are performing them alone. For example, in a historical study, Triplett (1898) showed that bicycle racing performance was better in social groups than alone. The presence of others has also been shown to decrease individual performance on a variety of tasks (Zajonc, 1965). Zajonc believed the presence of others is a factor that influences arousal. This experienced arousal in turn affects individual performance, with performance for well-learned responses improving with induced arousal and performance on, new tasks, or ones being learned, being impaired.

Therefore, arousal stemming from the presence of others augments dominant responses while undermining nondominant responses (Zajonc, 1965). Cottrell, Wack, Sekerak, and Rittle (1968) found that the presence of active observers had a positive influence on dominant responses but a hindering effect on nondominant behaviors. Previous studies have demonstrated that similar observer effects can be obtained with the presence of a video camera, which may help or hinder performance (Constantinou, Ashendorf, & McCaffrey, 2005). Other investigators have expanded upon this camera effect by demonstrating a benefit to performance similar to that seen with active observers (Putz, 1975). Our premise was that when participants learned and remembered in the presence of a camera, this would serve as a modern-day environmental stressor that might facilitate memory performance for survival information and/or processing while possibly diminishing performance for nonsurvival lists and/or processing. A recent study along these lines was conducted by Smeets, Otgaar, Raymaekers, Peters, and Merckelbach (2012). They had subjects initially take part in a psychosocial stressful activity involving a speech on a topic of their choosing as well as subsequently describing in detail their personality, both of which were carried out in front of a live audience (active observers) while being videotaped. In control participants, the speech was not followed by a description of personality and was delivered under nonstressful conditions in which they were alone (i.e., no experimenter and no active observers). Then all participants rated words for their relevance to survival or moving, followed by an incidental memory test showing that encoding for survivability and prior presence of stress independently aided memory performance. Smeets et al. (2012) actually assessed acute stress by measuring levels of cortisol. We did not take such measures. However, based on their success in elevating stress under conditions of videotaping participants, we believe it is reasonable to expect that our camera condition operated similarly in the present Experiment 2. Before we explored arousal, we conducted an initial study as it was important to establish our manipulations of interest within the DRM paradigm.

EXPERIMENT 1

The hypotheses for Experiment 1 were as follows. Generic survival processing, to be described below, and survival grasslands processing would lead to better veridical memory as well as greater false memory in line with the "more is less" pattern when compared to pleasantness and control conditions. Similarly, survival lists were expected to show higher levels of both true and false memory compared to nonsurvival lists. An interaction was hypothesized with the "more is less" pattern predicted to be most robust for those engaged in survival processing for survival-relevant lists.

Method

Participants

A total of 93 participants were run from two major universities. They were randomly assigned to one of four processing conditions with 22 participants in a generic survival processing group, 24 in the grasslands survival group (employing Nairne's grasslands scenario), 26 in the pleasantness processing group, and 21 participants in the control group given standard intentional-memory instructions but no specific processing task. Generic survival meant that subjects rated list words for their general relevance to survival without a scenario context or a definition of the term "survival." Regardless of condition, they were tested in small groups of five or six in size.

Materials

Six of the 24 lists compiled by Roediger and McDermott (1995) were selected for the encoding phase. Each list contained the 15 highest associates related to a critical nonpresented lure. The order of presentation of the study words was held constant, with the highest associate always presented first and followed by the remaining 14 items in descending order of association. For example, the list items related to the nonpresented word "sleep" were *bed, rest, awake, tired, dream, wake,* and *snooze*. The "sleep" list was one of three lists assumed to be survival-related, with the other two being "bread" and "doctor." Three lists selected on the belief they were nonsurvival in nature were "needle," "king," and "sweet." For the "needle" list the words included *thread, pin, eye, sewing, sharp, point,* and *prick*.

Participants who engaged in one of the two survival processing conditions rated each list word on a 1-to-5-point scale for survivability. Those in the generic survival condition rated each word on their interpretation of general survivability, whereas those in the survival grasslands condition rated each word on how it would help them survive in the classic Nairne et al. (2007) scenario.

Pleasantness processing required subjects to evaluate list items on a 1-to-5-point scale for degree of pleasantness, and the control group was asked to study each list in anticipation of a test of memory. In these four conditions, for both study and testing purposes, booklets were constructed with a cover page, followed by a series of rating pages and pages to free recall each list. In the control group's booklets blank pages replaced rating pages. Each booklet is more fully described below.

Design and Procedure

The main experiment employed a 2 × 4 mixed design. List type (survival-related and not survival-related) was varied within-subject and crossed with type of processing, a between-subjects variable noted above. The primary dependent measures were the proportion of study words correctly recalled, the proportion of critical nonpresented words falsely recalled, and assessment of total false recall, computed by adding critical item false memories and other intrusions related to the theme of the list. Other measures are described in the procedure subsection.

All subjects were informed that they would hear a series of word lists read by the experimenter, and that following each list they would be tested for free recall by receiving intentional learning and memory instructions. Participants in the survival encoding conditions were informed that after they heard each word they were to rate it for its relevance to survivability on a 5-point scale ranging from 1 (totally irrelevant to survival) to 5 (extremely relevant to survival). These subjects were encouraged to use the whole scale and to choose the rating that most accurately represented their survival judgment for each verbal item. Similar instructions were given to participants in the pleasantness processing condition as they rated each list item for its pleasantness on a 5-point scale in which 1 indicated unpleasant and 5 indicated pleasant. In the control encoding condition, subjects were asked to study each list for recall, but they did not engage in an experimenter-presented processing task.

Following initial instructions, the subjects were presented with the six study lists in one of three random orders. Each word was read aloud, with 2 seconds between items. In the survival and pleasantness conditions participants had their booklets open to a rating page as they evaluated each word per their group directions. Those in the control group listened to the words at a 2-second rate while their booklet was open to a blank page.

Once the first list had been presented, participants were asked to turn over their rating (or blank) page to one that contained a brief free recall instruction and lines on which to write all words they could remember. The use of intentional memory testing, combined with the manipulation of type of list (survival vs. nonsurvival), necessitated that the lists be tested one at a time. Ninety seconds were allotted to complete each free recall test. After a recall period had expired, participants were asked to stop and turn to the next page, which was blank for controls or displayed the appropriate rating scale for survivability or pleasantness. List 2 was then read at a 2-second rate and followed by a 90-second recall period. The same routine was followed for the remaining four lists. Once all six lists had been studied and tested, the booklets were collected and participants were debriefed and thanked.

Results

Findings are presented in subsections below. For all analyses reported the significance level was set at $\alpha = .05$, with *post hoc* comparisons conducted when relevant with Tukey's HSD.

Rating Analyses

The rating activities served primarily as orienting tasks at encoding. However, these judgments also provided normative information on our six lists and importantly were useful as a manipulation check (that is, were our intuitions about what constituted survival lists and nonsurvival lists correct?). For participants in the survival processing grasslands condition, we calculated the average survival ratings on the 1-to-5 scale for the doctor, bread, and sleep lists and then an overall mean across survival lists. The same computations were carried out for the nonsurvival lists of needle, king, and sweet. The mean survivability rating for survival lists was 2.94 ($SD = 0.68$), whereas a mean rating of 1.94 ($SD = 0.73$) was observed for nonsurvival lists. While you would expect nonsurvival lists to be low in perceived survivability, as in 1.94, there may be some concern that survival lists are viewed as middle of the road (2.94) regarding survivability. At least they are perceived to be of medium relevance to survival. To place our values in context, we examined a number of studies that have used the survival rating task and found that it was not uncommon for the rating data not to be reported. When such ratings were reported, they are strikingly similar to the current data. In the original Nairne et al. (2007) article, across experiments the survival ratings hovered around 3.00. Butler, Kang, and Roediger (2009) observed mean ratings for survival of 2.60 in both their moving and survival conditions, while Weinstein, Bugg, and Roediger (2008) found that on average survivability ratings were 2.62, 2.50, and 3.04 in their survival, moving, and pleasantness groups, respectively. We conducted a within-subject *t*-test that revealed a significant difference between list type ($t(17) = 12.85, p = .000$), with survival lists being rated as more relevant for survival than nonsurvival lists. This outcome confirmed that our chosen survival lists were in fact viewed as more survival-related than the lists that were deemed as nonsurvival in nature. As such, these findings served as a manipulation check for our type of list variable and therefore negated the need to conduct a pilot study. The same kind of analysis was carried out on the pleasantness ratings. For survival lists the mean pleasantness rating was 3.39 ($SD = 0.32$), whereas for nonsurvival lists it was 2.79 ($SD = 0.34$). Another within-subject t-test revealed a significant difference between list type ($t(17) = 6.75$, $p = .000$), with survival lists being rated as more pleasant than nonsurvival lists.

Correct Recall

These data were initially scored for the proportion of study words correctly recalled for each of the six lists, leading to a 6 (lists) × 4 (type of processing) mixed analysis of variance, with the list variable treated as a within-subject factor. The 6 × 4 ANOVA revealed that neither the main effect nor the interaction was significant. Thus, we collapsed across lists in the analysis below.

The 2 (list type) × 4 (processing type) mixed ANOVA for correct recall yielded no main effect for list type. Given the recall performance for the individual lists just noted above, a lack of an effect of list type is not surprising. Thus, survival lists were not better remembered than nonsurvival lists, contrary to our hypothesis. A significant main effect of processing was observed ($F(3,89) = 5.55$, $p = .002$). Unlike a number of studies by Nairne et al. (2007, 2008), processing for grasslands survival ($M = .60$, $SD = 0.07$) resulted in significantly worse correct recall than generic survival ($M = 0.68$, $SD = 0.11$) and pleasantness ($M = 0.69$, $SD = 0.06$), which did not differ significantly from each other. The interaction was not reliable.

False Recall for Critical Nonpresented Items

The 2 (list type) × 4 (processing type) mixed ANOVA for critical nonpresented items yielded no significant main effects or interaction. Survival lists ($M = .38$, $SD = 0.30$) did not produce more false memory than nonsurvival lists ($M = .33$, $SD = 0.28$), contrary to our hypothesis. Note, however, that substantial levels of false recall were observed in this experiment. Although the analysis was not significant ($F(3,89) = 1.70$, $p = .16$), it is instructive to see the false memory rates as a function of processing condition. The means are as follows: processing for survival grasslands ($M = .42$, $SD = 0.23$), generic survival ($M = .36$, $SD = 0.26$), pleasantness ($M = .28$, $SD = 0.15$), and the control ($M = .40$, $SD = 0.28$), which did not differ significantly from each other.

Total False Recall

A 2 (list type) × 4 (processing type) mixed ANOVA was computed for total false recall, measuring not only the critical nonpresented item but also any other intrusion related to the theme of the list presented items, such as remembering "blanket," a word not part of the *sleep* list. As in the analysis reported in the section above, there were neither significant main effects nor an interaction. The failure to find these effects contradicts our hypotheses. Even though we emphasized that there were substantial levels of false recall in this experiment, the presence of related intrusion errors that contributed to total false recall underscores the fact that traditional reporting of only critical nonpresented item analysis underestimates the false memory effect.

OVERALL ACCURACY

Following the lead of List (1986) and more recently several investigators, including Howe and Derbish (2010) and Otgaar and Smeets (2010), an overall accuracy statistic, ranging from 0.0 to 1.0, was calculated taking both true and false memory (critical nonpresented items only) into account. Higher scores, closer to 1.0, represent greater accuracy.

$$\text{Accuracy} = \frac{\text{Veridical memory}}{(\text{Veridical memory} + \text{False memory})}$$

A 4 (processing type) × 2 (list type) mixed design ANOVA, with repeated measures on the last factor, was used to compare participants' levels of overall accuracy. There was no significant main effect for list type, which was also the case in the correct recall analysis. A significant main effect for processing type was found ($F(3,89) = 2.72$, $p = .049$), with performance in the pleasantness processing condition ($M = .973$, $SD = 0.013$) exceeding performance in survival grasslands ($M = .955$, $SD = 0.023$) but not general survival ($M = .965$, $SD = 0.026$) or the control ($M = .961$, $SD = 0.027$). This result is somewhat different than what was observed in correct recall performance and is probably due to ceiling effects with this accuracy measure. Because this accuracy measure takes into account both true and false memory, the high levels of nearly equal accuracy across processing conditions are understandable given that type of processing did not influence the degree to which participants did not produce false memories. Even though pleasantness processing led to significantly greater accuracy, it is still the case that the survival conditions produced high accuracy. There were no significant interactions.

EXPERIMENT 2

The second experiment also addressed survival processing in the DRM paradigm using material from Roediger and McDermott's original lists. Continuing along the lines of our first study, we continued to investigate differences between survival and nonsurvival material. In addition to survival differences in processing, we manipulated participant awareness of performance evaluation via the presence or absence of a camera. Therefore, we likely (see Smeets et al., 2012) altered arousal for half the participants by video-recording their performance in the experiment. While no pretest for stress levels among participants was given, many other studies have shown the presence of a camera to yield similar behaviors and stress levels as performing tasks in front of an active observer (Constantinou et al., 2005; Putz, 1975; Smeets et al., 2012), which is well documented to increase arousal and alter performance.

The hypotheses for Experiment 2 were as follows. Survival processing would show no benefits to veridical (true) memory, although it would yield higher levels of false memory. Whereas differences were not found in Experiment 1, correct recall and false recall of survival lists were expected to be higher than nonsurvival lists. Also, to the extent that processing survival-related information is a dominant response, we hypothesized that the presence of an audience (camera) should facilitate true memory. To the extent that processing nonsurvival information is a nondominant response, we further hypothesized that the presence of an audience (camera) should interfere with true memory, as well as result in higher levels of false memory.

Method

Participants

A total of 91 participants recruited from a major university were randomly assigned to one of two camera conditions, with 46 participants in the camera group and 45 participants in the non-camera group. Participants were also randomly assigned to use one of three different types of processing, with 31 assigned to survival, 28 assigned to pleasantness, and 32 in the control group.

Materials

Four different DRM lists were selected for use in this experiment. As designated by their critical nonpresented item, the four lists were *chair, needle, smell,* and *doctor* (Roediger & McDermott, 1995).

The purpose of this study was to examine different survival-related aspects and their influence on memory in the DRM paradigm. As in Experiment 1, the word lists were chosen to reflect two differing relationships to survival. The chair and needle lists were thought to have little to do with survival, whereas the smell and doctor lists were seen as more survival-related. As with Experiment 1, list type was later confirmed by the rating analysis. As the previous study represented our first attempt to examine survival memory within the context of the DRM illusion, in Experiment 2 we decided to more carefully select the DRM lists. The survival lists were balanced against those in the nonsurvival category as the concreteness of the words in each list and the levels of false recall each list typically produces were held constant across the list conditions. Additionally, the forward associative strength, or the likelihood of the words in an individual list causing the participant to think about the critical non-presented item (CNP), was also balanced between survival and nonsurvival lists. All word lists were recorded and played back for participants to ensure no differences would be present during encoding due to fluctuations in tempo, pitch, and dialect from reading to reading between researchers. A tone was used to signify the start and stop of each list. Each list began with a tone, followed by 5 seconds of silence. Following this, the first list of words was read at exactly one word every 2 seconds. A visual metronome was used to ensure proper tempo while recording each list. Another tone was played immediately following the reading of the last word in list 1. Ninety seconds of silence were then provided for free recall. At the end of 90 seconds, another tone was played to let the participant know to stop recall. After a 10-second break to allow enough time to turn the page in their experimental booklet, another tone was played to signify the start of the next list, which followed 5 seconds later. This format continued until participants had recalled the last list.

As with Experiment 1, we provided all participants with test booklets that were used both in the processing and recall of each list. The control group did not engage in any type of instructed processing, as they viewed a blank booklet page while listening to the words before recall. Participants completed their rating tasks in the other two groups, pleasantness or survival, in the same manner described in the first study. All participants in the survival condition read the same grasslands scenario employed in the first experiment (Nairne et al., 2007).

While engaged in both the encoding and recall, half the participants were video-recorded as a stressor. The full-size VHS video recorder was placed on a tripod within constant line of sight of the participant.

Design and Procedure

Participants were run one at a time. We did not want the presence of one participant to potentially act as an active observer for other participants, their presence implying evaluation. We did this to ensure that the influence of the camera to create an awareness of social evaluation would be as salient as possible. It was also important for those not in the camera group to feel relaxed and not pressured or judged by anyone who may have been participating simultaneously All participants were presented with every word list, though the order in which the lists were presented was randomly assigned to each participant.

Participants in the camera condition then were informed that they would be video-recorded and that the researchers would watch the videos so their performance could be assessed at a later time. After confirming their understanding, the researcher then powered up the camera, checked proper focus and framing, and then began recording. After asking the participant if he or she was ready to begin, the researcher would then instruct the participant to turn to the first page of the booklet and then began playing the audio file, beginning with the first list. In all conditions the researcher would turn off the computer monitor so it would not distract the participant. Immediately following each list, participants were given 90 seconds to try and recall as many words as they could, in any order. Upon the completion of the last 90-second recall session, the researcher turned off the camera, if necessary, and collected the test booklet.

Results

Unless noted otherwise, all effects are evaluated at the alpha level of p <.05. Tukey's HSD was used to for all *post hoc* tests.

Rating Analyses

A paired samples t-test was calculated to compare the mean survival rating of the survival lists and nonsurvival lists. On a 1-to-5 scale the mean rating for all survival lists across all participants was 3.23 (SD = 0.64), and the mean rating for all nonsurvival lists was 2.60 (SD = 0.60). Survival lists were rated significantly higher than nonsurvival lists on how relevant they would be to helping a participant survive in the grasslands condition ($t(30) = 7.68$, p <.001). Similar to our first study, these findings served as a manipulation check for the type of list variable and again negated the need to conduct a pilot study.

A paired samples t-test was also calculated to compare the mean pleasantness rating of the survival and nonsurvival lists. The mean pleasantness rating for combined survival lists was 2.98 (SD = 0.28), and the mean rating for nonsurvival lists was 2.76 (SD = 0.37). Survival lists were rated significantly higher than nonsurvival lists on how pleasant participants found them to be ($t(27) = 3.10$, $p < .01$), thus replicating the Experiment 1 perception that survival lists are more pleasant.

Correct Recall

These data were initially scored for the proportion of study words correctly recalled for each of the four lists, leading to a 4 (lists) × 3 (type of processing) mixed analysis of variance with the list variable treated as a within-subjects factor. The 4 × 3 ANOVA revealed that neither thye main effect nor the interaction was significant. Thus, as in Study 1 we collapsed across lists in the analysis below.

A 3 (processing type) × 2 (camera presence) × 2 (list type) mixed design ANOVA, with repeated measures on the last factor, was used to compare participants' levels of correct recall. A significant main effect for processing type was found ($F(2,85) = 4.80$, $p = .011$). Participants who engaged in survival processing ($M = 0.58$, $SD = 0.08$) remembered significantly fewer words than both those using pleasantness processing ($M = 0.63$, $SD = 0.09$) and the no-rating control ($M = 0.64$, $SD = 0.07$). Table 8.1 shows correct recall means for the different processing and list types expressed as proportion correct. No significant main effect was found for list type ($F < 1.00$), with both survival and nonsurvival lists leading to similar levels of veridical memory. No significant main effect was found for camera condition ($F(1,85) = 1.63$, $p = .205$), indicating that the camera's presence did not influence a participant's level of true memory.

The main effect of processing type was qualified by a significant list type × processing type interaction ($F(2,85) = 3.19$, $p = .046$), with nonsurvival lists leading to higher levels of veridical memory than survival lists in the control condition. Follow-up tests revealed no other significant differences. This interaction is displayed in Table 8.1.

A significant list type × camera presence interaction was found ($F(1,85) = 6.43$, $p = .013$). Participants in the camera condition remembered significantly fewer words for nonsurvival lists than those in the non-camera condition. Recall of survival lists was not affected by the camera's presence. All other interactions were found to be not significant ($F(1,85) = 1.15$, $p = .321$). Table 8.2 shows the interaction between list type and processing type.

False Recall for Critical Nonpresented Items (CNPs)

A 3 (processing type) × 2 (camera presence) × 2 (list type) mixed design ANOVA, with repeated measures on the last factor, was used to compare

Table 8.1. Mean Proportion Correct Based on Processing Type by List Type

		Processing Type			
		Survival (mean)	Pleasantness (mean)	Control (mean)	Total (mean)
List Type	*Survival Lists*	.59	.64	.62	.62
	Nonsurvival Lists	.57	.62	.65	.62
	Total	.58	.63	.64	

Table 8.2. Mean Proportion Correct Based on Camera Presence by List Type

		Camera Presence		
		Camera (mean)	No Camera (mean)	Total (mean)
List Type	*Survival Lists*	.62	.61	.62
	Nonsurvival Lists	.59	.64	.62
	Mean	.61	.63	

participant's recall for CNPs. No significant main effect for processing type was found ($F(2,85) = 1.35$, $p = .265$). A significant main effect was found for list type ($F(1,85) = 10.10$, $p < .01$), with nonsurvival lists ($M = 0.61$, $SD = 0.38$) leading to significantly higher levels of false memory than survival lists ($M = .45$, $SD = 0.43$). Caution should be exercised in making too much of this "advantage" for survival lists, as they are still susceptible to producing illusory recall. Could false memories emanating from survival material be considered adaptive? In a similar vein, it has been suggested that greater levels of false memory following survival processing might be adaptive (Howe & Derbish, 2010). We will address this issue in the general discussion. This main effect is shown in Table 8.3. No significant main effect was found for camera condition ($F(1,85) = 1.60$, $p = .209$), with the camera's presence not influencing a participant's level of false memory for critical non-presented items.

The interaction between list type and processing was significant ($F(2,85) = 3.07$, $p = .052$). Further analysis found no significant list type difference for survival processing ($F < 1.00$). When asked to process information in terms of its pleasantness, participants had marginally lower levels of false recall ($t(27) = -1.91$, $p = .067$) of nonsurvival word lists ($M = 0.54$, $SD = 0.41$) than survival-related word lists ($M = 0.34$, $SD = 0.41$). When given no specific guidelines for processing (control), participants had significantly lower levels of false recall ($t(31) = -3.04$, $p < .01$) of nonsurvival word lists ($M = 0.71$, $SD = 0.36$) than survival-related word lists ($M = 0.42$, $SD = 0.43$). This interaction can be seen in Table 8.3.

Table 8.3. Proportion of False Memory for CNPs Based on Processing Type by List Type

		Processing Type			
		Survival (mean)	Pleasantness (mean)	Control (mean)	Total (mean)
List Type	*Survival Lists*	.56	.36	.42	.45
	Nonsurvival Lists	.56	.54	.70	.60
	Mean	.56	.45	.56	

Total False Recall

As with the first experiment, we calculated a measure called total false memory that reflects all thematically related intrusions produced by subjects. Total false memory can be as low as 0, with no upper limit for the measure; however, the grand mean for total intrusions across all conditions and lists in the experiment was 3.25. Thus, similar to Experiment 1, the intrusion rate for related words other than those designated as critical averaged slightly more than one additional false memory per participant.

$$\text{Total false memory} = \text{Critical non-presented items} + \text{Other related intrusions}$$

A 3 (processing type) × 2 (camera presence) × 2 (list type) mixed design ANOVA, with repeated measures on the last factor, was used to compare participants' levels of total false recall. No significant main effect for processing type was found ($F(2,85) = 1.40$, $p = .252$). List type also showed no significant main effect ($F < 1.00$). Those in the camera condition ($M = 3.74$, $SD = 2.36$) were more likely to recall items not heard during encoding than those in the non-camera condition ($M = 2.76$, $SD = 1.91$), showing a significant main effect for camera presence ($F(1,85) = 4.655$, $p = .034$).

Overall Accuracy

As with Experiment 1, the calculation for overall accuracy was:

$$\text{Accuracy} = \frac{\text{Veridical memory}}{(\text{Veridical memory} + \text{False memory})}$$

A 3 (processing type) × 2 (camera presence) × 2 (list type) mixed design ANOVA, with repeated measures on the last factor, was used to compare participants' levels of overall accuracy. No significant main effect for processing type was found ($F(2,85) = 2.23$, $p = .114$); however, the trend was consistent with the true memory findings. There was a significant main effect for list type ($F(1,85) = 10.70$, $p = .002$), with survival lists ($M = 0.67$, $SD = 0.27$) having better overall accuracy than non-survival lists ($M = 0.57$, $SD = 0.23$). No significant differences were found between camera conditions, although the trend was consistent with the total false memory findings showing detriments to memory performance with the camera present ($F(1,85) = 1.60$, $p = .21$). There were no significant interactions.

GENERAL DISCUSSION

Throughout this volume there has been much discussion of Nairne and colleagues' methodology to examine the notion that human memory evolved to be particularly tuned to survival. While their approach may be new, the idea of such tuning is not a recent issue; the study of the role of cognition in survival has a long history (see Cosmides & Tooby, 1992, for a review, and Jerison, 1973, for an earlier review). Nairne and colleagues' methodology, entailing the introduction of a survival processing paradigm involving the grasslands scenario, sparked considerable attention: since Nairne's first studies were published just some 6 or so years ago, a flourish of papers have appeared that extend the original work, expand the paradigm, and test the generalizability of the results. Perhaps it is not too surprising that these extensions across many laboratories have produced some mixed results, with many studies replicating the superiority in memory afforded by encoding in a survival-related context but a growing number of other investigations showing the opposite effect when stimuli or conditions are varied. Indeed, the chapters in this volume, including ours, add to the variety of findings.

Some of this mixture surely can be traced to differences among the study materials, differences across survival scenarios, and the nature of the retrieval task. For example, in the present experiments we extended, as did Howe and Derbish (2010) and Otgaar and Smeets (2010), the encoding material to DRM lists while also examining the degree to which the DRM material itself was related to survival. We were also interested in more than veridical memory and joined a growing focus on how survivability might be related to false recollection and the possibility that illusory memory may actually be adaptive in an evolutionary sense (Howe & Derbish, 2010; Nairne & Pandeirada, 2008a). Indeed, as discussed earlier, our initial foray into the study of adaptive memory can be traced to Nairne et al.'s (2007) finding of an advantage of survival processing (within the grasslands context) in accurate recall, while also showing a trend toward survival encoding creating increased false memory, as indexed by greater rates of intrusion errors. This "more is less" pattern of higher true memory accompanied by higher false memory is fairly common (Brainerd & Reyna, 2005; Toglia et al., 1999) and is now being observed in adaptive memory investigations (Howe & Derbish, 2010; Otgaar & Smeets, 2010). Experiment 1 was mainly designed to further examine this pattern using materials, DRM lists, that Toglia et al. (1999) had employed in their work. Somewhat surprisingly, we did not replicate Nairne's results of survival processing producing better veridical memory; in fact, we found that true memory was lowest in the survival grasslands condition. Similarly, Howe and Derbish (2010) showed that in terms of total accuracy, there was no survival processing advantage, with encoding for survival relevance producing the poorest level of retention. A major reason for this was that survival processing led to higher levels of false memory. This state of affairs is consistent with the notion that illusory memory may be adaptive, a suggestion convincingly made by Howe and Derbish.

Because semantic and/or elaborative encoding (Brainerd, Yang, Reyna, Howe, & Mills, 2008) is often theorized to drive the illusory memory effect, if survival processing is a super-level semantic task, as suggested by the survival processing advantage in veridical memory over other well-established semantic forms of processing (e.g., self-reference; pleasantness; see Nairne & Pandeirada, 2008a), it is understandable that adaptive memory-oriented encoding would boost false memory. But why did we not see a bump in veridical memory? This lack of increase in true memory seems real enough given that we observed a relative decrease in veridical recall in the survival grasslands condition compared to pleasantness processing in Experiment 1. We replicated this decrease in Experiment 2.

The key to understanding this seems to be list structure and how it relates to item-specific and relational processing. When unrelated lists are presented at study, as in many of Nairne and colleagues' experiments, participants focus on the specific items, and a robust survival processing advantage is observed. This facilitation based on study items being oriented to survival is reduced when categorized lists are employed (Nairne & Pandeirada, 2008a) that promote the encoding of relationships. When DRM lists are used, as in the present experiments, clearly they permit considerable relational processing, which leads to no advantage or even a disadvantage. DRM lists would also require some degree of item-specific processing regardless of the encoding task. Support for this idea was provided by Burns et al. (2011), who showed that survival processing engaged participants in both item-specific and relational processing while encoding such words. When they constructed situations wherein nonsurvival encoding conditions promoted both item-specific and relational processing, the survival processing advantage was eliminated. While pleasantness ratings clearly involve item-specific processing for unrelated word lists, when DRM lists are used, even some relational processing is inescapable when participants rate the individual words for pleasantness. Thus, based on Burns et al.'s (2011) results, it is not surprising that we did not observe a survival processing benefit.

Thus, the present findings regarding the influence of encoding scenarios stand in contrast to Nairne's original work. Much of this contrast turns on our argument that stimulus materials do make a difference, in fact a big difference, in whether or not a survival processing advantage is observed, and therefore questions can be raised about the generalizations of the grasslands scenario to memory in general. One might expect broad applications because survival processing should be effective to the extent it was an important element at the time when memory modules in humans evolved and were honed over our ancestral history. Thus, the environmental factors in the grasslands scenario are the same theorized to be present during the formation of these specific memory modules. That being said, the benefits of survival processing are not confined to the grasslands scenario but have been demonstrated using scenarios such as outer space (Kostic, McFarlan, & Cleary, 2012) and predators, zombies, not possibly present when humans evolved (Soderstrom & McCabe, 2011). Although the effectiveness of survival processing is not global in the sense that it is effective in all memory tasks, memory modules that evolved under constraints of ancestral situations should be global in that they are still effective when enlisted for processing material under conditions that could not have been present when they evolved (Soderstrom & Cleary, Chapter 6 in this volume).

Whereas, as reiterated above, survival processing has demonstrated its superiority across different forms of encoding (Kang, McDermott, & Cohen, 2008; Nairne et al., 2007; Weinstein et al., 2008), these benefits have not been replicated in the DRM paradigm, as demonstrated in recognition memory by Howe and Derbish (2010) and the current experiments. Why might this be the case? One possibility is that a majority of the experiments demonstrating survival processing's superiority, including Nairne's initial 2007 study, employed incidental memory as opposed to intentional memory. Mnemonic situations where survival memory mechanisms excel may mirror the context in which they evolved. As early humans most likely did not develop all memory mechanisms while intentionally learning about their surroundings, it makes sense that the benefits found using survival processing do not necessarily extend to the intentional memory tasks used in both of our experiments. It is doubtful that our ancestors intentionally dedicated most of what they learned to memory. It is more likely that the day-to-day memories vital to survival were initially encoded incidentally, in survival scenarios, therefore strengthening that specific memory module. Of course, beyond initial encoding, when a particular survival context was re-experienced, intent to remember and elaborating upon a memory module can be thought of as adaptive strategies activated by the learner. Although our findings raise questions regarding the ability to generalize survival processing, they support, we believe, the idea that the memory modules have evolved into their current states. As a product of evolution, perhaps the human mind is "customized" for the retention of survival-relevant information.

In addition to the intentionality issue, many experiments supporting the survival processing advantage also employed nonthematic word lists, as opposed to categorically oriented lists. In recognizing this, Nairne and Pandeirada (2008a) conducted survival processing experiments in which the encoding materials were categorized lists. As we noted earlier, with these lists they still found a survival processing advantage, though it was only marginally reliable in their Experiment 1, and thus categorical word lists led to an attenuated survival processing advantage compared to unrelated lists. A more "salient" form of categorized material would be DRM lists, in part because DRM lists often produce greater levels of false memory in adult participants (see Howe, Wimmer, Gagnon, & Plumpton, 2009, for examples). Prior to the present studies, Howe and Derbish (2010) set out to replicate the survival grasslands findings using DRM lists. Staying true to Nairne's original procedure, they randomized DRM lists of words for presentation, made the task incidental (Howe & Derbish, 2010), included a distractor task after study, and administered a recognition test. In line with Nairne, their participants in the survival condition produced the best level of correct recall. Although this demonstrates benefits to survival processing in the DRM paradigm, circumstances not typical of the DRM paradigm were required. Furthermore, this benefit disappeared when the data were scored for accuracy using the ratio that takes into account false memory (we also employed this in both of our experiments). Howe and Derbish found not only that participants in the survival condition produced greater veridical memory, as noted above, but also found significantly greater recognition of critical lures compared with their other two conditions. This effect was present in all three types of word lists (negative, neutral, and survival), and is consistent with the "more is less" pattern (Toglia et al., 1999). Interestingly, this composite accuracy ratio showed that accuracy rates were significantly higher in the pleasantness condition compared with the survival condition (Howe & Derbish, 2010).

The composite accuracy ratio results in the present experiments were generally in line with Howe and Derbish's (2010) composite accuracy data as well as similar findings reported by Otgaar and Smeets (2010). Furthermore, in the current studies conducted within a traditional DRM format, survival grasslands processing led to the lowest levels of uncorrected recall in both experiments. Interestingly, in our first study, encoding in the survival grasslands context also led to lower recall than processing in the generic survival condition. It is unclear why this was the case, though generic survival processing may have served as another control condition in that our generic processing did not engage participants in a scenario environment.

Turning to the survival characteristics of the materials, we hypothesized that recall for survival-relevant lists would be better than for nonsurvival lists. In Experiment 1 we found no differences. In Experiment 2 all lists were normed for levels of correct recall (see Roediger, Watson, McDermott, & Gallo, 2001), and in the control group only, nonsurvival lists yielded higher true memory than survival lists. This finding is limited to a subsection of our participants in the second experiment, with no significant differences found among all processing groups. Still, the trend is toward detriments in correct recall for survival-related information. This conclusion must be tempered by the fact that this pattern did not hold up when analyzing overall memory performance, as the high levels of intrusions in *all* conditions overshadowed any benefits found in the control condition for true memory for nonsurvival material. In general nonsurvival lists had a high proclivity for intrusions compared to survival lists, especially in the control group. Although this does not align with the "more is less" pattern (Toglia et al., 1999), showing high levels of true memory and high levels of false memory, found by Howe and Derbish (2010) and Experiment 1, it does support the notion that memory may be tuned to better remember survival-relevant information more accurately. This survival benefit is most noticeable in high arousal situations, perhaps similar to those in which they evolved, as described next.

The arousal generated by the presence (or implied presence) of others has been shown to facilitate performance of dominant responses while hindering nondominant responses (Zajonc, 1965). As our brain mechanisms have evolved over time, survival-relevant material ought to have been processed more often, become familiar, and therefore be a dominant response. Nonsurvival material, on the other hand, may be more novel, less processed, and therefore nondominant. For recall of nonsurvival-related information there were reliable decrements in veridical memory as a function of a camera present during the recall task. This makes sense, as memory for nonsurvival material would be less beneficial in high-stress (survival) conditions, while survival material would remain unaffected.

In addition to hindering veridical memory performance for nonsurvival lists, presumed arousal caused by the camera, as in Smeets et al. (2012), led to higher levels of intrusions. Here, social facilitation seems to have a positive effect on dominant responses and a negative effect on nondominant responses for survival-relevant material and situations. Flawless memory is unlikely to be a practiced behavior (intentionally or not) and therefore likely to be a nondominant response. As such, we saw high levels of intrusions in the camera condition compared to the non-camera condition, which can likely be traced to poor performance for nondominant functions. While hindering overall memory performance, the presence of false memories may be beneficial (see Schacter, Guerin, & St. Jacques, 2011, for a discussion of adaptivity and memory distortions). When evaluating the array of dangers presented by predators, confusion about which predators have or have not been encountered might be quite advantageous. As such, surviving is predicated on learning from past experiences, but maybe more so in being able to anticipate future situations and the behaviors necessary for survival in these circumstances. This was nicely summarized by Nairne and Pandeirada (2008b, p. 240) when they stated, "The fact that memory is fundamentally constructive rather than reproductive, often laced with relevant but 'false' recollections, provides prima facie evidence for this claim that memory evolved to help predict the future, not remember the past."

REFERENCES

Anderson, J. R., & Schooler, L. J. (2000). The adaptive nature of memory. In E. Tulving & F. I. Craik (Eds.), *The Oxford handbook of memory* (pp. 557–570). New York: Oxford University Press.

Brainerd, C. J., Holliday, R. E., Reyna, V. F., Yang, Y., & Toglia, M. P. (2010). Developmental reversals in false memory: Effects of emotional valence and arousal. *Journal of Experimental Child Psychology, 107*, 137–154.

Brainerd, C. J., & Reyna, V. F. (2005). *The science of false memory*. New York: Oxford Press.

Brainerd, C. J., Yang, Y., Reyna, V. F., Howe, M. L., & Mills B. A. (2008). Semantic processing in "associative" false memory. *Psychonomic Bulletin & Review, 15*, 1035–1053.

Brown, S. C., & Craik, F. I. (2000). Encoding and retrieval of information. In E. Tulving & F. I. Craik (Eds.), *The Oxford handbook of memory* (pp. 93–107). New York: Oxford University Press.

Budson, A. E., Todman, R. W., Chong, H., Adams, E. H., Kensinger, E. A., Krangel, T. S., & Wright, C. I. (2006). False recognition of emotional word lists in aging and Alzheimer disease. *Cognitive and Behavioral Neurology, 19*, 71–78.

Burns, D. J., Burns, S. A., & Hwang, A. J. (2011). Adaptive memory: Determining the proximate mechanisms responsible for the memorial advantages of survival processing. *Journal of Experimental Psychology: Learning, Memory, and Cognition, 37*, 206–218.

Butler, A. C., Kang, S. H. K., & Roediger, H. L. (2009). Congruity effects between materials and processing tasks in the survival processing paradigm. *Journal of Experimental Psychology: Learning, Memory, and Cognition, 35*(6), 1477–1486.

Constantinou, M., Ashendorf, L., & McCaffrey, R. J. (2005). Effects of a third party observer during neuropsychological assessment: When the observer is a video camera. *Journal of Forensic Neuropsychology, 4*(2), 39–47.

Corson, Y., & Verrier, N. (2007). Emotions and false memories: Valence or arousal? *Psychology Science, 18*, 208–211.

Cosmides, L., & Tooby, J. (1992). Cognitive adaptations in social exchange. In J. H. Barkow, L. Cosmides, & J. Tooby (Eds.), *The adapted mind: Evolutionary psychology and the generation of culture* (pp. 163–228). New York: Oxford University Press.

Cottrell, N. B., Wack, D. L., Sekerak, G. J., & Rittle, R. H. (1968). Social facilitation of dominant responses by the presence of an audience and the mere presence of others. *Journal of Personality and Social Psychology, 9*(3), 245–250.

Darwin, C. (1859). *On the origin of species*. London: John Murray.

Deese, J. (1959). On the prediction of occurrence of particular verbal intrusions in immediate recall. *Journal of Experimental Psychology, 58*, 17–22.

Howe, M. L., Candel, I., Otgaar, H., Malone, C., & Wimmer, M. C. (2010). Valence and the development of immediate and long-term false memory illusions. *Memory, 18*(1), 58–75.

Howe, M. L., & Derbish, M. H. (2010). On the susceptibility of adaptive memory to false memory illusions. *Cognition, 115*(2), 252–267.

Howe, M. L., Wimmer, M. C., Gagnon, N., & Plumpton, S. (2009). An associative-activation theory of children's and adults' memory illusions. *Journal of Memory and Language, 60*, 229–251.

Hunt, R. R., & Einstein, G. O. (1981). Relational and item-specific information in memory. *Journal of Verbal Learning & Verbal Behavior, 20*, 497–514.

Jerison, H. J. (1973). *Evolution of the brain and intelligence*. New York: Academic Press.

Kang, S. H. K., McDermott, K. B., & Cohen, S. M. (2008).The mnemonic advantage of processing fitness-relevant information. *Memory & Cognition, 36*(6), 1151–1156.

Kostic, B., McFarlan, C. C., & Cleary, A.M. (2012). Extensions of the survival advantage in memory: Examining the role of ancestral context and implied social isolation. *Journal of Experimental Psychology: Learning, Memory, and Cognition, 38*, 1091–1098.

List, J. A. (1986). Age and schematic differences in the reliability of eyewitness testimony. *Developmental Psychology, 22*, 50–57.

Mandler, G. (2007). *History of modern experimental psychology: From James and Wundt to cognitive science.* Cambridge: MIT Press.

Miller, G. A. (1956). The magic number seven, plus or minus two: Some limits on our capacity for processing information. *The Psychological Review, 63*, 81–97.

Nairne, J. S., & Pandeirada, J. N. S. (2008a). Adaptive memory: Is survival processing special? *Journal of Memory and Language, 59*(3), 377–385.

Nairne, J. S., & Pandeirada, J. N. S. (2008b). Adaptive memory: Remembering with a stone-age brain. *Current Directions in Psychological Science, 17*, 239–243.

Nairne, J. S., Pandeirada, J. N. S., & Thompson, S. R. (2008). Adaptive memory: The comparative value of survival processing. *Psychological Science, 19*, 176–180.

Nairne, J. S., Thompson, S. R., & Pandeirada, J. N. (2007). Adaptive memory: Survival processing enhances retention. *Journal of Experimental Psychology: Learning, Memory, and Cognition, 33*(2), 263–273.

Neisser, U. (1967). *Cognitive psychology.* East Norwalk, CT: Appleton-Century-Crofts.

Otgaar, H., & Smeets, T. (2010). Adaptive memory: Survival processing increases both true and false memory in adults and children. *Journal of Experimental Psychology: Learning, Memory, and Cognition, 36*(4), 1010–1016.

Pesta, B. J., Murphy, M. D., & Sanders, R. E. (2001). Are emotionally charged lures immune to false memory? *Journal of Experimental Psychology: Learning, Memory, and Cognition, 27*(2), 328–338.

Putz, V. (1975). The effects of different modes of supervision on vigilance behavior. *British Journal of Psychology, 66*, 157–160.

Roediger, H. L., & McDermott, K. B. (1995). Creating false memories: Remembering words not presented in lists. *Journal of Experimental Psychology, 21*, 803–814.

Roediger, H. L., Watson, J. M., McDermott, K. B., & Gallo, D. A. (2001). Factors that determine false recall: A multiple regression analysis. *Psychonomic Bulletin & Review, 8*, 385–407.

Schacter, D. L., Guerin, S. A., & St. Jacques, P. L. (2011). Memory distortion: an adaptive perspective. *Trends in Cognitive Science, 15*, 467–474.

Smeets, T., Otgaar, H., Raymaekers, L., Peters, M. J. V., & Merckelbaxh, H. (2012). Survival processing under stress. *Psychonomic Bulletin & Review, 19*, 113–118.

Soderstrom, N. C., & McCabe, D. M. (2011). Are survival processing memory advantages based on ancestral priorities? *Psychonomic Bulletin & Review, 18*(3), 564–569.

Symons, C. S., & Johnson, B. T. (1997). The self-reference effect in memory: A meta-analysis. *Psychological Bulletin, 121*, 371–394.

Toglia, M. P., Neuschatz, J. S., & Goodwin, K. A. (1999). Recall accuracy and illusory memories: When more is less. *Memory, 7*(2), 233–256.

Triplett, N. (1898). The dynamic factors in pacemaking and competition. *American Journal of Psychology, 9*, 507–533.

Weinstein, Y., Bugg, J. M., & Roediger, H. L., III. (2008). Can the survival recall advantage be explained by basic memory process? *Memory & Cognition, 36*(5), 913–919.

Zajonc, R. B. (1965). Social facilitation. *Science, 149*(3681), 269–274.

9

Survival Processing Does Not Improve Paired-Associate Learning

BENNETT L. SCHWARTZ AND BROCK R. BROTHERS

In a variety of nonhuman species, there is evidence to support the notion that memory systems have evolved to allow species to adapt to particular niches (Adachi, Chou, & Hampton, 2009; Sherry & Schacter, 1987). It is therefore logical to suspect that human evolution has shaped our memory processes, given the range and power of human memory systems (Nairne & Pandeirada, 2008). Evolutionary processes are difficult to address in studies of human cognition because cognition does not leave fossils, nor can contemporary cognitive psychologists use any aspect of natural selection as an independent variable in experiments. Thus, evolutionary psychology requires research to look at variables that may have been shaped by natural selection in the distant past and to examine how they affect memory in modern humans. To accomplish this, research must be creatively designed to examine the effects of memory variables that might have influenced the evolution of human memory.

One such methodology of evaluating possible evolutionary effects on memory is the survival processing paradigm, developed by James Nairne and his colleagues. In the seminal study, Nairne, Thompson, and Pandeirada (2007) asked participants to rate a list of unrelated words in terms of their relevance to surviving in the grasslands of a foreign land. Participants were later given a free recall test for the unrelated words. Nairne et al. found that participants who made survival-relevance judgments recalled more words on a free recall test than did participants instructed to use other well-established encoding techniques, such as making pleasantness judgments or judging the words in terms of self-relevance. Other researchers have shown survival processing to produce better free recall than other scenarios, such as moving to a new house, planning for a bank robbery, and even surviving in a hostile city (Butler, Kang, & Roediger, 2009; Kang, McDermott, & Cohen, 2008; Kostic, McFarlan, & Cleary, in press; Nairne & Pandeirada, 2011; Nairne, Pandeirada, & Thompson, 2008; Seamon et al., 2012; Weinstein, Bugg, & Roediger, 2008).

Nairne and Pandeirada (2010) suggested that survival processing is likely the result of a memory module tuned to ancestral conditions. That is, embedded in our cognitive architecture are mechanisms that are tuned specifically to those stimuli that are related to survival in our ancestral environment. For a prehistoric human, remembering information vital to survival may have led to natural selection for this ability. If natural selection is applicable to memory abilities, we should see evidence of it in memory in modern humans. Nairne and Pandeirada reason that modern stimuli that most closely match this ancestral tuning will be remembered better than those that match less closely.

If this argument is correct, then the advantages of survival processing should be seen across various tests of memory and should not be restricted to the specific methodology employed by Nairne and Pandeirada in their various experiments (see Nairne & Pandeirada, 2011). That is, if an evolutionarily based survival memory system exists, it ought to be observable across many situations in which information is cast against a survival background. Therefore, we should see survival processing effects in many different scenarios that call on mnemonic processing, in both incidental and intentional learning, and with a variety of memory tests. In the current study, we examined the effects of survival processing on paired-associate learning with both incidental and intentional learning.

We choose paired-associate learning for three reasons. First, to our knowledge, survival processing had not yet been tested with paired associates when we started the project (but see Seamon et al., 2012). Associative learning is a major form of human learning, presumably shaped by natural selection (Platt, Brannon, Briese, & French, 1996). Because so much of human memory is cue-dependent (e.g., Tulving & Pearlstone, 1966), it is important to examine if survival processing is effective in memory situations that call on cue–target associations. Second, paired-associate learning is frequently used in memory experiments and clearly benefits from elaborative processing, such as pleasantness judgments and imagery instructions (Wang & Thomas, 1995). If survival processing has the generalizability expected if it is an evolutionary module, we should see its effects in cued recall tests of paired associates as well as free recall of items from unrelated lists. Third, we can design experiments that are identical except for one respect, whether or not the participants know if their memories will be tested. That is, in one condition, we can tell people to focus on the survival processing task because we are interested in those judgments without making any reference to a later memory test. In another condition, we can inform the participants that a memory test will follow. Thus, we can examine intentional versus incidental learning.

On another level, paired-associate learning models many forms of real-world learning, such as learning vocabulary in a new language and learning difficult words for standardized tests (Kornell, 2009). If paired-associate learning benefited from survival processing, there would be clear practical implications of survival processing research. We could encourage students and other learners to employ survival processing when engaged in learning activities.

To date, only one other experiment has examined survival processing manipulations with a cued recall test. Seamon et al. (2012) presented participants with a 14-paragraph story read by an experimenter. Participants were instructed to make a survival-relevance judgment or a meaning-relevance judgment for each paragraph in the story. Following the presentation of the story, the participants answered cued-recall questions concerning factual information in the story. Seamon et al. found no evidence that the survival-relevance judgments led to greater levels of correct cued recalls than did the meaning-relevance judgments. Indeed, performance was equal to that when evaluating the meaning relevance of the story. Thus, the Seamon et al. data add to the recent spate of studies finding limitations to the extent to which survival processing generalizes (e.g., Soderstrom & McCabe, 2011; Tse & Altarriba, 2010).

In our paradigm, participants make specific judgments on discrete paired associate items, so we contend that our paradigm more closely matches the Nairne paradigm than Seamon et al. (2012). To examine paired-associate learning, we used language translation equivalents in Swahili–English (Experiments 1 and 2) and Lithuanian–English (Experiments 3 and 4). Participants encoded the pairs using survival processing or other instructions, such as pleasantness judgments. Later, a cued-recall test was administered in which the new-language word was presented and participants were asked to recall the English translation.

The current study was designed with two goals in mind. First, we wished to determine if survival processing improves the memory for paired associates in a language learning paradigm, tested by cued recall. Our initial hypothesis was that survival processing would benefit paired-associate learning. Given the proposed mechanisms for the survival processing effect in free recall, it should generalize to other kinds of learning and memory testing (Nairne & Pandeirada, 2008). However, as will be seen shortly, we found no evidence to support this hypothesis. In Experiments 2, 3, and 4, we were also interested to see if survival processing would affect meta-memory judgments for unrecalled items. We looked at both feeling-of-forgetting judgments (Halamish et al., 2011) and feeling-of-knowing judgments (Thomas, Bulevich, & Dubois, 2011) to examine meta-memory judgments. If meta-memory judgments are sensitive to variables that affect memory, the judgments would be sensitive to survival processing as well. However, some research suggests that meta-memory judgments are sensitive to cues that are linked with memorability but are not linked directly to memorability itself (e.g., Koriat, 1993). Thus, the inclusion of the meta-memory judgments allowed us to determine if meta-memory judgments are sensitive to survival processing or if our meta-memory is not attuned to this factor (Palmore, Garcia, Bacon, Johnson, & Keleman, 2012).

EXPERIMENT 1

Experiment 1 was conducted as a preliminary experiment, exploring the application of survival processing and paired-associate learning. We conducted it in a large classroom of students in an undergraduate Psychology class. As such, all students received the same list of words in an identical order, though the order itself was created randomly. This creates a confound with order effects but also allowed us to test many participants.

Method

Participants

A total of 138 Florida International University students participated in the first experiment in return for research credit. This and all experiments described here were approved by Florida International University's Institutional Review Board (IRB).

Materials and Procedure

A total of 20 Swahili words, each paired with their English translation, were shown on PowerPoint slides from a screen in the middle of a class auditorium. The Swahili words used for Experiments 1 and 2 were chosen from the Nelson and Dunlosky (1994) norms and were of moderate difficulty. All participants were run simultaneously in a large auditorium. The experiment was a between-subject design with three levels (pleasantness judgments, savannah-survival judgments, and city-survival judgments).

The experimenter (BRB) introduced the experiment and distributed informed consent forms to each participant. He then asked them to participate in a memory experiment and told them they would view translation pairs in a PowerPoint presentation. He then distributed instruction sheets concerning how each participant was to encode the stimuli. Each participant read instructions that were given on individual sheets of paper. The participants were randomly assigned to one of the three different conditions, which were adapted from Kang et al. (2008). Participants did not discuss their instructions with each other or recall collaboratively. The conditions were as follows:

> *Word pleasantness.* "I am going to show you a list of word pairs on a PowerPoint slide show. Each slide will display a single word in Swahili on the left side, and its English translation on the right side, for 5 seconds. Your task is to rate the pleasantness of each word (on a scale of 1 to 5; 1 being very unpleasant and 5 being very pleasant) Some of the words on these slides may sound or look pleasant to you while some words may not—it is up to you to decide."
>
> *Survival—grassland setting.* "Imagine as best as possible that you are stranded in foreign grassland, such as the plains of Africa. You are the only one in this environment and you have no survival materials. Over the course of a few months, you will need to find steady supplies of food and water and protect yourself from predators. We are going to show you a list of word pairs on a PowerPoint slide show. Each slide will display a single word in Swahili on the left side, and its English translation on the right side, for 5 seconds. Your task is to rate how relevant each of these words would be for you in this survival situation (on a scale from 1 to 5; 1 being barely relevant and 5 being highly relevant). Some of the words on these slides may be relevant and some may not—it is up to you to decide."
>
> *Survival—city setting.* "Imagine as best as possible that you are stranded in a foreign city. You are the only one in this environment and you have no survival materials. Over the course of a few months, you will need to find steady supplies of food and water and protect yourself from attackers. We are going to show you a list of word pairs on a PowerPoint slide show. Each slide will display a single word in Swahili on the left side, and its English translation on the right side, for 5 seconds. Your task is to rate how relevant each of these words would be for you in this survival situation (on a scale of 1 to 5; 1 being barely relevant and 5 being highly relevant). Some of the words on these slides may be relevant and some may not—it is up to you to decide."

After reading instructions, we showed participants a list of 20 Swahili words and their English translation. Each word pair was shown for 5 seconds. During this time, we instructed the participants to rate each word pair's relevance (according to their particular condition) on a Likert scale with values from 1 to 5. After the participants had seen all 20 words, we showed the list again in the same order, with each word being visible for 3 seconds. Ratings were made during the second presentation as well. Following the second trial of learning, participants engaged in an arithmetic task for 5 minutes to prevent rehearsal. Following the 5-minute distractor task, the participants received a recall test. The participants were shown a slide with a Swahili word on the overhead screen and were instructed to write down the English translation on the answer sheet we provided to them. The order at recall was randomly selected and was different from the order of presentation during learning. The recall sheets were collected from each participant, and then finally the participants were debriefed and thanked for their participation.

Results

Participants' mean ratings did not differ across conditions, $F < 1$. The mean pleasantness was 3.1, the mean grassland-survival rating was 2.9, and the mean city-survival rating was 2.8. The mean difference in rating hovered around 3.0 for both pleasantness and survival in Experiments 2 through 4 as well and therefore will not be reported further for the sake of brevity.

We did not find any recall advantage of the grasslands scenario relative to the other conditions, $F < 1$. The mean level of recall was 21 percent in the pleasantness condition, 21 percent in the grasslands-survival condition, and 18 percent in the city-survival condition. Subsequent analyses found a congruity effect (e.g., Butler et al., 2009). Words that were rated high (above the mean) on survival relevance were more likely to be remembered in the survival processing condition than words that were rated less high on survival relevance. Congruity effects are common in survival processing experiments (see Palmore et al, 2012).

Discussion

We initially expected the survival processing conditions (grasslands) to produce better recall than the control conditions. However, we found no differences. Before drawing any conclusions, we were concerned about a number of factors in this experiment. First, because all participants received the same order of the same stimuli, we were concerned that order effects may have swamped any survival processing effect. Thus, in subsequent experiments, we tested participants on an individual basis and randomized the order of presentation during encoding and during testing. Second, students were aware that a memory test would follow. Therefore, they may have ignored our instructions and used their own study strategies. As a result, in all subsequent studies, we focused participants' attention to the instructional tasks but did not inform them that a memory test would ensue.

EXPERIMENT 2

In Experiment 2, we tested participants individually. We did not give participants instructions that memory would be tested after they finished rating the pairs. In this way, at least instructionally, the experiment involved incidental learning. In addition, we were originally interested in testing how different processing scenarios would affect meta-memory judgments. We anticipated that survival processing would lead to better recall, but that participants would not show metacognitive awareness. In fact, in Experiment 2, we found the opposite pattern.

Method

Participants and Materials

A total of 61 Florida International University students participated. All of these participants were Introductory Psychology students and did not overlap with the participants involved in Experiment 1. We used the same Swahili–English word pairs as we used in Experiment 1.

Procedure

The experiment was a between-subject design with two levels (pleasantness judgments and savannah-survival judgments).

Participants were tested individually on Macintosh computers. We randomized the order of presentation for all participants during both encoding trials and at the time of retrieval. On arriving at the lab, each participant was randomly assigned to the survival processing condition (30 participants) or the pleasantness judgment condition (31 participants). Participants then read the instructions to evaluate each translation pair either in terms of its pleasantness or in terms of whether or not the concept represented by the pair would be useful to survival. Participants were instructed to evaluate the word pairs in terms of survival processing or pleasantness ratings but were not informed that there would be a subsequent memory test. The instructions were as follows.

> *Survival processing.* "You will be presented with a short series of Swahili words and their translation in English. The Swahili word will be presented on the left and its translation on the right. In this task, please imagine that you are stranded in the grasslands of a foreign land, without any basic survival materials. You have recently received word that a dangerous predator has been seen in the area. You will need to avoid and/or escape from the predator to ensure your survival. You will see each word pair for 10 seconds. During that time, think how relevant each of these word pairs would be in your attempt to avoid the predator."
>
> *Pleasantness judgments.* "You will be presented with a short series of Swahili words and their translation in English. The Swahili word will be presented on the left and its translation on the right. Some of the words may be pleasant in meaning, sound, or appearance, and some may not. Please evaluate each word in terms of its pleasantness as you read each pair. You will see each word pair for 10 seconds. During that time, think about whether or not the word pair is pleasant or unpleasant."

After reading instructions, we showed participants a list of 20 Swahili words and their English translation. Each word pair was shown for 10 seconds. During this time, the participants were instructed to rate each word pair's relevance (according to their particular condition) on a Likert scale with values from 1 to 5. After the participants had read all 20 words, we showed the list again in a new random order, with each word being visible for 10 seconds. Participants made ratings again during the second presentation.

Following the second trial of learning, participants were given instructions concerning recall. They would again see the Swahili word, and we instructed them to recall the English translation. We discouraged participants from guessing and told them to type in an answer only if they were fairly confident that they were correct. They were then shown the Swahili cues in a random order and asked to recall the English translation. If they knew the English word, they typed it in. If they did not, we asked them if they had learned the word and forgotten it (henceforth, a feeling-of-forgetting judgment). The participants were then given a recognition test for unrecalled items. For the recognition test, the Swahili word was again presented, followed by four alternatives. All four of the alternatives were English words used as targets in the experiment, one of which was the correct translation. This means that each target word was used four times in the experiment, three times as a distractor and once as a target. This meant that participants could do some process of elimination in recognition. However, the recognition test showed no effects of survival processing and was mainly included to assess the accuracy of the meta-memory judgments.

Results and Discussion

Participants recalled more words in the pleasantness judgment condition (34 percent) than in the survival processing condition (20 percent), $t(59) = 2.02$, $p < .05$ (Fig. 9.1).

In this experiment, we found a statistically significant advantage in cued recall for those items that had been encoded with pleasantness judgments over survival

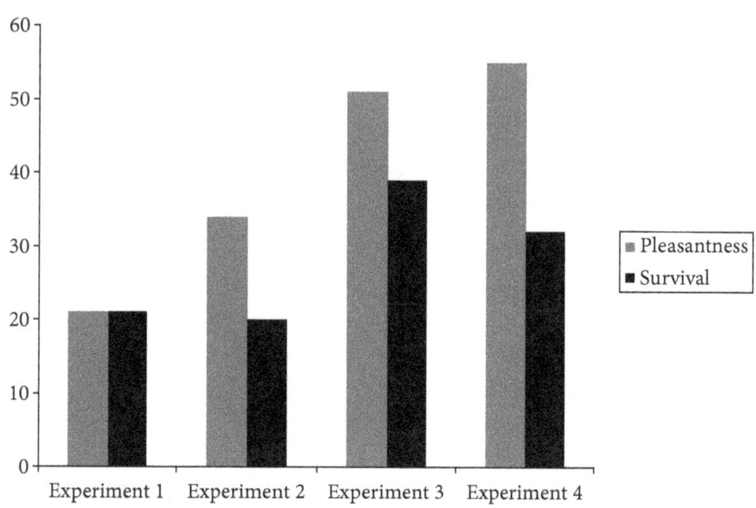

Figure 9.1 Mean recall as a function of survival processing vs. pleasantness judgments in Experiment 1, 2, 3, and 4.

processing in this experiment. Because in this experiment we tested participants one by one with order randomized for each participant, the effect cannot be dismissed as an order effect. Moreover, because the instructions did not specify that memory would be tested later, we also followed the usual survival processing procedures in which encoding is incidental, thus following the original paradigms of Nairne et al. (2007). However, what we found was a reverse survival processing effect: survival processing led to *worse* memory than the pleasantness judgments.

We also found that for unrecalled items, feeling-of-forgetting judgments were higher in the survival processing condition than in the pleasantness judgment condition, $t(59) = 2.04$. However, this effect did not replicate in the next two experiments, so we report it here only to be thorough. Recognition of unrecalled items was not affected by the experimental manipulation ($t = 1.2$). The overall mean recognition was 52 percent.

In this experiment, we found that pleasantness judgments instructions led to cued-recall advantages compared to survival processing instructions. Because few studies have shown that survival processing led to a deficit in performance (but see Klein, Robertson, & Delton, 2011), because the effect size was quite small, and because we wondered if including the feeling-of-forgetting judgments might have altered the manner in which participants approached retrieval, we conducted a replication in Experiment 3. In this experiment, we used a condition with meta-memory judgments and another condition without meta-memory judgments.

EXPERIMENT 3

Experiment 3 involved a conceptual replication of Experiment 2. The differences were that we used new stimuli, Lithuanian–English pairs instead of Swahili–English pairs. This allowed us to determine if the negative survival processing effect would replicate with other materials. Experiment 3 was identical in other ways to Experiment 2. In Experiment 3, we tested to determine if including the meta-memory judgments was affecting recall of the targets. Thus, in Experiment 3, we included one condition in which we included meta-memory judgments and a recognition test and we included one condition in which we did not include either the feeling-of-forgetting judgments or the final recognition test.

Method

Participants and Materials

A total of 124 Florida International University students participated. In this experiment, we switched from the Swahili–English pairs to Lithuanian–English translation equivalents, drawn from the norms of Grimaldi, Pyc, and Rawson (2010). Twenty-one translations were chosen from the top of the list in Appendix A of Grimaldi et al. (2010). In Experiments 1 and 2, our lists were 20 words long, but in this one we used a 21-word list.

The experiment was a between-subject design with two levels (pleasantness judgments and savannah-survival judgments). We also compared the presence or absence of feeling-of-forgetting judgments as a second between-subject variable.

Procedure

The procedure for Experiment 3 was identical to Experiment 2, except for the stimuli. We also ran a condition in which we did not ask for feeling-of-forgetting judgments or recognition of unrecalled items.

Results and Discussion

In Experiment 3, participants recalled more words in the pleasantness judgment condition (52 percent) than in the survival processing condition (40 percent), $t(58) = 2.02$, $p < .05$ (see Fig. 9.1). When feeling-of-forgetting and recognition were not collected, participants recalled more words in the pleasantness judgment condition (50 percent) than in the survival processing condition (37 percent), $t(62) = 2.68$, $p < .05$ (see Fig. 9.1).

It is also worth noting that the Lithuanian–English translations were easier to learn than the Swahili–English translations.

In Experiment 3, the instruction manipulation did not affect the magnitude of the feeling-of-forgetting judgments made on unrecalled paired associates, $t = 0.67$. Recognition of unrecalled targets did not differ between the conditions, $t = 1.21$. The overall recognition level was 55 percent.

EXPERIMENT 4

Experiment 4 was methodologically similar to Experiment 3 with one exception: we used a standard meta-memory measure, feeling-of-knowing judgments, instead of a nonstandard measure, feeling-of-forgetting judgments. Because most studies do not use feeling-of-forgetting judgments, we were not convinced that participants understood them in the way we intended. Thus, we switched to the more standard measure.

Method

Participants, Materials, and Procedure

A total of 61 Florida International University students participated. The method was identical to Experiment 3 except that participants made feeling-of-knowing judgments instead of feeling-of-forgetting judgments. The experiment was a between-subject design with two levels (pleasantness judgments and savannah-survival judgments).

Results and Discussion

In Experiment 4, participants recalled more words in the pleasantness judgment condition (55 percent) than in the survival processing condition (32 percent), $t(59) = 3.70$, $p < .05$ (see Fig. 9.1).

In this experiment, the instruction manipulation did not affect the magnitude of the feeling-of-forgetting judgments made on unrecalled paired associates, $t = 1.78$. Recognition of unrecalled targets did not differ between the conditions.

GENERAL DISCUSSION

Our initial hypothesis was incorrect: survival processing does *not* benefit learning of paired associates. In all of the experiments, survival processing did not result in benefits to learning relative to all other conditions. However, in four experiments, we found that pleasantness judgments led to better recall of paired associates than did survival processing judgments. In the first two experiments, we used Swahili–English translation pairs, and in the last two experiments, we used Lithuanian–English word pairs. It did not matter; for both sets of stimuli, pleasantness judgments led to better recall than did survival processing judgments (see Klein et al., 2011, for similar findings).

These findings are consistent with several recent findings of limited generalizability to the survival processing paradigm. For example, Tse and Altarriba (2010) showed that survival processing did not increase the sensitivity of items to implicit memory tests relative to control items. Burns, Hwang, and Burns (2011) found that survival processing was mostly eliminated for categorized lists (but see Otgaar & Smeets, 2010). Palmore et al. (2012) found that when congruity effects were accounted for, judgments of learning were not sensitive to survival processing. Savine, Scullen, and Roediger (2011) found that survival processing does not benefit the learning of faces (but see Otgaar, Smeets, & van Bergen, 2010, for positive results with other visual stimuli). Seamon et al. (2012) found that meaning-relevance instructions resulted in the same memory performance as survival-relevance instructions. Finally, Klein et al. (2011) found that processing for future planning led to better recall than survival processing. Our finding that survival processing does not affect paired-associate learning adds to this list. Thus, there is a growing surge of research that replicates the basic survival processing effect but finds that it does not generalize greatly beyond its initial scope (i.e., Nairne et al., 2007), much of it reviewed in the chapters in this volume.

We assert that the limited generalizability of survival processing argues against its inclusion as a set of cognitive modules shaped by natural selection. Indeed, Klein et al. (2011) have argued that "survival" itself is too broad a term to think of in terms of natural selection. That is, natural selection favors those traits that do survive, not whether those traits concern survival. This has led other researchers to search for more specific evolutionarily based mechanism, such as planning for the future (Klein et al., 2011) or learning relevant social information (Klein et al., 2009). In the current context, if remembering survival-relevant information were shaped by natural selection, one would expect its benefits to include many situations in which people are considering material in terms of its relevance to survival. Therefore, what can best explain the survival processing effect?

Burns et al. (2011) looked at the proximate mechanisms that cause the survival processing effect. They found that survival processing led to the use of both item-specific and relational processing while encoding words. When they manipulated conditions, such that other encoding conditions also benefited from both item-specific and relational processing, the advantages of survival processing over those conditions disappeared. Thus, Burns et al.'s data suggest that survival processing advantages may simply reflect the combined use of two processing strategies, and that other paradigms, such as learning faces or learning paired associates, may also draw on both processing strategies, thereby eliminating the advantages that survival processing might have over other learning strategies.

What then drives the reverse survival processing effect observed here? Following Burns et al. (2011), we suspect that survival processing leads to the increased use of relational processing across items in free recall of word lists. That is, it helps participants construct stories to connect items together. However, in the current studies, using cued recall, relational processing across items is not relevant. What is important is meaning-based associative processing between the novel item (the word in Swahili or Lithuanian) and its familiar translation equivalent. In this case, what benefits encoding are processes that create that associative processing between the items. Pleasantness judgments may focus learning on associating the items, whereas survival processing may focus the learning on relations among items. Thus, it is possible that we would have found a survival processing effect if we had asked participants to free recall English translations (or Swahili or Lithuanian words) rather than recall specific English words when presented with specific other-language words.

Human beings are biological organisms, and human cognition is a function of biological processes in the human brain. So it is certain that human cognition evolved in the distant past. However, what the processes were that drove human cognition are still poorly understood (Klein et al., 2009), and how those processes affect memory performance in modern populations is likely complex and difficult to study. We applaud Nairne and colleagues (e.g., Nairne & Pandeirada, 2011) for attempting to crack this difficult nut. However, it does not seem that the survival processing effect in free recall of single-word lists is necessarily a paradigm that will shed light on how natural selection drove the evolution of human memory. Nonetheless, we are optimistic that the study of the adaptive nature of human memory will blossom and will eventually lead us to both better questions and better answers.

AUTHORS' NOTE

The authors thank Andrew Laginess and Marie Annoual for their assistance with data collection. We thank Henry Otgaar, Stanley Klein, Lisa Son, and Michael Toglia for comments on earlier drafts of this chapter. The research was approved by the Florida International University Institutional Review Board.

REFERENCES

Adachi, I., Chou, D. P., & Hampton, R. R. (2009). Thatcher effect in monkeys demonstrates conservation of face perception across primates. *Current Biology, 19*, 1273–1279.

Burns, D. J., Hwang, A. J., & Burns, S. A. (2011). Adaptive memory: Determining the proximate mechanisms responsible for the memorial advantages of survival processing. *Journal of Experimental Psychology: Learning, Memory, and Cognition, 37*, 206–218.

Butler, A. C., Kang, S. H., & Roediger, H. L., III. (2009). Congruity effects between materials and processing tasks in the survival processing paradigm. *Journal of Experimental Psychology: Learning, Memory, and Cognition, 35*, 1477–1486. doi:10.1037/a0017024

Grimaldi, P. J., Pyc, M. A., & Rawson, K. A. (2010). Normative multitrial recall performance, metacognitive judgments, and retrieval latencies for Lithuanian-English paired associates. *Behavior Research Methods, 42*, 634–642.

Halamish, V., McGillivray, S., & Castel, A. D. (2011). Monitoring one's own forgetting in younger and older adults. *Psychology and Aging, 26*, 631–635.

Kang, S. H. K., McDermott, K. B., & Cohen, S. M. (2008). The mnemonic advantage of processing fitness-relevant information. *Memory & Cognition, 36*, 1151–1156. doi:10.3758/MC.36.6.1151

Klein, S. B., Cosmides, L., Gangi, C. E., Jackson, B., Tooby, J., & Costabile, K. A. (2009). Evolution and episodic memory: An analysis and demonstration of a social function of episodic recollection. *Social Cognition, 27*, 283–319.

Klein, S. B., Roberson, T. E., & Delton, A. W. (2011). The future-orientation of memory: Planning as a key component mediating the high levels of recall found with survival processing. *Memory, 19*, 121–139.

Koriat, A. (1993). How doe we know that we know? The accessibility account of the feeling of knowing. *Psychological Review, 100*, 609–639.

Kornell, N. (2009). Optimizing learning using flashcards: Spacing is more effective than cramming. *Applied Cognitive Psychology, 23*, 1297–1317.

Kostic, B., McFarlan, C. C., & Cleary, A.M. (2012). Extensions of the survival advantage in memory: Examining the role of ancestral context and implied social isolation. *Journal of Experimental Psychology: Learning, Memory, and Cognition, 38*, 1091–1098.

Nairne, J. S., & Pandeirada, J. N. S. (2008). Adaptive memory: Remembering with a stone-age brain. *Current Directions in Psychological Science, 17*, 239–243. doi:10.1111/j.1467-8721.2008.00582.x

Nairne, J. S., & Pandeirada, J. N. S. (2010). Adaptive memory: Nature's criterion and the functionalist agenda. *American Journal of Psychology, 123*, 381–390.

Nairne, J. S., & Pandeirada, J. N. S. (2011). Congruity effects in the survival processing paradigm. *Journal of Experimental Psychology: Learning, Memory, and Cognition, 37*, 539–549. doi: 10.1037/a0021960

Nairne, J. S., Pandeirada, J. N. S., & Thompson, S. R. (2008). Adaptive memory: the comparative value of survival processing. *Psychological Science, 19*, 176–180. doi:10.111/j.1467-9280.2008.02064.x

Nairne, J. S., Thompson, S. R., & Pandeirada, J. N. S. (2007). Adaptive memory: survival processing enhances retention. *Journal of Experimental Psychology: Learning, Memory, and Cognition, 33*, 263–273. doi:10.1037/0278-7393.33.2.263

Nelson, T. O., & Dunlosky, J. (1994). Norms of paired-associate recall during multitrial learning of Swahili-English translation equivalents. *Memory, 2*, 325–335.

Otgaar, H., & Smeets, T. (2010). Adaptive memory: Survival processing increases both true and false memory in adults and children. *Journal of Experimental Psychology: Learning, Memory, and Cognition, 36*, 1010–1016.

Otgaar, H., Smeets, T., & van Bergen, S. (2010). Picturing survival memories: enhanced memory after fitness-relevant processing occurs for verbal and visual stimuli. *Memory & Cognition, 38*, 23–28. doi:10.3758/MC.38.1.23

Palmore, C. C., Garcia, A. D., Paige Bacon, L., Johnson, C. A., & Keleman, W. L. (2012). Congruity influences memory and judgments of learning during survival processing. *Psychonomic Bulletin & Review, 19*, 119–125. DOI 10.3758/s13423-011-0186-6

Platt, M. L., Brannon, E. M., Briese, T. L., & French, J. A. (1996) Differences in feeding ecology predict differences in performance between golden lion tamarins (*Leontopithecus rosalia*) and Wied's marmosets (*Callithrix Kuhli*) on spatial and visual memory tasks. *Animal Learning & Behavior, 24*, 384–393.

Savine, A. C., Scullen, M. K., & Roediger, H. L. (2011). Survival processing of faces. *Memory & Cognition, 39*, 1359–1373. doi:10.3758/ s13421-011-0121-0

Seamon, J. G., Bohn, J. M., Coddington, I. E., Ebling, M. C., Grund, E. M., Haring, C. T., Jang, S.-J., Kim, D., Liong, C., Paley, F. M., Pang, L. K., & Siddique, A. H. (2012, January 30). Can survival processing enhance story memory? Testing the generalizability of the adaptive memory framework. *Journal of Experimental Psychology: Learning, Memory, and Cognition*. Advance online publication. doi: 10.1037/a0027090

Sherry, D. F., & Schacter, D. L. (1987). The evolution of multiple memory systems. *Psychological Review,94*, 439–454.

Soderstrom, N. C., & McCabe, D. P. (2011). Are survival processing memory advantages based on ancestral priorities. *Psychonomic Bulletin & Review, 18*, 564–569.

Thomas, A. K., Bulevich, J. B., & Dubois, S. (2011) The role of contextual information in episodic feeling of knowing. *Journal of Experimental Psychology: Learning, Memory & Cognition, 38*, 96–108.

Tse, C.-S., & Altarriba, J. (2010). Does survival processing enhance implicit memory? *Memory & Cognition, 38*, 1110–1121.

Tulving, E., & Pearlstone, Z. (1966). Availability versus accessibility of information in memory for words. *Journal of Verbal Learning and Verbal Behavior, 5*, 381–391.

Wang, A. Y., & Thomas, M. H. (1995). The effect of keywords on long-term retention: Help or hindrance. *Journal of Educational Psychology, 87*, 468–475.

Weinstein, Y., Bugg, J. M., & Roediger, H.L. (2008). Can the survival recall advantage be explained by basic memory processes? *Memory & Cognition, 36*, 913–919.

10

Proximate Cognitive Mechanisms Underlying the Survival Processing Effect

EDGAR ERDFELDER AND MEIKE KRONEISEN

What are the adaptive problems human episodic memory was designed to solve? One answer that has received considerable support in recent years is that nature "tuned" our memory systems to process and remember fitness-relevant information in general and survival-relevant information in particular (e.g., Nairne, Thompson, & Pandeirada, 2007). Given the premium placed on long-term memory for survival-relevant information, our ancestors had survival advantages and were more likely to reproduce. In line with this proposition, several recent studies have shown that information processed in the context of a hypothetical ancestral survival scenario is recalled and recognized much better than information encoded using alternative procedures and techniques known to be highly effective (e.g., visual imagery, pleasantness ratings, self-relevance ratings). This mnemonic benefit of survival-related processing has been called the survival processing effect (e.g., Kroneisen & Erdfelder, 2011).

The survival processing effect proves to be replicable and robust (for reviews see Nairne, 2010, and Nairne & Pandeirada, 2008b). It generalizes to different word materials, presentation modalities, experimental designs, age groups, and populations and different measures of episodic long-term memory (e.g., Nairne et al., 2007; Otgaar & Smeets 2010; Otgaar, Smeets, & van Bergen, 2010; Weinstein, Bugg, & Roediger, 2008). How can this striking effect be explained? In evolutionary accounts of human behavior, a complete explanation of a psychological phenomenon always entails both an ultimate and a proximate explanation. As outlined by Scott-Phillips, Dickins, and West (2011, p. 38), "ultimate explanations are concerned with the fitness consequences of a trait or behavior and whether it is (or is not) selected. In contrast, proximate explanations are concerned with the mechanisms that underpin the trait or behavior—that is, how it works. Put another way, ultimate explanations address evolutionary function (the 'why' question), and proximate explanations address the way in which that function is achieved (the 'how' question)."

In research on the survival processing effect, ultimate explanations have dominated the scientific arena so far. According to Nairne and collaborators (Nairne, 2010; Nairne & Pandeirada, 2008a, 2008b, 2010, 2011; Nairne et al., 2007; Nairne, Pandeirada, & Thompson, 2008), survival processing taps into the ultimate function of human episodic memory, namely to enable encoding, storage, and retrieval of survival-relevant information in the first place. In other words, typically humans easily remember information previously evaluated with respect to its survival relevance because this memorial skill facilitates adaptive behaviors that help them survive. In contrast, Klein and collaborators (Klein, Robertson, & Delten, 2010, 2011) maintain that the ultimate function of human episodic memory "is intrinsically prospective: It is used to support future decisions and judgments, which cannot be known in advance" (Klein, Cosmides, Tooby, & Chance, 2002, p. 313). According to this account, the memorial advantage emerges because survival processing encourages future-oriented thoughts of planning more strongly than control conditions do, thus serving the ultimate function of facilitating future decisions. Both explanations have influenced much of the research on the survival processing effect. However, for a complete understanding of this effect, neither of these (and possible other) ultimate explanations suffices. Knowing why human episodic memory contributes to inclusive fitness maximization does not imply anything with regard to how memory works in specific encoding and retrieval contexts. In the present chapter, we therefore focus on proximate explanations of the survival processing advantage that aim at uncovering the memory mechanisms driving this effect.

Surprisingly, systematic work on the proximate mechanisms underlying the mnemonic benefit of survival processing has begun only recently (see Howe & Otgaar, in press). Hence, we still lack clear and unequivocal answers to a number of important scientific questions: How does the cognitive system implement this benefit? What are the relevant cognitive moderators influencing the strength of the survival processing effect? Does the effect vanish under certain conditions or is it robust against variations of the standard conditions? Which cognitive mechanisms mediate the effect of survival processing on memory performance? Is it possible to explain the observed results in terms of well-established theories of episodic memory? In the present chapter, we attempt to answer these questions by isolating the cognitive mechanisms primarily responsible for the survival processing benefit. In the first section, we briefly review past research on the survival processing effect, thereby providing boundary conditions for appropriate explanations of this effect. Section 2 outlines the proximate explanatory accounts proposed in the relevant literature so far. Next, based on a review of the available empirical evidence, we evaluate these accounts in Section 3, primarily focusing on moderators and potential mediators of the survival processing effect. Finally, in Section 4, we summarize the conclusions that can be drawn from past research and discuss their implications for evolutionary accounts of human episodic memory in general and the survival processing effect in particular.

THE MNEMONIC BENEFIT OF SURVIVAL PROCESSING

In the standard survival processing paradigm, participants process a list of unrelated words (typically 30 concrete words), not knowing that their memory for these words will be tested later on (i.e., incidental learning conditions). They are instructed to imagine being stranded in the grasslands of a foreign land, trying to find steady supplies of food and water and trying to protect themselves from predators. Next, participants are asked to rate each of the words with respect to its relevance for the imagined survival scenario, using a 5-point Likert scale ranging from "totally irrelevant" (1) to "extremely relevant" (5) (Nairne et al., 2007, p. 264). After a brief distractor task lasting for a few minutes, a surprise memory test follows. To establish the retention benefit of survival processing, this survival processing task is usually compared with one or two control tasks, most frequently the moving control task or the pleasantness rating task (Nairne et al., 2007, Exp. 1). The moving control task is structurally isomorphic to the survival processing task but does not include any survival aspects. Specifically, participants are instructed to imagine moving to a foreign land. Thus, they would need to locate and purchase a new home and to transport their belongings. In line with the instructions for the survival scenario, the participants then have to rate each word with respect to its relevance for the moving problem. In contrast, the pleasantness rating task does not involve a scenario participants are asked to imagine. Rather, each word in the list is simply rated for its pleasantness on a 5-point rating scale ranging from "totally unpleasant" (1) to "extremely pleasant" (5). Pleasantness ratings are known to induce high levels of retention by fostering item-specific processing (e.g., Engelkamp, Biegelmann, & McDaniel, 1998; Hunt & Einstein, 1981) and thus provide a useful benchmark for establishing a survival processing benefit.

Compared with these control conditions, Nairne et al. (2007, Exp. 1, p. 267) found a clear benefit of survival processing in the final recall test. Whereas participants recalled about 52 percent of the words both in the moving and in the pleasantness rating condition, they recalled about 60 percent in the survival processing condition—that is, an increase of about 15 percent in average free recall rates. Thus, the effect is rather strong when measured on absolute performance scales. In terms of standardized effect sizes such as Cohen's d, the sample effect size $d \approx .68$[1] observed in the original study by Nairne et al. (2007, Exp. 1) corresponds to a medium to large effect according to Cohen's (1988, p. 26) effect size conventions.

The survival processing benefit shows up both in between-subjects designs (e.g., Burns, Hart, Griffith, & Burns, in press, Exp. 1; Nairne et al., 2007, Exp. 1) and within-subject designs (e.g., Burns et al., in press, Exp. 2; Nairne et al., 2007, Exp. 2) and also for different word materials (e.g., Weinstein et al., 2008, Exp. 1). Moreover, the effect is not limited to free recall but is also found in recognition tests (e.g., Burns et al., in press; Kang, McDermott, & Cohen, 2008; Nairne et al., 2007, Exp. 3). Across experiments, condition differences in average ratings or processing times per items cannot explain the survival processing advantage. Sometimes rating latencies are longer in the survival processing condition than in control conditions; sometimes they are not. However, the survival processing effect is significant in both cases. Similar results hold for differences in average relevance ratings or pleasantness ratings between conditions. In addition, there is initial evidence that items processed in the survival processing paradigm are resistant against directed forgetting instructions (Pinho & Pandeirada, 2011). Moreover, participants appear to be aware of the memorial benefits of survival processing during learning, because the survival processing effect also shows up in judgments of learning (Palmore, Garcia, Bacon, Johnson, & Kelemen, 2012).

Importantly, the survival processing advantage is not limited to the moving and the pleasantness rating control tasks. Replacing these tasks by others that typically induce high levels of retention—such as self-reference ratings for words, imagery ratings, self-generation of words, and intentional learning—produces comparable results (Nairne et al., 2007, 2008). In fact, the survival processing advantage tends to be even larger when compared to intentional learning or self-reference ratings. Also, replacing the moving control scenario by other presumably more interesting and appealing control scenarios does not diminish the effect (Kang et al., 2008; Nairne et al., 2008).

Likewise, the survival processing effect is not restricted to specific populations. It replicates in various countries and different languages (e.g., Dutch, English, German, Japanese, and Portuguese). Moreover, it shows up not only in young adults but also in children aged 4 to 10 (Aslan & Bäuml, 2012) and 8 to 11 years (Otgaar & Smeets, 2010, Exp. 2). Nouchi (2012) found the effect also in older adults (i.e., adults with a mean age of about 70 years).

Although most studies assessed memory for concrete words, the survival processing effect generalizes to other materials as well. Otgaar et al. (2010), for example, found comparable survival processing benefits for pictures and words (see also Albuquerque, Carvalho, Oliveira, & Capelo, 2011). Moreover, Savine, Scullin, and Roediger (2011, Exp. 5) reported a survival processing effect for neutral statements, and Nairne, VanArsdall, Pandeirada, and Blunt (2012) showed that there may also be a survival processing benefit in memory for locations, provided the relevance-rating task is modified to an ease-of-collecting or ease-of-capturing rating task.

To summarize, the survival processing benefit is a very robust and typically rather strong effect that is found not only under the standard conditions originally investigated by Nairne et al. (2007) but also for different experimental designs, memory tests, materials, and populations and different types of control conditions. Naturally, the question arises how the cognitive system produces this remarkably stable effect. We address this question in the next section.

POSSIBLE PROXIMATE EXPLANATIONS

Right from the beginning of research on the survival processing effect, possible proximate mechanisms underlying this effect have been discussed, though mostly not in detail (see, e.g., Nairne et al., 2007, pp. 270–271). In this section, we summarize and systematize the existing explanatory accounts. To begin with, we confine ourselves to outlines of the basic ideas. We will not evaluate these ideas at this point. A more detailed discussion of the pros and cons of the different hypotheses in light of the available empirical evidence will follow in the subsequent section.

Proximate explanations can be roughly classified in affective, structural, and process explanations. Affective explanations maintain that the survival processing task influences memory via motivational or emotional reactions to the task's characteristics, whereas structural and process explanations presume that the task affects (or is mediated by) the structure of the memory representation or the process of information encoding, respectively. We describe these three different approaches in turn.

Affective Explanations

Media exposure and motivation. A motivational explanation already considered by Nairne et al. (2007) maintains that the survival processing task is more engaging than control tasks and thus prompts more intense and persistent forms of information processing—that is, more cognitive effort during encoding. Cognitive effort was shown to be a powerful predictor of memory performance quite some time ago (e.g., Tyler, Hertel, McCallum, & Ellis, 1979). In the context of the survival scenario, this motivational effect might be triggered by the fact that survival is an important and reoccurring topic in different types of popular media, such as books, movies, or TV shows.

Arousal hypothesis. Another possible mediating mechanism is emotional arousal. Survival processing addresses an exciting topic of high relevance and thus can be argued to cause more arousal and emotional processing of words (Nairne et al., 2007). As is well known, arousal and emotional processing can boost memory, thus providing a possible explanation for the survival processing effect (e.g., Levine & Pizarro, 2004; McGaugh, 2003).

Stress hypothesis. Although related to the arousal hypothesis, the stress hypothesis is unique in focusing on the physiological (i.e., hormonal) processes induced by stressful events and their memorial consequences. Stress is invoked whenever survival is critical. Moreover, it is well known that physiological stress mediated via adrenalin and noradrenalin release activates the hypothalamic-pituitary-adrenalin axis and, in turn, causes secretion of the stress hormone cortisol. Cortisol has been shown to impair memory retrieval (e.g., Smeets, 2011) but to foster storage when released in the encoding and consolidation phase of information processing (e.g., Smeets, Otgaar, Candel, & Wolf, 2008). Thus, it is conceivable in principle that physiological stress responses contribute to the survival processing effect, particularly under stress-eliciting context conditions.

Structural Explanations

Survival modules. Evolutionary psychologists maintain that the cognitive system includes evolutionarily acquired "cognitive modules" designed to solve specific adaptive problems of humans in ancestral environments (e.g., Tooby & Cosmides, 1992, 2005). Several of these modules are presumably tailored to detection and retention of special types of nutrition, predators, and toxins. As suggested by Nairne et al. (2007), the survival processing task might activate one or several of these specialized survival modules, thereby producing the mnemonic benefit.

Planning modules. According to Klein et al. (2010, 2011), the crucial component in the survival processing task is future-oriented thought and planning rather than thoughts about survival. Generation of survival-relevance ratings for a number of objects may in fact involve planning how to use these objects in the respective environment. Thus, survival processing might tap into an evolutionarily acquired planning module—that is, "an evolved capacity to imagine and plan for personal future contingencies, especially plans not tied to current drives and needs" (Klein et al., 2010, p. 13).

Schema-related encoding. Compared to the pleasantness control condition, the grasslands survival scenario can be argued to provide a clearer and more differentiated schema, script, or stereotype for information encoding (e.g., Otgaar et al., 2011). Schema-related processing is known to facilitate elaborated encoding of information and to boost retrieval, especially when using free recall as a memory measure (e.g., Bransford & Johnson, 1972; Mandler, 1984). For these reasons, Nairne and colleagues included the moving scenario as a control condition in many of their studies. Since the moving scenario involves schema- and script-based information processing as well, the survival processing advantage should diminish if schema-related encoding is the only factor relevant for the survival processing effect. However, as outlined by Weinstein et al. (2008), the moving scenario is perhaps less suitable as a control condition because the survival scenario provides an unusual and novel encoding context whereas the moving scenario does not. Moreover, the scenarios differ on several dimensions, not only with regard to the survival aspect. Thus, to evaluate the schema-related encoding explanation properly, schematic control conditions resembling the survival scenario on all these dimensions seem to be mandatory. In addition, an interesting prediction of the schema-related encoding account is that survival processing should enhance not only true memories but also false recall and recognition of schema-consistent information that was not part of the study episode (e.g., Brewer & Treyens, 1981). By implication, survival processing may also have detrimental effects on memory performance when false memories are taken into account.

Process Explanations

Congruity. Items that fit into the conceptual framework in which they are processed are recalled and recognized better than items incongruent with this framework. This phenomenon has been called the "congruity effect" (Craik & Tulving, 1975; Schulman, 1974). For example, words classified as compatible with a given sentence frame are later remembered better than words classified as incompatible, although deep semantic processing is involved in either case. The memorial benefit of congruity is usually explained by elaboration: Congruent items, in contrast to incongruent items, can be integrated in or tied to coherent preexisting knowledge structures that provide effective frameworks for retrieval later on (e.g., Craik, 2002, p. 307).

Congruity may play an important role in the survival processing effect for two reasons. First, words congruent with the survival processing task (i.e., words with high survival-relevance ratings) should be easier to remember than incongruent words (i.e., words with low survival-relevance ratings). Thus, congruity might act as a powerful within-list moderator of the survival processing advantage. Second, if the average congruity of the word material differs between the survival processing task and control tasks, this could artificially inflate or reduce the survival processing effect. Thus, the survival processing effect could at least partly be due to a confound between survival processing and item congruity (Butler, Kang, & Roediger, 2009; Nairne et al., 2007).

Richness of encoding. As already discussed by Nairne et al. (2007), the survival processing task might induce particularly effective elaboration and distinctive processing of information. From this perspective, "survival processing is simply another form of 'deep processing,' albeit a particularly good one" (Nairne & Pandeirada, 2008a, p. 378). Using somewhat diverse terminology, related explanations focusing on conceptual elaboration, on distinctive processing, or on both types of encoding have been suggested by Howe and Derbish (2010), Kroneisen and Erdfelder (2011), Nouchi (2012), and Burns et al. (in press). The common root of these accounts is the richness-of-encoding perspective advanced by Craik and Tulving (1975; see also Craik, 2002). In contrast to its predecessor, the levels-of-processing approach (Craik & Lockhart, 1972), the richness-of-encoding perspective does not distinguish just two homogeneous types of encoding, "deep" (semantic) versus "shallow" (non-semantic) processing. Rather, "deep processing" is conceived as a heterogeneous class of cognitive processes that differ along dimensions of processing quality and processing quantity (Craik, 2002). In fact, because both the survival and the moving processing tasks involve deep semantic processing, the original levels-of-processing approach cannot explain the well-established finding that the former task reliably elicits higher levels of recall and recognition than the latter. In contrast, richness of encoding can account for this result, assuming that the survival processing task provides a richer, more complex framework of encoding than the moving processing task does. As already shown by Craik and Tulving (1975, Exp. 7), words judged with respect to their fit into "complex" sentences produce higher levels of recall than words assessed with respect to their fit into "simple" sentences. Complex encoding frameworks enable more richly elaborated memory representations than simple ones, resulting in two memorial consequences (see Craik, 2002, pp. 306–307). First, memory representations of items will be more distinctive (i.e., more differentiated) and thus better distinguishable from other items in memory at retrieval (e.g., Schacter & Wiseman, 2006). Second, items are linked to or integrated in complex knowledge structures that guide retrieval later on. Thus, to the degree that retrieval conditions activate the memory representations established during encoding (Moscovitch & Craik, 1976), the richness-of-encoding hypothesis predicts both the congruity effect as outlined in the previous section (because of knowledge integration) and, in addition, better retention of richly encoded information even after controlling for congruity (because of higher distinctiveness).

Single-item and relational processing. A particularly effective strategy to encode a list of items is to combine single-item and relational processing (Einstein & Hunt, 1980; Hunt & Einstein, 1981; Hunt & McDaniel, 1993). Single-item processing involves encoding of features and specific characteristics of individual items. In contrast, relational processing focuses on relations between different items—for example, semantic relations between different words of a list. Nairne and Pandeirada (2008a) suggested that, by requiring participants to categorize words in survival-relevant versus survival-irrelevant words, survival processing perhaps induces relational processing of semantically heterogeneous words. Burns, Burns, and Hwang (2011) refined this idea by proposing both single-item and relational processing to be involved in survival processing, producing a retention benefit relative to control conditions that are largely limited to single-item processing (e.g., pleasantness ratings). In addition, Burns et al. (in press) recently proposed that survival processing also fosters item-specific processing, resulting in better memory for list items compared to other control scenarios that also involve relational processing (e.g., the moving scenario).

Interactive imagery. As shown by Nairne and collaborators (Nairne et al., 2008, Nairne & Pandeirada, 2010), both simple imagery of words and ease-of-imagery of the scenario in general are not likely to contribute to the survival processing effect because the memorial benefit persists even if these influences are taken into account.

However, interactive imagery has not been assessed in these studies. Interactive imagery is a special form of mental imagery known to be highly beneficial particularly in associative learning tasks (Bower, 1970). One way to foster interactive imagery for words to be associated (e.g., "cat" and "car") is to link these words by imagining self-generated actor–action–object phrases (e.g., "The cat jumps into the car"). Subjects instructed to visualize concepts interacting with each other are able to learn associations between these words much faster than control groups. According to Bower (1970), interactive imagery causes stronger associations between concepts. Applied to the survival processing task, interactive imagery of using certain objects in a survival scenario (e.g., "I use a stone as a missile") could foster stronger associations between each of the words and the scenario, thus enabling later recall of these words when the scenario is activated. To account for the available evidence, it must be assumed that the survival processing paradigm prompts spontaneous interactive imagery more than control conditions do. Given the unusual and threatening context induced by the paradigm, this seems indeed plausible.

Self-referential processing. The self-reference effect refers to the memorial benefit of information processed in reference to one's self (see Symons & Johnson, 1997, for a review). Because the standard survival processing task involves assessments of the personal relevance of objects in a threatening situation, it seems plausible that self-referential processing may be an important factor contributing to the survival processing effect. However, as outlined in the previous section, Nairne and collaborators included self-reference ratings as a control condition in several of their experiments. They always found that survival-relevance ratings produce significantly higher levels of retention than self-reference ratings. Does this rule out a role of self-referential processing in the survival processing effect? As recently argued by Klein (2012), the self-reference rating task used by Nairne et al. (2007, 2008) may not have been fully effective because it did not activate retrieval of episodic memories from one's life. However, such autobiographical retrieval experiences are crucial for obtaining a strong self-reference effect (Klein, 2012). Hence, it seems mandatory to have a closer look at the role of self-referential processing in the survival processing paradigm.

Gist processing. Still another mechanism possibly involved in the survival processing effect is gist processing (Brainerd, Reyna, & Ceci, 2008; Otgaar & Smets, 2010). Gist processing could be enhanced because survival-relevance ratings effectively link up all list items with the common theme of survival in the grasslands rather than directing attention toward verbatim aspects of the words. Since gist processing can be seen as a form of relational processing (Burns et al., in press), the gist-processing hypothesis resembles the single-item and relational processing account discussed before. However, in contrast to Burns et al. (in press), Otgaar and Smeets (2010) maintain that the task fosters a specific type of relational processing (i.e., gist processing) rather than item-specific processing. An interesting prediction of this account is that survival processing should enhance not only true memories but also false memories for lure items that fit the gist of the encoding episode.

EVALUATION OF PROXIMATE EXPLANATIONS

In this section, we evaluate the explanatory accounts outlined in the previous section based on an exhaustive review of the empirical results on survival processing available so far. The structure of this section matches the structure of the previous one. Thus, affective explanations will be assessed first, followed by structural explanations and process explanations.

Affective Explanations

Media exposure and motivation. Kang et al. (2008) designed an experiment that addresses several affective explanations of the survival processing effect. The authors reasoned that if emotional arousal, motivational engagement, or media exposure of survival topics plays a predominant role, then the effect should vanish when the original survival scenario is compared with isomorphic control scenarios that resemble the survival scenario in terms of arousal, excitement, novelty, and media exposure. To this end, they designed a bank heist control scenario similar in structure to the grassland survival scenario. Surprisingly, and in contrast to the prediction of affective theories, the bank heist control scenario neither eliminated nor mitigated the survival processing benefit. Probably as a consequence of the longer word list used by Kang et al. (2008), their recall rates were generally somewhat lower than those observed in the original experiments of Nairne et al. (2007). However, the overall result pattern was almost identical to the one obtained in studies based on the moving control scenario. These results were replicated when the recall test was exchanged for a recognition test (Kang et al., 2008, Exp. 2) and when the survival and bank heist scenarios were replaced by 90-second video clips taken from popular movies involving survival on a desert island (i.e., *Cast Away*) and a bank heist (i.e., *Inside Man*), respectively. Thus, given the fact that Kang et al.'s bank heist scenario is comparable to the survival scenario in terms of media presence and motivational engagement, there is good reason to believe that these factors cannot account for the survival processing effect.

Arousal hypothesis. Based on the assumption of Kang et al. (2008) that the bank heist scenario matches the grasslands survival scenario not only in media exposure but also in novelty and excitement, the results outlined in the previous paragraph also provide evidence against the arousal explanation. However, mortal danger in an unknown grassland environment might be perceived as more threatening and thus also as more arousing than exposure to a bank heist. Similar objections can be raised with respect to other control scenarios designed to match the survival scenario in terms of arousal and excitement (e.g., the vacation scenario, Nairne et al., 2008).

To cope with these objections, it seems advisable to test the arousal hypothesis more thoroughly. To this end, Soderstrom and McCabe (2011) constructed three control scenarios identical to the grasslands survival scenario except that two words were exchanged. In control scenarios 3 and 4, the word "grasslands" was replaced by "city," thus transforming the "ancestral" survival scenario into a "modern" survival scenario. Moreover, in control scenarios 2 and 3, the word "predator" was exchanged for "zombie." This modification transforms the realistic threat scenario of Nairne et al. (2007) into an unrealistic threat scenario. All four scenarios were rated with respect to arousal and valence. As expected, results indicated that the two scenarios involving zombies were associated with higher arousal levels and were evaluated more negatively than the other two conditions. In line with the arousal hypothesis and irrespective of the context provided in the scenario (i.e., grasslands vs. city), the zombie scenarios produced the highest levels of recall, even higher than those achieved with Nairne et al.'s (2007) original grassland survival scenario. This result suggests an important role of arousal and/or valence in the survival processing effect. However, as also shown by Soderstrom and McCabe (2011), differences in arousal or valence ratings cannot fully account for the observed pattern in recall rates: The retention benefit for the two zombie processing tasks remained significant even after controlling for effects of arousal and valence ratings statistically. By implication, properties other than emotional content contribute to the fact that replacing "predators" by "zombies" boosts the survival processing benefit.

In their study on survival processing effects for pictures, Otgaar et al. (2010) also assessed effects of picture valence and arousal elicited by the pictorial stimuli. Although high-arousal and high-pleasure pictures were recalled better than low-arousal and low-pleasure pictures, respectively (Otgaar et al., 2010, Table 1, p. 26), this effect did not interact with the processing condition: The advantage of the survival condition compared to moving and pleasantness control conditions was equally strong for each of the arousal and valence classes of stimuli. This is in line with results previously reported for word stimuli (Nairne et al., 2007, p. 271). Thus, arousal and valence improve retention independently from survival processing, a result that also speaks against a role of arousal in the survival processing effect.

Although arousal in general cannot explain the survival processing effect, some authors have argued that a specific arousal aspect, mortality salience, might play a key role in the effect. According to Hart and Burns (2012, p. 265), for example, "death is an abstract, concerning, personally relevant problem that seems likely to activate a unique cognitive context oriented toward processing information more deeply or complexly." In three experiments, Hart and Burns (2012) were indeed able to show that, compared to two different control conditions unrelated to death, priming of death-related thoughts prior to providing pleasantness ratings enhanced later recall significantly. Unfortunately, Hart and Burns did not include the standard grassland survival scenario of Nairne et al. (2007) as a comparison condition. It remains unclear, therefore, whether mortality salience can fully account for the survival processing effect. Bell, Röer, and Buchner (2013) recently closed this gap by comparing the standard grassland survival scenario with both the moving scenario and a clearly death-related control scenario (floating in outer space with dwindling oxygen supplies). In contrast to the mortality salience hypothesis, the death-related scenario did not improve recall. Recall performance in this condition was similar to performance in the moving and the pleasantness control conditions and significantly worse than in the standard survival condition. Similar results were recently observed by Klein (in press). Using a between-subjects design, he compared a modified death scenario lacking contextual detail with both the standard survival scenario and a pleasantness rating control condition. Again, recall rates in the death scenario condition did not differ significantly from the control condition but were significantly worse than in the survival scenario condition. In sum, as shown by both Bell et al. (2013) and Klein (in press), scenario negativity or mortality salience cannot account for the survival processing advantage.

Stress hypothesis. Assuming that the survival processing task elicits physiological stress and thereby improves information storage, Smeets, Otgaar, Raymaekers, Peters, and Merckelbach (2012) reasoned that past results could have underestimated the survival processing effect because the neutral lab context dampens stress reactions. Therefore, they assessed the survival processing effect under conditions of realistic social stress as elicited by the Trier Social Stress Test (TSST) and control conditions without social stress. If the stress hypothesis holds, then the TSST condition should enhance the survival processing effect compared to no-stress control conditions. However, in contrast to this prediction, the same survival processing benefit was observed relative to the moving scenario, irrespective of whether social stress was induced during encoding or not. Although the TSST manipulation enhanced cortisol secretion and recall memory as expected, it did not interact with the processing task (survival vs. moving). Moreover, cortisol levels did not differ significantly between the survival and the moving scenario. Descriptively, there was a slight tendency for higher cortisol levels in the moving condition. Hence, Smeets et al.'s (2012) results clearly rule out the physiological stress hypothesis.

Structural Explanations

Survival modules. There is agreement among evolutionary psychologists that cognitive modules evolved under Pleistocene conditions that dominated the largest part of the phylogenesis of mankind. If the survival processing advantage is due to an evolutionarily acquired "survival module," one would therefore expect the effect to be stronger for ancestral than for modern survival scenarios. As shown by Weinstein et al. (2008) and Nairne and Pandeirada (2010), this appears to be true in some cases. Modern city survival scenarios of different types show weaker survival processing advantages than ancestral survival scenarios. However, based on the evolutionary framework, one would also expect a gender-specific effect. Men should outperform women when tested with survival scenarios related to hunting, whereas women should outperform men when using survival scenarios related to typical problems of gatherers such as finding edible fruits and vegetables. However, Nairne, Pandeirada, Gregory, and Van Arsdall (2009) found no interaction of gender and type of survival scenario, inconsistent with this evolutionary gender-specialization hypothesis.

Even more problematic is the result of Soderstrom and McCabe (2011) discussed previously. Their study shows that there is no general superiority of ancestral over modern encoding contexts (see also Kostic, McFarlan, & Cleary, 2012). In Soderstrom and McCabe's (2011) experiment, being threatened by zombies in a city produced higher recall rates than being threatened by predators in the grasslands (i.e., the original survival scenario). This result is very difficult to reconcile with the survival modules account.

An additional problem is the finding of Howe and Derbish (2010) and Otgaar and Smeets (2010) that survival processing enhances not only true memories but also false memories (i.e., intrusions and false alarms), often resulting in a null effect or even a negative overall effect when recall and recognition measures are corrected for intrusions and false alarms, respectively. Moreover, the survival processing effect does not generalize to source memory tests (Bröder, Krüger, & Schütte, 2011), indirect memory tests (Tse & Altarriba, 2010), and memory for faces (Savine et al., 2011), findings that are also unexpected under the survival module account. Evolutionarily acquired memory modules should boost memory performance generally and automatically, in contrast to the findings summarized here. In sum, although some results are in line with the survival module account, the overall pattern appears to be at odds with this explanation.

Planning modules. To assess the effect of future orientation and planning on retention, Klein et al. (2010), prepared different "camping in the woods" scenarios and asked three groups of participants to rate information in a past-oriented perspective (How likely was it that each of the objects was at the camping trip you remember?), in a nontemporal perspective (How likely is it that each of the objects is at the campsite that you have imagined?), and in a future-oriented perspective (How likely is it that you would take each of the objects as you plan your camping trip?). These three groups were compared with a group receiving a forest survival scenario along with the instruction to rate how much the objects will increase the chances for survival. As predicted, the future-oriented planning group achieved the highest levels of retention in a surprise free recall test, followed by the survival processing group. The remaining two groups performed worst. Subsequently, in a more direct comparison of future-oriented thinking and survival processing, Klein et al. (2011) compared three encoding tasks designed to induce (1) survival processing with planning, (2) survival processing without planning, and (3) planning without survival. As expected, the two groups with planning instructions achieved highest levels of recall; the survival-without-planning group (2) performed worst.

Clearly, the results of Klein et al. (2010, 2011) are in line with their theoretical framework. However, given the substantial changes in the scenarios and the lack of control conditions corresponding to those investigated by Nairne et al. (2007, 2008), they are open to alternative interpretations as well. A major problem is that Klein et al.'s (2010, 2011) instructions for planning conditions implicitly foster thoughts about possible uses of objects in the respective context, whereas the instructions for nonplanning control conditions do not. Thus, the experimental conditions appear to be confounded with elaboration and distinctiveness of encoding. In Klein et al.'s (2011, p. 127) survival-without-planning scenario, for example, participants are instructed "to rate how likely each of these items is something you can eat to increase your chances of survival," an instruction that resembles a simple classification task. This instruction clearly does not foster thoughts about different possible uses and functions of objects in the same way as Nairne's relevance-rating instruction does. Unfortunately, processing times per item were not measured so that we cannot assess possible differences in rating times. Thus, given the results available so far, it remains unclear whether it is the planning *per se* or the thoughts about different uses of objects *vis-à-vis* a specific goal that is crucial for the memorial benefit. Moreover, since several of the control conditions investigated by Nairne and collaborators also involved future orientation and planning explicitly (e.g., the moving and the vacation scenarios) but failed to produce recall rates comparable to survival processing, the evidence suggesting a dominant role of planning in the survival processing effect is still weak (see, however, Klein et al., 2010, pp. 18–19; Klein, in press).

Schema-related encoding. To assess the role of schema-based encoding and online schematic processing in the survival processing effect, several predictions of this approach can be tested. One prediction is that schema-based encoding should enhance not only correct recall but also false memories—that is, schema-related intrusions in recall tests and false alarms in recognition tests. In fact, Howe and Derbish (2010) found more false recognitions in the Deese-Roediger-McDermott

(DRM) paradigm, and Otgaar and Smeets (2010) observed more intrusions triggered by survival processing both in children and adults. However, it is not clear so far whether these false memories conform to schema-consistent intrusions as predicted by schema theories. Additional research on this question is needed.

More conclusive evidence comes from comparisons between the standard survival processing task and well-designed control conditions involving equivalent degrees of schematic processing. Several researchers followed this route. To reiterate, Weinstein et al. (2008, Exp. 2) compared the original grasslands survival scenario with a city survival scenario using almost identical wording. Rating instructions and the scenario wording were identical in both conditions except that two words were exchanged in the city survival scenario: The word "grasslands" was replaced by "city," and the word "predators" was replaced by "attackers." Despite the high degree of similarity between these scenarios, there was a significant retention benefit for the ancestral survival scenario in comparison to the modern survival scenario. In a conceptual replication of this study, Nairne and Pandeirada (2010) observed almost identical results.

In sum, the grasslands survival scenario produces higher levels of retention than the city survival scenario despite the fact that these scenarios differ in only two words. Since both scenarios match on almost all dimensions, this result is clearly problematic for the schematic processing account. If combined with the fact (1) that several other schema-based control scenarios (e.g., the vacation scenario, cf. Nairne et al., 2008, Exp. 2; the bank heist scenario, cf. Kang et al., 2008; and the survival-stereotype scenario, cf. Otgaar et al., 2011) failed to produce recall rates comparable to the grasslands survival scenario and (2) that ancestral contexts provide higher levels of retention also for modified survival scenarios focusing on infection avoidance and food collection (Nairne & Pandeirada, 2010, Exp. 2 and 3), the conclusion seems justified that schematic processing cannot explain the survival processing benefit. Nevertheless, the schematic processing account—much like the gist hypothesis—may be useful in understanding the disadvantages of survival processing in terms of increases in false memories as observed by Howe and Derbish (2010) and Otgaar and Smeets (2010).

Process Explanations

Congruity. Nairne and collaborators discussed the possible role of item congruity in their first study on survival processing (Nairne et al., 2007, Exp. 1). To avoid congruity-biased item pools, they used random word lists from a variety of different semantic categories in most of their experiments. Moreover, they controlled for possible differences in word congruity by comparing average ratings between experimental conditions. Importantly, the survival processing advantage was observed even when control conditions did not differ in average ratings from the survival processing condition.

On the other hand, Nairne et al. (2007, p. 265) observed that, in the survival processing paradigm, "retention increased monotonically with average rating…, with recall averaging .47 for items given a relevance rating of 1 and averaging .72 for items given a rating of 5." This could indeed reflect a within-list congruity effect that moderates the size of the survival processing effect. However, this result could also be due to the fact that episodic memory is sensitive to the degree of fitness relevance and thus prioritizes information highly relevant for survival.

Butler et al. (2009) provided strong evidence for the former explanation. In their first experiment, they replicated both the survival processing effect with random word lists and the correlation between relevance ratings and recall rates. Like others (e.g., Kroneisen & Erdfelder, 2011; Kroneisen, Erdfelder, & Buchner, 2013; Palmore et al., 2012) they found a significant correlation not only for survival-relevance ratings but also for moving-relevance ratings, suggesting that within-list congruity effects are involved in both conditions. In addition, they designed three new word lists (1) congruent with the survival scenario, (2) congruent with a bank heist scenario, and (3) incongruent with both scenarios (irrelevant words). Half of the participants rated these three word lists with respect to their relevance for the survival scenario and the other half with respect to their relevance for the bank heist scenario as introduced by Kang et al. (2008). Irrespective of whether the three word lists were presented in a blocked or an unblocked (i.e., random) fashion, they observed a survival processing memory advantage for survival-related word lists, a survival processing disadvantage for robbery-related words, and no differences between scenarios for irrelevant words (Butler et al., 2009, Exp. 2 and Exp. 3, respectively). Moreover, the recall of robbery-congruent words in the bank heist condition was not worse than the recall of survival-congruent words in the survival condition, showing that the survival processing advantage was effectively eliminated for scenario-congruent word lists (for similar results, see Palmore et al., 2012).

Does this mean that congruity can explain the survival processing effect exhaustively? For two reasons, the answer is no. First, several studies showed that controlling statistically for differences in item congruity does not eliminate the effect. For example, looking only at items that receive lowest relevance ratings either for the survival or moving scenario, a pronounced survival processing advantage is still observed in many studies (Kroneisen & Erdfelder, 2011; Kroneisen et al., 2013; Nairne & Pandeirada, 2011). Second, not all experimental results of Butler et al. (2009) proved to be conceptually replicable (cf. Albuquerque et al., 2011; Nairne & Pandeirada, 2011). For example, using the word lists of Butler et al. (2009), Nairne and Pandeirada (2011) still found survival processing advantages when controlling the congruity of word lists experimentally. This was observed for word lists including only incongruent (or irrelevant) items (Nairne & Pandeirada, 2011, Exp. 2), for word lists including only congruent items (Nairne & Pandeirada, 2011, Exp. 3), and for word lists composed of equal amounts of congruent and incongruent items (Nairne & Pandeirada, 2011, Exp. 3). Hence, although congruity clearly is an important source of variance in survival processing data, it "is only a partial explanation of a robust phenomenon" (Palmore et al., 2012, p. 124).

Richness of encoding. A number of predictions can be derived from the richness-of-encoding hypothesis. Because this hypothesis includes conceptual elaboration as one basic mechanism, it predicts the congruity effect that was found to be influential in survival processing research. Moreover, because this hypothesis also claims higher distinctiveness of memory representations as a consequence of survival processing, it additionally predicts survival processing advantages even if item congruity is controlled statistically or experimentally. As shown in the previous section, both predictions appear to be in line with the data.

The distinctiveness prediction derives from the assumption that, in order to come up with a relevance rating for an object, people implicitly attempt to generate a number of relevance arguments (i.e., ideas about possible uses of objects) *vis-à-vis* a certain scenario. Higher distinctiveness of memory representations is assumed to result from the enhanced complexity of the survival processing context (and thus more options for unique relevance arguments) compared to control contexts such as the moving, the bank heist, or the vacation scenario. One method to test this assumption is to manipulate the complexity of the survival processing scenario. If the complexity is reduced, the survival processing effect should diminish or even vanish. Following this reasoning, Kroneisen and Erdfelder (2011) designed a reduced survival scenario identical in wording to the original scenario except that only a single survival problem (i.e., lack of potable water) was mentioned instead of the three survival problems addressed in the original scenario (i.e., predators, lack of food, and lack of potable water). As predicted by the richness-of-encoding hypothesis, Kroneisen and Erdfelder (2011, Exp. 1 and 2) replicated the survival processing effect for the original survival scenario but found no such effect for the reduced scenario.

An additional assumption inherent in the richness-of-encoding hypothesis is that not only the quality but also the quantity of processing matters. Quality of processing is largely determined by the complexity of the encoding context (e.g., type of scenario) whereas quantity of processing depends on the number of relevance arguments generated for a specific object *vis-à-vis* a specific scenario. Quantity of processing has a much smaller (if any) effect for encoding contexts low in complexity. The higher the context's complexity the stronger the effect of the number of relevance arguments generated by a participant on both conceptual elaboration and distinctiveness. In other words, the memorial benefit of survival processing should correlate with the number of relevance arguments generated by a participant. To test this prediction, Kroneisen and Erdfelder (2011, Exp. 3) asked half of their participants to generate four relevance arguments for each word. The other participants were asked to generate one argument only. Each group was split into one subgroup that generated relevance arguments for the survival scenario and an additional subgroup that worked on the moving scenario. In line with the predictions, they observed a strong survival processing effect in the four-arguments condition and a null effect in the one-argument condition.

In a very direct test of the richness-of-encoding hypothesis, Röer, Bell, and Buchner (in press, Exp. 2) recently asked their participants to write down all ideas that came to mind with respect to the usefulness of an object in the previously presented scenario. As predicted by the richness-of-encoding hypothesis, they found larger numbers of generated ideas in the standard survival scenario compared to several fitness-irrelevant control scenarios. Moreover, they replicated the survival processing advantage in a surprise free recall test and showed that, for all scenarios investigated, recall rates increased with the number of previously generated ideas, again corroborating the richness-of-encoding hypothesis. In their third experiment, Röer et al. (in press) provided evidence that the congruity effects reported in the literature may also be due to pragmatic distinctiveness of memory representations as indicated by the number of ideas generated for a word with respect to a specific scenario. Congruent word–scenario combinations were associated with larger numbers of generated ideas than incongruent word–scenario combinations. What is more, they replicated the typical congruity effects in a surprise free recall test and showed that the overall level of recall was significantly correlated with the number of previously generated ideas, in line with the result of their second experiment.

An additional prediction was recently tested by Bell et al. (2013, Exp. 3): According to the richness-of-recoding hypothesis, survival processing benefits should be limited to concrete words and vanish (or at least diminish) for abstract words. The reason is that, in contrast to the former, the latter words do not allow for generation of many relevance arguments *vis-à-vis* the survival scenario. The authors observed that survival-relevance ratings were associated with higher levels of recall in a subsequent surprise free recall test than death-relevance ratings. However, in line with the prediction, this effect was found only for concrete words and not for abstract words.

Importantly, the richness-of-encoding hypothesis predicts an advantage of survival processing only for memory tests that benefit from the conceptually integrated and distinctive memory representations of objects established during encoding. Whereas free recall and recognition of words and pictures conform to this constraint, implicit memory tests such as word stem completion, source memory tests for the location of a word on the computer screen, recognition memory tests for faces of speakers, and recall tests for story content, among others, do not. Consistent with these expectations, word stem completion (Tse & Altaribba, 2010), source memory for location (Bröder et al., 2011; Nairne, VanArsdall, Pandeirada, & Blunt, 2012, Fig. 1), face recognition (Savine et al., 2011), and both cued and free recall of factual story content (Seamon et al., 2012, Exp. 3–5) do not show memorial benefits when information is processed with respect to its relevance for survival. Results may change, however, when the relevance rating task is modified to an ease-of-collecting or ease-of-capturing rating task that fosters relational processing of location information and intrinsic features of objects on the computer screen. As shown by Nairne et al. (2012, Exp. 1 and 2), under these conditions there may be a survival processing benefit in source memory for locations.

There are a number of additional predictions of the richness-of-encoding hypothesis that have not been tested so far. For example, because relevance argument generation and cognitive elaboration require cognitive resources, the survival processing effect should diminish when cognitive resources are scarce—for example, under working memory load. So far there is only indirect evidence supporting this prediction, namely that older adults (presumably characterized by reduced working memory capacity) show a smaller survival processing effect compared with young adults (Nouchi, 2012).

Although richness of encoding can explain a multitude of different findings, some open questions have not been satisfactorily answered so far. One potential problem is the larger number of intrusion errors and false alarms found under survival processing conditions in some studies (Howe & Derbish, 2010; Otgaar & Smeets, 2010). Although the richness-of-encoding hypothesis predicts more intrusion errors as a consequence of elaboration and integration in preexisting knowledge structures (Howe & Derbish, 2010), one would perhaps expect that enhanced distinctiveness of memory representations counteracts intrusion errors and false alarms (as predicted by the distinctiveness heuristic framework, see Schacter & Wiseman, 2006). Obviously, as shown by Howe, Derbish, Otgaar, and Smeets, this is not always the case. A possible explanation is that distinctiveness resulting from survival processing is pragmatic rather than semantic in nature. In other words, different functions or uses of objects come to mind during encoding (resulting in "pragmatic distinctiveness" of memory representations; cf. Kroneisen & Erdfelder, 2011, p. 1554), not necessarily different semantic or perceptual features of objects. Thus, pragmatic distinctiveness does not necessarily prevent intrusions of semantic associates or schema-typical lures in recall or false alarms in DRM-recognition tests (Howe & Derbish, 2010), provided these lures resemble target items in functional or pragmatic aspects.

A second potential problem relates to Savine et al.'s (2011) claim that the lack of a survival processing advantage for faces is inconsistent with a levels-of-processing explanation. They argue that face memory has been shown to benefit from deep processing. Hence, because survival processing does not improve face recognition, it cannot involve deep processing. The problem with this argument is that it does not distinguish between different types of deep processing. According to the richness-of-encoding hypothesis, the memorial advantage of survival processing is neither due to depth of processing in general nor to the survival scenario considered in isolation. Rather, the specific process of generating relevance arguments *vis-à-vis* the survival scenario is assumed to be crucial for obtaining the effect. If the relevance rating task is replaced by a different task that does not encourage generation of relevance arguments (as was the case in Savine et al.'s face processing task), then the survival processing advantage is expected to vanish even if the survival scenario remains otherwise unchanged.

A third problem concerns the purported "complexity" of the grassland survival scenario. Admittedly, the factors rendering this scenario more complex than control contexts are still not well defined. The number of different survival problems activated by the encoding context likely contributes to its complexity (Kroneisen & Erdfelder, 2011, Exp. 1 and 2). In fact, it seems plausible that processing information with respect to its relevance for finding food, locating potable water, and defending against predators (survival scenario) is more complex than assessing relevance with respect to locating and purchasing a new home and transporting one's belongings (moving scenario). However, it is more difficult to account for differences between scenarios that vary in one or two words only (Nairne & Pandeirada, 2010; Soderstrom & McCabe, 2011; Weinstein et al., 2008). A powerful moderator perhaps involved in these studies is the creativity and engagement stimulated by the processing task to think about unusual, idiosyncratic, and unique uses of objects. Such thoughts would necessarily result in very distinctive and salient memory representations and thus boost later recall.[2] This mechanism could also explain why the survival processing advantage generalizes to many different survival processing contexts, not just those relevant for human evolution (Kostic et al., 2012). Moreover, it could account for the puzzling result that zombie survival scenarios enhance memory performance even more than the original grassland survival scenario does (Soderstrom & McCabe, 2011). However, for the time being, this is a *post hoc* hypothesis that awaits further research.

Single-item and relational processing. There are different ways to assess the role of single-item and relational processing in the survival processing effect. Nairne and Pandeirada (2008a) used a straightforward method to test whether the benefit of survival processing is largely due to relational processing of words from different semantic categories. They reasoned that if this hypothesis is correct, then the survival processing benefit should vanish or at least diminish when semantically unrelated words are replaced by categorized word lists from different semantic categories. Relational processing tasks are redundant when study items invite relational processing by default. In fact, previous research has often shown that relational processing tasks like sorting items into categories fail to enhance retention for obviously categorized word lists (e.g., Einstein & Hunt, 1980; Hunt & Einstein, 1981). Does the same hold when survival processing is applied to semantically categorized word lists? In two experiments using word lists consisting of eight exemplars from each of four unique semantic categories, Nairne and Pandeirada (2008a) still found statistically significant survival processing advantages when compared with a pleasantness rating control condition. This was observed for both between-subjects and within-subject designs (Nairne & Pandeirada, 2008a, Exp. 1 and Exp. 2, respectively). However, the survival processing effect was slightly weaker than usually obtained in experiments with semantically unrelated words (i.e., an increase of about 10–15 percent correct recall due to survival processing compared to up to 30 percent found for lists of unrelated words). Nevertheless, given the fact that the control condition in the experiments of Nairne and Pandeirada (2008a) effectively combines single-item processing (induced by pleasantness ratings) and relational processing (induced by categorized word lists presented in a blocked fashion)—a combination that is known to boost recall memory—the persistence of the survival processing benefit is remarkable.

Burns et al. (2011) criticized two aspects of Nairne and Pandeirada's (2008a) experiments. First, most if not all semantic categories used in these experiments were inherently related to survival (i.e., fruits, vegetables, four-footed animals, and types of human dwellings). Second, the type of relational processing induced by survival processing might deviate from the relational processing implied by the semantic category structure. Both problems could promote performance in the survival processing condition. Therefore, Burns et al. (2011) used categorized word lists consisting of four words from each of 12 unique semantic categories that bear less obvious thematic relations to the survival processing scenario. Compared with the pleasantness rating and category sorting control conditions, the standard survival processing task again produced highest levels of recall (Burns et al., 2011, Exp. 1 and 2). This time, however, the survival processing advantage was statistically significant only with respect to the latter but not with respect to the former control condition. Because the category sorting task involves relational processing only whereas the pleasantness rating task involves both single-item and relational processing for categorized word lists, this result is in line with the hypothesis that survival processing encourages both types of processing and thus produces memorial effects similar to pleasantness ratings applied to categorized word lists. Moreover, when replacing the categorized word lists by lists of words from different *ad hoc* categories without obvious categorical structure, the picture reverses: Pleasantness ratings of unrelated words produce the lowest level of recall, whereas both the survival processing task and the category sorting task enhance recall and do not differ significantly (Burns et al., 2011, Exp. 3 and 4). For lists of unrelated words, pleasantness ratings are limited to one type of processing (i.e., single-item processing). The category sorting task, in contrast, combines both types of processing, thus presumably producing memorial benefits resembling those of survival processing.

These results—including additional analyses of clustering scores, cumulative recall curves, and recognition scores—suggest an important role of single-item and relational processing in the survival processing effect. It is less clear, however, whether these two proximate mechanisms can fully explain the effect. One problem acknowledged by Burns et al. (2011, p. 211) is the rather low statistical power of their critical *t*-tests assessing the survival processing benefit. Even more importantly, all four experiments of Burns et al. (2011) produced average survival-relevance ratings, clearly smaller than the relevance ratings typically obtained in survival processing experiments. Average ratings ranged between 2.26 and 2.39 in the experiments of Burns and collaborators, whereas they usually range between 2.7 to 3.3. As outlined before, relevance ratings and recall rates are positively correlated (e.g., Butler et al., 2009; Kroneisen & Erdfelder, 2011; Nairne et al., 2007). Thus, if the word material of Nairne and Pandeirada (2008a) is argued to invoke a bias favoring the survival processing effect, then it seems legitimate to argue that the material of Burns et al. (2011) invokes a bias against this effect. In addition, it remains unclear why fitness-irrelevant relevance rating tasks such as the moving control task produce significantly lower levels of retention than Nairne et al.'s (2007) survival task, although both types of tasks provide coherent thematic frameworks for relational processing of words from different semantic categories. Does the moving task promote less single-item processing, less relational processing, or both? Burns et al. (in press) recently provided evidence favoring an explanation in terms of single-item processing. However, Burns et al. (in press, p. 8) also acknowledged "that survival processing may sometimes enhance the encoding of relational information." In addition, a number of important questions have not been answered satisfactorily so far. Why does the survival processing task promote item-specific or relational processing more than control tasks do? Which factors moderate the contributions of the two types of processes? Is it possible to modify the survival processing task so that single-item or relational processing increases or decreases selectively? Little is known about these issues. For the time being, therefore, it seems too early for a final assessment of the importance of single-item and relational processing in the survival processing effect.

Interactive imagery. One way to evaluate the interactive imagery hypothesis is to compare the standard relevance rating conditions with interactive imagery conditions in which participants are explicitly instructed to imagine using each object in the respective scenario and then rate the ease of interactive imagery. According to the interactive imagery hypothesis, explicit instructions should promote using this method to strengthen the association between words and the scenario, thus improving memory performance in both survival and control conditions. Kroneisen et al. (2013) followed this rationale and found no benefit of interactive imagery instructions in either scenario. In contrast, interactive imagery reduced recall performance particularly in the survival condition, causing the survival processing advantage to disappear under these conditions. Clearly, interactive imagery is not involved in the survival processing effect. Obviously, forcing participants to engage in interactive imagery distracts them from more efficient types of information processing that normally produce the survival processing advantage in relevance rating tasks.

Self-referential processing. To assess the possible role of self-referential processing in the survival processing paradigm more rigorously, Klein (2012) replicated the survival and self-reference rating tasks used by Nairne et al. (2007, 2008) and compared them to an improved self-reference condition in which participants were asked to "first bring to mind an important personal experience involving each of the items and then rate how easily the item brought to mind an important experience" (Klein, 2012, p. 1236). This wording resembles standard instructions in typical self-reference studies more closely than Nairne's instruction does. As predicted, reliably higher recall rates were observed in the improved self-reference rating condition compared with the self-relevance condition investigated by Nairne and collaborators. Moreover, *post hoc* Tukey tests revealed no significant differences between the improved self-reference condition and the survival processing task in two experiments, a finding that was interpreted as showing "that self-referential recall tasks are not necessarily inferior to tasks that promote thoughts of survival, provided they are self tasks of the right type" (Klein, 2012, p. 1240). However, given the fact that statistical power was not controlled, it cannot be ruled out that these null effects are actually due to low sensitivity of the *post hoc* Tukey tests. Descriptively, the survival processing task led to higher recall rates than the improved self-reference task in both experiments (i.e., .45 vs. .40 in Exp. 1 and .42 vs. .38 in Exp. 2), suggesting that the self-reference effect tends to be smaller than the survival processing effect even when optimized instructions are used. In addition, other studies have shown that experimental manipulations of the first-person versus third-person perspective in the survival processing task do not have significant effects on performance (Kang et al., 2008, Exp. 3; Weinstein et al., 2008). If self-referential processing is of crucial importance for the emergence of the survival processing effect, then one would expect this effect to vanish or at least diminish when processed from a third-person perspective. Thus, although Klein's (2012) results suggest that the benefit of survival-relevance ratings relative to self-relevance ratings may have been overestimated in past studies, the evidence for a causal role of self-referential processing in the survival processing effect still appears to be weak. We do agree, however, with Klein's (2012, p. 1241) conclusion that "additional research clearly is necessary." The present results do not rule out that the survival processing task automatically triggers more self-referential processing than, say, the moving task does, irrespective of whether a survival scenario is viewed from the first-person or the third-person perspective. They also do not rule out that the strength of the survival processing effect is moderated by the amount of self-referential processing. If self-referential processing indeed plays a causal role in the survival processing effect, then both hypotheses should hold. At this point, we simply lack supportive evidence for these hypotheses and therefore have to await further research.

Gist processing. The gist hypothesis maintains that (1) survival processing enhances gist processing in the first place and (2) items that fit the gist traces established during encoding are more likely recalled, irrespective of whether they were actually part of the encoding episode or not. This hypothesis is supported by a number of experiments showing higher levels of false recall and false recognition under survival processing conditions (Howe & Derbish, 2010; Otgaar & Smets, 2010). Also, the lack of a survival processing advantage in face recognition is in line with this hypothesis (Savine et al., 2011). However, a number of results on the role of item-specific processes and item distinctiveness in the survival processing effect (e.g., Burns et al., in press; Kroneisen & Erdfelder, 2011; Röer et al., in press) are not easily reconciled with the gist hypothesis. In fact, some studies (e.g., Kroneisen & Erdfelder, 2011; Kroneisen et al., 2013; Nairne et al., 2007) showed that the survival processing effect in free recall tends to be stronger for items receiving low relevance ratings (i.e., items not easily linked up with the common theme of survival in the grasslands) than items receiving high relevance ratings. Hence, although gist processing likely is involved in survival processing and can account for several findings, the gist hypothesis falls short of explaining strong survival processing effects for survival-irrelevant items.

CONCLUSIONS

Although research on the cognitive mechanisms involved in the survival processing effect is still in its infancy, the available empirical evidence allows for a preliminary assessment of the proximate explanations suggested in the literature. We distinguished affective, structural, and process explanations. Among the affective explanations, the arousal hypothesis has received most support, primarily because of the intriguing results obtained by Soderstrom and McCabe (2011) using arousal-stimulating zombie scenarios. However, although arousal appears to contribute to the survival processing effect, it also seems clear that this affective mechanism cannot fully account for the memorial benefit of survival processing (Soderstrom & McCabe, 2011). Importantly, research has also shown that motivational effects triggered by media exposure of survival topics are likely to be negligible (Kang et al., 2008). Furthermore, mortality salience (Bell et al., 2013) and physiological stress induced by survival processing do not mediate arousal effects on the survival processing effect (Smeets et al., 2012).

The evidence for structural explanations is generally weak. Trying in a variety of studies to trace back the survival processing effect to beneficial effects of schematic processing or the operation of evolutionarily acquired "survival modules" has produced a number of inconsistent results. Currently, the evidence is less clear with respect to the role of planning in the survival processing effect. However, although the results of Klein et al. (2010, 2011) can be attributed to evolutionarily acquired "planning modules," they are open to alternative explanations as well. More research is needed before a final assessment can be made.

Overall, process explanations have received most empirical support so far. Although interactive imagery can be ruled out as a factor contributing to the survival processing effect and a predominant role of self-referential processing is unlikely given the available results, there is strong evidence for a dominant role of congruity, richness of encoding, and the combined effect of single-item and relational processing in the survival processing effect. Congruity has been shown to explain a large portion of the variance in survival processing data. However, it is also clear that the survival processing advantage persists even when congruity effects are controlled statistically or experimentally. The same holds for the combined effect of single-item and relational processing. The richness-of-encoding hypothesis, based on the revised depth-of-processing framework and dating back to the work of Craik and Tulving (1975), can account for most of the available results. The strength of this hypothesis is primarily due to the fact that it predicts more elaboration and conceptual integration (and, thus, the congruity effect) and, in addition, higher distinctiveness of items processed with respect to their possible survival relevance (see Craik, 2002). It can thus explain why other hypotheses can account for only part of the variance in survival processing data.

An additional strength of the richness-of-encoding hypothesis is its heuristic potential. Richness of encoding predicts a number of additional effects that have not been rigorously tested so far—for example, detrimental effects of (1) concurrent working memory load during encoding, (2) aging, and (3) modifications of the relevance rating task on the size of the survival processing effect. However, despite these advantages, some questions remain. Currently, the richness-of-encoding framework does not uniquely and precisely identify the crucial components of the encoding context (i.e., the scenario and—to some degree—the processing task) that are necessary and sufficient for obtaining the effect. Although the richness-of-encoding account shares this weakness with other successful proximate explanations (i.e., the arousal hypothesis and the single-item-relational-processing account), this clearly is an important topic for future research.

We conclude that four well-established domain-general mechanisms of human memory—memorial benefits caused by (1) arousal, (2) item congruity, (3) richness of encoding, and (4) the combination and single-item and relational processing—have proven successful in accounting for many results on moderators and mediators of the survival processing effect (for a similar conclusion, see Howe & Otgaar, 2013). In contrast, hypotheses attributing the survival processing advantage to the operation of domain-specific, evolutionarily acquired "cognitive modules" tailored to solve specific adaptive problems our ancestors faced in Pleistocene environments (i.e., survival modules or planning modules) were comparatively less successful. Does this imply that the evidence summarized here is inconsistent with an evolutionary explanation of the survival processing benefit in human memory? Indeed, several evolutionary psychologists have argued that proper evolutionary theories must focus on domain-specific proximate explanations, mainly because domain-general mechanisms serving several adaptive goals simultaneously are unlikely to evolve (e.g., Buss, 1995; Tooby & Cosmides, 1992, 2005). From this perspective, assuming that the survival processing effect reflects an evolved adaption, the proper domain-specific mechanism producing it has yet to be found. However, there are alternative perspectives on evolutionary psychology that deny a necessary link between evolutionary theorizing and domain-specific proximate explanations (see, e.g., Buller, 2005a, 2005b; Over, 2003). Buller (2005c, p. 884) explained this perspective nicely: "Evolutionary Psychologists argue that distinct adaptive problems throughout our evolutionary history would have required distinct solutions, and they conclude that these distinct solutions must be distinct cognitive mechanisms. But all that follows is that distinct adaptive problems would have required distinct *behavioral solutions*, not that they would have required distinct modules. Indeed, domain-general cognitive mechanisms could operate on domain-specific inputs, and make use of information 'specific to' those inputs, in order to generate domain-specific behavioral solutions to the problems they encounter. The need for behavioral solutions specific to adaptive problems in our evolutionary history wouldn't have necessarily selected for modules."

Given the robustness and generality of the survival processing advantage and converging evidence based on other memory paradigms (e.g., Wilson, Darling, & Skyes, 2011), we do not consider it likely that the survival processing effect is accidental. Rather, as argued by Nairne and collaborators, this effect appears to tap into an evolved adaptive function of human episodic memory that contributes to inclusive fitness maximization. However, following Buller (2005a, 2005c), we believe this ultimate explanation does not necessarily imply a domain-specific proximate explanation of the mechanisms driving this effect (see Bell & Buchner, 2012, for a similar conclusion concerning memory for cheaters). As shown in this chapter, evidence is accumulating that a few well-known domain-general mechanisms of human episodic memory—most importantly, beneficial effects of richness of encoding—can account for variations in the strength of the survival processing effect within and between experiments. Hence, they are promising candidates for a complete proximate explanation of survival processing benefits in episodic memory.

NOTES

1. We estimated d as follows. The mean recall rate increase due to survival processing was about .60 –.52 = .08. MSE = .014 (Nairne et al., 2007, p. 265) implies that the standard deviation within groups was SQR(MSE·147/150) = .117. Hence, $d \approx .08/.117 = .68$.
2. Consider the Apollo 13 disaster as an example. It seems likely that most who saw the Apollo 13 movie are able to recall the objects that were used to rescue the astronauts from death by suffocation.

REFERENCES

Albuquerque, P. B., Carvalho, P. F., Oliveira, H., & Capelo, A. M. (2011). *Word and visual stimuli recall in the survival processing paradigm*. Poster presented at the 5th International Conference on Memory, York, UK.

Aslan, A., & Bäuml, K.-H. (2012). Adaptive memory: Young children show enhanced retention of fitness-related information. *Cognition, 122*, 118–122.

Bell, R., & Buchner, A. (2012). How adaptive is memory for cheaters? *Current Directions in Psychological Science, 21*, 403–408.

Bell, R., Röer, J. P., & Buchner, A. (2013). Adaptive memory: The survival-processing memory advantage is not due to negativity or mortality salience. *Memory & Cognition, 41*, 490–502.

Bower, G. H. (1970). Imagery as a relational organizer in associative learning. *Journal of Verbal Learning and Verbal Behavior, 9*, 529–533.

Brainerd, C. J., Reyna, V. F., & Ceci, S. J. (2008). Developmental reversals in false memory: A review of data and theory. *Psychological Bulletin, 134*, 343–382.

Bransford, J. D., & Johnson, M. K. (1972). Contextual prerequisites for understanding: Some investigations of understanding and recall. *Journal of Verbal Learning and Verbal Behavior, 11*, 717–726.

Brewer, W. F. & Treyens, J. C. (1981). Role of schemata in memory for places. *Cognitive Psychology, 13*, 207–230.

Bröder, A., Krüger, N., & Schütte, S. (2011). The survival processing effect should generalize to source memory, but it doesn't. *Psychology, 2*, 896–901.

Buller, D. J. (2005a). *Adapting minds: Evolutionary psychology and the persistent quest for human nature*. Cambridge, MA: MIT Press.

Buller, D. J. (2005b). Evolutionary psychology: The emperor's new paradigm. *Trends in Cognitive Sciences, 9*, 277–283.

Buller, D. J. (2005c). Get Over: Massive modularity. *Biology and Philosophy, 20*, 881–891.

Burns, D. J., Burns, S. A., & Hwang, A. J. (2011). Adaptive memory: Determining the proximate mechanisms responsible for the memorial advantages of survival processing. *Journal of Experimental Psychology: Learning, Memory, and Cognition, 37*, 206–218.

Burns, D. J., Hart, J., Griffith, S. E., & Burns, A. D. (in press). Adaptive memory: The survival scenario enhances item-specific processing relative to a moving scenario. *Memory*.

Buss, D. M. (1995). Evolutionary psychology: A new paradigm for psychological science. *Psychological Inquiry, 6*, 1–30.

Butler, A. C., Kang, S. H. K., & Roediger, H. L. III. (2009). Congruity effects between materials and processing tasks in the survival processing paradigm. *Journal of Experimental Psychology: Learning, Memory, and Cognition, 35*, 1477–1486.

Cohen, J. (1988). *Statistical power analysis for the behavioral sciences* (2nd ed.). Hillsdale, NJ: Erlbaum.

Craik, F. I. M. (2002). Levels of processing: Past, present…and future? *Memory, 10*, 305–318.

Craik, F. I. M., & Lockhart, R. S. (1972). Levels of processing: A framework for memory research. *Journal of Verbal Learning and Verbal Behavior, 11*, 671–684.

Craik, F. I. M., & Tulving, E. (1975). Depth of processing and the retention of words in episodic memory. *Journal of Experimental Psychology: General, 104*, 268–294.

Engelkamp, H., Biegelman, U., & McDaniel, M. A. (1998). Relational and item-specific information: Trade-off and redundancy. *Memory, 6*, 307–333.

Einstein, G. O., & Hunt, R. R. (1980). Levels of processing and organization: Additive effects of individual item and relational processing. *Journal of Experimental Psychology: Human Learning and Memory, 6*, 588–598.

Hart, J., & Burns, D. J. (2012). Nothing concentrates the mind: Thoughts of death improve recall. *Psychonomic Bulletin & Review, 19*, 264–269.

Howe, M. L., & Derbish, M. H. (2010). On the susceptibility of adaptive memory to false memory illusions. *Cognition, 115*, 252–267.

Howe, M. L., & Otgaar, H. (2013). Proximate mechanisms and the development of adaptive memory. *Current Directions in Psychological Science, 22*, 16–22.

Hunt, R. R., & Einstein, G. O. (1981). Relational and item-specific information in memory. *Journal of Memory and Language, 20*, 497–514.

Hunt, R. R., & McDaniel, M. A. (1993). The enigma of organization and distinctiveness. *Journal of Memory and Language, 20*, 497–514.

Kang, S. H. K., McDermott, K. B., & Cohen, S. M. (2008). The mnemonic advantage of processing fitness-relevant information. *Memory & Cognition, 36*, 1151–1156.

Klein, S. B. (2012). A role for self-referential processing in tasks requiring participants to imagine survival on the savannah. *Journal of Experimental Psychology: Learning, Memory, and Cognition, 38*, 1234–1242.

Klein, S. B. (in press). The effects of thoughts of survival and thoughts of death on recall in the adaptive memory paradigm. *Memory*.

Klein, S. B., Cosmides, L., Tooby, J., & Chance, S. (2002). Decisions and the evolution of memory: Multiple systems, multiple functions. *Psychological Review, 109*, 306–329.

Klein, S. B., Robertson, T. E., & Delton, A. W. (2010). Facing the future: Memory as an evolved system for planning future acts. *Memory & Cognition, 38*, 13–22.

Klein, S. B., Robertson, T. E., & Delton, A. W. (2011). The future-orientation of memory: Planning as a key component mediating the high levels of recall found with survival processing. *Memory, 19*, 121–139.

Kostic, B., McFarlan, C. C., & Cleary, A. M. (2012). Extensions of the survival advantage in memory: Examining the role of ancestral context and implied social isolation. *Journal of Experimental Psychology: Learning, Memory, and Cognition, 38*, 1091–1098.

Kroneisen, M., & Erdfelder, E. (2011). On the plasticity of the survival processing effect. *Journal of Experimental Psychology: Learning, Memory, and Cognition, 37*, 1552–1563.

Kroneisen, M., Erdfelder, E., & Buchner, A. (2013). The proximate memory mechanism underlying the survival processing effect: Richness of encoding or interactive imagery? *Memory, 21*, 494–502.

Levine, L. J., & Pizarro, D. A. (2004). Emotion and memory research: A grumpy overview. *Social Cognition, 22*, 530–554.

Mandler, J. M. (1984). *Stories, scripts, and scenes: Aspects of schema theory*. Hillsdale, NJ: Lawrence Erlbaum Associates.

Moscovitch, M., & Craik, F. I. M (1976). Depth of processing, retrieval cues, and uniqueness of encoding as factors in recall. *Journal of Verbal Learning and Verbal Behavior, 15*, 447–458.

McGaugh, J. L. (2003). *Memory and emotion: The making of lasting memories.* New York: Columbia University Press.

Nairne, J. S. (2010). Adaptive memory: Evolutionary constraints on remembering. In B. H. Ross (Ed.), *The psychology of learning and motivation: Advances in research and behavior* (Vol. 53, pp. 1–32). San Diego, CA: Academic Press.

Nairne, J. S., & Pandeirada, J. N. S. (2008a). Adaptive memory: Is survival processing special? *Journal of Memory and Language, 59*, 377–385.

Nairne, J. S., & Pandeirada, J. N. S. (2008b). Adaptive memory: Remembering with a stone-age brain. *Current Directions in Psychological Science, 17*, 239–243.

Nairne, J. S., & Pandeirada, J. N. S. (2010). Adaptive memory: Ancestral priorities and the mnemonic value of survival processing. *Cognitive Psychology, 61*, 1–22.

Nairne, J. S., & Pandeirada, J. N. S. (2011). Congruity effects in the survival processing paradigm. *Journal of Experimental Psychology: Learning, Memory, and Cognition, 37*, 539–549.

Nairne, J. S., Pandeirada, J. N. S., Gregory, K. J., & Van Arsdall, J. E. (2009). Adaptive memory: Fitness relevance and the hunter-gatherer mind. *Psychological Science, 20*, 740–746.

Nairne, J. S., Pandeirada, J. N. S., & Thompson, S. (2008). Adaptive memory: The comparative value of survival processing. *Psychological Science, 19*, 176–180.

Nairne, J. S., Thompson, S. R., & Pandeirada, J. N. S. (2007). Adaptive memory: Survival processing enhances retention. *Journal of Experimental Psychology: Learning, Memory, and Cognition, 33*, 263–273.

Nairne, J. S., VanArsdall, J. E., Pandeirada, J. N. S., & Blunt, J. R. (2012). Adaptive memory: enhanced location memory after survival processing. *Journal of Experimental Psychology: Learning, Memory, and Cognition, 38*, 495–501.

Nouchi, R. (2012). The effect of aging on the memory enhancement of the survival judgment task. *Japanese Psychological Research, 54*, 210–217.

Otgaar, H., & Smeets, T. (2010). Adaptive memory: Survival processing increases both true and false memory in adults and children. *Journal of Experimental Psychology: Learning Memory and Cognition*, *36*, 1010–1016.

Otgaar, H., Smeets, T., Merckelbach, H., Jelicic, M., Verschuere, B., Galliot, A.-M., & van Riel, L. (2011). Adaptive memory: Stereotype activation is not enough. *Memory & Cognition*, *39*, 1033–1041.

Otgaar, H., Smeets, T., & van Bergen, S. (2010). Picturing survival memories: Enhanced memory after fitness-relevant processing occurs for verbal and visual stimuli. *Memory & Cognition*, *38*, 23–28.

Over, D. E. (Ed.) (2003). *Evolution and the psychology of thinking: The debate.* New York: Psychology Press.

Palmore, C. C., Garcia A. D., Bacon L. P., Johnson, C. A., & Kelemen W. L. (2012). Congruity influences memory and judgments of learning during survival processing. *Psychonomic Bulletin & Review*, *19*, 119–125.

Pinho, M. S., & Pandeirada, J. N. S. (2011, July). *Directed forgetting in the survival processing paradigm.* Poster presented at the 5th International Conference on Memory, York, UK.

Röer, J. P., Bell, R., & Buchner, A. (in press). Is the survival processing memory advantage due to richness of encoding? *Journal of Experimental Psychology: Learning, Memory & Cognition.*

Savine, A. C., Scullin, M. K., & Roediger, H. L. (2011). Survival processing of faces. *Memory & Cognition*, *39*, 1359–1373.

Schacter, D. L., & Wiseman, A. L. (2006). Reducing memory errors: The distinctiveness heuristic. In R. R. Hunt & J. B. Worthen (Eds.), *Distinctiveness and memory* (pp. 89–108). Oxford, UK: Oxford University Press.

Schulman, A. I. (1974). Memory for words recently classified. *Memory & Cognition*, *2*, 47–52.

Scott-Phillips, T. C., Dickins, T. E., & West, S. A. (2011). Evolutionary theory and the ultimate-proximate distinction in the human behavioral sciences. *Perspectives on Psychological Science*, *6*, 38–47.

Seamon, J. G., Bohn, J. M., Coddington, I. E., Ebling, M. C., Grund, E. M., Haring, C. T., Jang, S.-J., Kim, D., Liong, C., Paley, F. M., Pang, L. K., & Siddique, A. H. (2012). Can survival processing enhance story memory? Testing the generalizability of the adaptive memory framework. *Journal of Experimental Psychology: Learning Memory and Cognition*, *38*, 1045–1056.

Smeets, T. (2011). Acute stress impairs memory retrieval independent of time of day. *Psychoneuroendocrinology*, *36*, 495–501.

Smeets, T., Otgaar, H., Candel, I., & Wolf, O. T. (2008). True or false? Memory is differently affected by stress-induced cortisol elevations and sympathetic activity at consolidation and retrieval. *Psychoneuroendocrinology*, *36*, 495–501.

Smeets, T., Otgaar, H., Raymaekers, L., Peters, M. J. V., & Merckelbach, H. (2012). Survival processing in times of stress. *Psychonomic Bulletin & Review*, *19*, 113–118.

Soderstrom, N. C., & McCabe, D. P. (2011). Are survival processing memory advantages based on ancestral priorities? *Psychonomic Bulletin & Review*, *18*, 564–569.

Symons, C. S. & Johnson, B. T. (1997). The self-reference effect in memory: A meta-analysis. *Psychological Bulletin*, *121*, 371–394.

Tooby, J., & Cosmides, L. (1992). The psychological foundations of culture. In J. H. Barkow, L. Cosmides, & J. Tooby (Eds.), *The adapted mind: Evolutionary psychology and the generation of culture* (pp. 19–136). New York: Oxford University Press.

Tooby, J., & Cosmides, L. (2005). Conceptual foundations of evolutionary psychology. In D. Buss (Ed.), *The handbook of evolutionary psychology* (pp. 5–67). New York: Oxford University Press.

Tse, C.-S., & Altarriba, J. (2010). Does survival processing enhance implicit memory? *Memory & Cognition*, *38*, 1110–1121.

Tyler, S. W., Hertel, P. T., McCallum, M. C., & Ellis, A. C. (1979). Cognitive effort and memory. *Journal of Experimental Psychology: Human Learning and Memory*, *5*, 607–617.

Weinstein, Y., Bugg, J. M., & Roediger, H. L. (2008). Can the survival recall advantage be explained by basic memory process? *Memory & Cognition*, *36*, 913–919.

Wilson, S., Darling, S., & Skyes, J. (2011). Adaptive memory: fitness relevant stimuli show a memory advantage in a game of pelmanism. *Psychonomic Bulletin & Review*, *18*, 781–786.

SECTION 3

Adaptive Memory in Distinctive Populations

11

Memory Errors in Adaptive Recollections

HENRY OTGAAR, MARK L. HOWE, TOM SMEETS,
LINSEY RAYMAEKERS, AND JOHAN VAN BEERS

Memory scholars are increasingly embracing the idea that our memory can be studied from an evolutionary perspective, thereby making inferences about which processes were shaped by natural selection (see Nairne & Pandeirada, 2010). Broadly speaking, this idea is borrowed from the suggestion that the lion's share of the development of our cognitive machinery happened during the Pleistocene era (approximately from 1.8 million to 10,000 years ago) and that as a result, memory processes are especially sensitive to fitness-relevant stimuli. Furthermore, this idea indicates that our memory will prioritize the processing of information that might increase survival success (e.g., Klein, Cosmides, Tooby, & Chance, 2002).

Experimental work in this area has mainly been hampered because of the impossibility to determine the precise circumstances that affected our ancestors' cognitive development, thereby leading to *just so* explanations of psychological phenomena (Gould & Lewontin, 1979). However, recently, Nairne, Thompson, and Pandeirada (2007) developed an ingenious memory paradigm rooted in the idea that memory performance should be boosted when fitness-relevant stimuli are processed. In this paradigm, participants are instructed to imagine being stranded on the grasslands of a foreign country and facing the potential danger of nearby predators. After this instruction, participants have to rate unrelated words for their relevance to this survival scenario. Intriguingly, surprise recall and recognition tests show that words are better remembered when they are processed according to a survival context relative to control contexts (e.g., moving to a foreign country, pleasantness). According to Nairne and colleagues (2007, 2008), this finding suggests that our memory system has been crafted to preferentially process survival-related information, presumably by forces of natural selection.

Following the original study, researchers started to study the boundary conditions of this survival recall advantage. Most studies have found that the survival recall advantage could be replicated using a wide variety of control conditions and stimuli (e.g., Kang, McDermott, & Cohen, 2008; Nairne & Pandeirada, 2010; Nairne et al.,

2008; Otgaar et al., 2011; Otgaar, Smeets, & van Bergen, 2010; Weinstein, Bug, & Roediger, 2008; Wilson, Darling, & Sykes, 2010; but see Savine, Scullin, & Roediger, 2011). To provide some examples, Kang et al. (2008) were interested in learning if the survival recall advantage was mediated by the survival scenario being more arousing and novel than the control scenarios. To test this idea, they compared the original survival scenario with a control condition that was equally high on arousal and novelty (i.e., a bank heist). Results showed that the survival recall effect remained intact. Nairne et al. (2008) additionally compared the survival scenario with a host of control conditions known to enhance memory retention (e.g., imagery, intentional learning). They found that even in this situation, participants in the survival condition outperformed the participants in the other memory-enhancing conditions in terms of memory retention. Besides these studies, other control conditions have also been pitted against the survival scenario yet have failed to eliminate the survival recall advantage (e.g., city, Weinstein et al., 2008; university professor, Otgaar et al., 2011).

For example, Weinstein et al. (2008) compared the original survival scenario with a scenario designed to induce comparable schematic processing as the survival scenario. That is, they used a scenario in which participants had to imagine being stranded in a city in a foreign land in danger of attackers. Otgaar et al. (2011) compared the original survival scenario with a scenario known to benefit memory performance (i.e., imagining being a college professor specializing in memory). In both studies, the survival recall advantage was present.

Researchers have also examined whether other classes of stimuli (e.g., pictures, location) could show a survival recall advantage. Otgaar et al. (2010) examined whether the survival recall effect was present when pictures were used as stimuli. In two experiments, they found that for pictorial stimuli and words, superior memory retention was evident when stimuli had to be processed for survival relevance (see also Wilson et al., 2010). Nairne, VanArsdall, Pandeirada, and Blunt (2012) found similar results. They showed that people were better in remembering the location of pictorial stimuli when the stimuli were relevant for survival. Taken together, the above-mentioned studies evidently yield reliable support for the idea that memory has been sculpted to retain fitness-relevant information and can be examined via the survival processing effect.

However, although studies have found the survival recall effect being robust across a number of dimensions, there are also studies that could not reproduce the effect. Soderstrom and McCabe (2011), for example, could not detect words being superiorly recalled after survival processing when the survival group was compared with a group in which participants had to imagine being attacked by zombies. No survival processing advantage also emerged when faces that likely stimulate item-specific rather than gist processing were used as stimuli (Savine et al., 2011). The latter finding was probably due to the fact that face recognition relies more on the recognition of details and not on the processing of gist, which is likely involved in the survival processing effect (Otgaar & Smeets, 2010).

Part of the failure to detect survival increases in memory retention stems from the fact that recently a few investigators have reported studies wherein they tested several proximate mechanisms that might mitigate the survival processing advantage. For example, no survival processing advantage was evident when groups were equated on item-specific and relational processing (Burns, Burns, & Hwang, 2011) or when the survival scenario was less distinctive and had fewer opportunities for elaboration (Howe & Derbish, Chapter 5 in this volume; Kroneisen & Erdfelder, 2011). Furthermore, Klein (2012) found that part of the survival processing advantage could be explained by relying on the self-reference effect. That is, Klein showed that when participants in the survival processing paradigm were asked to retrieve an episodic memory and then rate how easily words bring to mind a personal experience, the survival processing effect is eliminated. Therefore, self-referential processing might underpin the survival processing effect.

Of importance in the current study is the finding that survival processing is linked to an increase in relational processing and elaboration. What this implies is that survival processing would not only prioritize true memories but would also affect the formation of false memories. The underlying idea behind this is that an increase in relational processing would lead to more spreading activation in memory, which in turn would have ramifications for the development of false memories (Howe, Wimmer, Gagnon, & Plumpton, 2009). This proposition is derived from associative activation theory (Howe et al., 2009), which stipulates that true and false memories arise out of associative activation among related concepts in an individual's knowledge base.

Evidence for this claim comes from studies showing that survival processing not only boosts true memory performance but likewise also amplifies the risk for memory errors. For example, in the original study by Nairne et al. (2007; Experiment 1), it was demonstrated that both the survival group and a control group (moving to another country) remembered more incorrect words than a group that had to indicate the pleasantness of the words. Otgaar et al. (2010) showed that participants in the survival group remembered more incorrect details in pictures than participants in the pleasantness group (no differences between survival and moving).

Two studies that have systematically tested whether survival processing would perpetuate false memory production were carried out by Otgaar and Smeets (2010) and Howe and Derbish (2010). In these studies, participants received semantically related word lists known to reliably elicit spontaneous false memories (Roediger & McDermott, 1995). Both studies found that false memory rates were elevated when children (Otgaar & Smeets, 2010) and adults (Howe & Derbish, 2010; Otgaar & Smeets, 2010) were asked to process words for survival relevance compared to when participants were asked to process words in control conditions (see also Smeets, Otgaar, Raymaekers, Peters, & Merckelbach, 2012). Although one might argue that memory would be truly adaptive if memory illusions would not slip into our experiences, memory illusions can also be seen as a byproduct of a flexible memory system (Howe, 2011; Schacter, Guirin, & St. Jacques, 2011). For example, Howe, Garner, Charlesworth, and Garner (2011) showed that participants who were vulnerable to false memories were more likely to generate correct answers on subsequent problem-solving tasks. Also, Dewhurst, Thorley, Hammond, and Ormerod (2011) demonstrated that creativity tasks that rely on convergent thinking are positively associated with an increased susceptibility to false memories. These studies suggest that memory illusions not only have to be regarded as malignant errors but can also be seen as memories that in some contexts can have positive consequences.

In the studies conducted by Otgaar and Smeets (2010) and Howe and Derbish (2010), survival processing had an impact on the development of *spontaneous* false memories. These are the types of false memories that arise without external pressure and are the result of endogenous processes (spreading activation, Howe et al., 2009; gist extraction, Brainerd, Reyna, & Ceci, 2008). However, there also exists another category of false memories called *suggestion-based* false memories (Loftus, 2005). These false memories are the result of external suggestive pressure and have been subjected to a plethora of empirical scrutiny because of the legal consequences that these false memories might have (e.g., prosecution of innocent suspects). Importantly, like spontaneous false memories, there is evidence showing that in some conditions (i.e., strong suggestive pressure, imagination), suggestion-based false memories rely on the construction of new, but incorrect, memory traces and are not always the byproduct of social demands (Otgaar, Verschuere, Meijer, & Van Oorsouw, 2012). For example, in the Otgaar et al. (2012) study, children who developed false memories of being on a hot air balloon ride were faster to confirm than deny that this occurred on a response latency deception computer task. This pattern was also evident for true memories. Specifically, in this experiment, 8-year-old children received false suggestions that they were on a hot air balloon ride and were also asked about their first school day. Then, they received a response latency deception task in which they had to confirm or deny statements related to the false (e.g., "I have flown in a hot air balloon ride") and true (e.g., "I am in the third grade"). Confirming statements that refer to actual memories typically lead to faster response times and lower error rates than when statements are denied (i.e., when participants have to lie). Interestingly, for both true and false memories this pattern was present, thereby suggesting that false memories were processed as genuine memories.

Suggestion-based false memories are elicited by use of two main false memory paradigms, misinformation and implantation paradigm (Otgaar et al., 2012). In general, in both paradigms, participants are presented with suggestive questions and/or assertions about false details (misinformation paradigm) or false events (implantation paradigm). Studies have shown that these manipulations cause participants to accept the suggestions and incorporate them into their prospective memory reports (Loftus, 2005).

Although debate exists about whether similar mechanisms play a role in the development of both types (spontaneous and suggestion-based) of false memories (Brainerd et al., 2008), memory scholars have noted that both types of memory illusions could be adaptive in nature (see Schacter et al., 2011). Thus, one could expect that like spontaneous false memories, suggestion-based false memories are more likely to occur when people are presented with a survival context. The purpose of the current study is to address this issue. More specifically, in the present study, we examined whether suggestion-based false memories are more likely to occur when children and adults are situated in a survival context.

The reason for including children in the current study is twofold. First, one could expect that memory would be adaptive only if the survival processing effect would not occur solely in adults but also in other age groups (e.g., children). Indeed, studies show that the survival recall advantage is also present in child samples (Aslan & Bäuml, 2012; Otgaar & Smeets, 2010). The current study was designed to examine whether this effect can be replicated. Second, suggestion-based false memories are more easily elicited in children, especially very young children, than in adults (Ceci & Bruck, 1993), and thus it may be expected that children are also more likely to develop suggestion-based false memories than adults when a survival context is presented.

To elicit suggestion-based false memories, we let 11- and 12-year-old children and adults take part in an *intentional* suggestion-based false memory procedure. Although the original survival paradigm uses an *incidental* memory paradigm, other studies have found the survival processing effect also to occur in intentional memory paradigms (i.e., participants knew that their memory was being tested; Howe & Derbish, 2010). Also, we decided to use a customary paradigm to evoke suggestion-based false memories (Loftus, 2005). More specifically, participants were presented with a story that contained either a survival or control (moving to a foreign country) connotation. After this, participants received suggestive and memory questions about the story (Seamon et al., 2012). Our predictions were the following. First, we hypothesized that memory performance would be better for the survival story than for the control story. Second, we predicted that in the survival story, both children and adults were more likely to falsely assent to suggestive pressure than in the control story.

METHOD

Participants

Participants were 40 undergraduates (mean age = 25.28, *SD* = 4.70) at Maastricht University and 40 children (mean age = 11.43, *SD* = 0.50). Children were recruited from elementary schools in the southern part of the Netherlands. Parents provided written consent for their children's participation in the study. Children assented to their participation and received a small present in return for their involvement. Adult participants also provided written consent and received financial compensation (5 euro) or course credits for their participation. The study was approved by the standing ethical committee of the Faculty of Psychology and Neuroscience, Maastricht University.

Materials

Suggestion-based memory paradigm. For the current study, we constructed a memory paradigm involving suggestive questions that is built upon well-known standard suggestibility paradigms such as the Gudjonsson Suggestibility Scale (Gudjonsson, 1984) and the Bonn Test of Statement Suggestibility (Endres, 1997). Specifically, we developed two stories (i.e., survival and moving) that participants had to read. These two stories were similar in gist in the sense that they were about a boy going to Hawaii on a boat. In the survival story, however, the boat crashes and the boy is stranded on a foreign island and is in danger from predators and without any food and water (see Appendix). As can be seen, this story is in line with the survival instruction that is normally presented to participants in the survival processing paradigm. In the moving story, the boy moves to Hawaii and has to find a new house and make new friends. In this paradigm, participants received a free recall phase and were presented with suggestive questions about the story. Suggestive questions could be divided into *Yield* and *Shift* questions (see below). Yield questions refer to suggestive questions in which in an incorrect answer is suggested to participants (e.g., "The boy was 19 years old, wasn't he?"). Shift questions involve questions that suggest that a previous response was incorrect (e.g., "Think again. Was Sven 19 or 20 years old?"; Scullin & Ceci, 2001).

To explore whether the stories differed in survival relevance, we conducted a pilot study in which we asked 18 students (mean age = 24.28, $SD = 2.54$; 16 female) to indicate on a 5-point Likert scale (1 = not survival-related, 5 = extremely survival-related) how survival-relevant the two stories were. Paired samples t-tests showed that the survival story ($M = 4.78$, $SD = 0.43$) was indeed rated as significantly more survival-related than the moving story ($M = 2.01$, $SD = 0.54$; $t(17) = 17.26$, $p < .001$). We also conducted a pilot study with 20 children (mean age = 11.20, $SD = 0.41$) and asked them which story they found survival-related; they had to provide a categorical judgment (survival-related vs. not survival-related). All children indicated that the survival story was the more survival-related.

Design and Procedure

The current study made use of a 2 (age: children vs. adults) × 2 (condition: survival vs. moving) factorial design. Participants were randomly allocated to the survival versus moving scenario (20 in each condition).

Adult participants were tested individually in a quiet testing room and children were tested in separate rooms at their elementary school. Participants were told that they had to read a story and that they would receive several other tasks. When they had read the story, they were asked to report everything they could remember about the story. After the free recall phase, participants were provided with a 10-minute filler task (Tetris). Next, participants received both suggestive and explicit memory questions about the story (see examples in the next section).

Scoring

For the free recall phase, 23 specific details (e.g., that the boy went to Hawaii) for each story could be scored. For the suggestive and memory questions, participants received six Memory questions for each story (e.g., "Did Sven go to Hawaii?"). The suggestive questions could be divided in nine Yield ("Sven, wasn't he 19 years old?") and five Shift questions ("Are you sure? I will ask you once more. Did Sven find mangos or bananas?"). The Yield questions implied that a certain detail was present when in fact it was not. The Shift questions were questions in which it was suggested that a previous answer was incorrect. We also computed a total suggestibility score by summing up the Yield and Shift scores. Importantly, we considered participants to have a false memory when they falsely assented to either type of suggestive question.

RESULTS

We conducted a series of 2 (age group: children vs. adults) × 2 (condition: survival vs. moving) univariate ANOVAs on our dependent variables (free recall, Memory questions, Shift questions, Yield questions, and total suggestibility score). No

Table 11.1. Memory Questions, Suggestive Questions, and Accuracy Scores as a Function of Condition (Survival vs. Moving) and Age (Children vs. Adults)

	Survival		Moving	
	Children	**Adults**	**Children**	**Adults**
Free recall (max. 23)	12.75 (3.04)	17.70 (2.75)	9.45 (3.32)	14.50 (3.02)
Memory questions (max. 6)	5.65 (0.49)	5.75 (0.55)	5.25 (0.72)	5.70 (0.47)
Yield questions (max. 9)	3.45 (1.36)	1.30 (1.30)	2.50 (1.54)	0.90 (1.48)
Shift questions (max. 5)	2.15 (1.46)	0.15 (0.49)	1.95 (1.47)	0.15 (0.47)
Total suggestibility	5.60 (2.37)	1.45 (1.50)	4.45 (2.27)	1.05 (1.50)
Net accuracy	0.71 (0.09)	0.91 (0.09)	0.75 (0.13)	0.94 (0.08)

Data are given as mean (SD).

significant interactions were found for any of these analyses (all *F*s < 1; all *p*s >.05). For the free recall, there was a significant main effect of condition ($F(1,77) = 23.17$, $p < .001$, partial $\eta^2 = 0.23$), revealing that the survival group had significantly higher true memory scores than the moving group (Table 11.1). We also found a significant main effect of age group ($F(1,77) = 54.85$, $p < .001$, partial $\eta^2 = 0.42$), indicating that the adult participants had higher free recall scores than the children.

For the Memory questions, although the survival group had higher true memory levels than the moving group this difference was not significant ($F(1,77) = 3.14$, $p = .08$). We did find a significant age group effect ($F(1,77) = 4.68$, $p = .034$, partial $\eta^2 = 0.06$), with adults remembering more items correctly than children.

Interestingly, our analysis also showed that for both age groups, the survival group ($F(1,76) = 4.50$, $p = .037$, partial $\eta^2 = 0.06$; 1 missing value) had significantly higher false memory rates on the Yield questions relative to the moving group. Thus, our analysis showed that both children and adults had higher false memory rates when they were presented with a survival-related story than with a control story. We also found a significant age group effect ($F(1,76) = 34.71$, $p < .001$, partial $\eta^2 = 0.31$), with children being at higher risk for false memories than adults. However, there was no significant interaction ($F < 1$).

When we focused our analyses on the Shift questions, we did not find a significant condition × age group interaction or a significant main effect of dondition (both *F*s < 1). Our results did demonstrate a significant age group effect ($F(1,76) = 61.93$, $p < .001$, partial $\eta^2 = 0.45$), with children shifting more in their answers than adults. With respect to the total suggestibility scores, we found a significant main effect of age group only ($F(1,76) = 64.92$, $p < .001$, partial $\eta^2 = 0.46$), showing that children had higher total suggestibility scores than adults. All other effects were not significant (*F*s < 1).

Finally, we computed net accuracy scores to examine whether the survival group had higher or lower accuracy scores [proportion memory questions/proportion (memory questions + total suggestibility)] (see Howe & Derbish, 2010; Otgaar & Smeets, 2010) than other participants in the study. Our analysis showed that adult participants had higher accuracy scores than children (effect of age group: $F(1,76) = 77.27$, $p < .001$, partial $\eta^2 = 0.50$). However, our analysis did not yield a significant main effect of condition or a significant age group × condition interaction (both *ps* >.05).

DISCUSSION

According to recent research, the *sine qua non* of memory is that it has been shaped to retain fitness-relevant information (Nairne & Pandeirada, 2010). The aim of the current experiment was to assess whether the superior processing of survival information also amplifies the vulnerability to memory errors. To investigate this, we asked adults and children to participate in a suggestion-based memory paradigm in which they were included in either a survival or a moving group.

Our most important findings were the following. First, we found that suggestion-based false memories were more likely to occur when a survival story was provided than when participants received a moving story. Second, we showed that for both children and adults, true memories were higher in the survival condition than in the moving condition. That is, when participants were presented with a survival story, they correctly remembered more about this survival story than when they received a moving story, replicating and extending the basic survival processing effect. Third, our study demonstrated that overall, children had lower true memory levels yet higher false memory rates than adults, also replicating literature on memory development (e.g., Ceci & Bruck, 1993). Indeed, when we calculated net accuracy rates, this same pattern was confirmed inasmuch as adults were more accurate than children. This pattern of accuracy rates runs contrary to the usual developmental finding for spontaneous false memories, in which adults tend to generate more false memories (having lower net accuracy rates) than children (Brainerd et al., 2008).

Our finding that the survival story resulted in superior retention for both children and adults extends earlier work on adaptive memory (Nairne & Pandeirada, 2008). This work has mainly used an incidental memory paradigm and found that survival processing resulted in excellent memory retention. To date, only a few studies have used intentional memory paradigms (Howe & Derbish, 2010; Seamon et al., 2012). Howe and Derbish, for example, showed that the survival processing effect remained intact even using intentional memory paradigms. Our study extends this by using stories as stimuli, and we showed that for both age groups, the survival group exhibited superior memory performance (but see Seamon et al., 2012).

Importantly, the fact that we also found the survival recall advantage being present in child samples is interesting because up until now, there have been only two studies that examined whether this effect occurs in other samples (Aslan & Bäuml, 2012; Otgaar & Smeets, 2010; see also Raymaekers, Smeets, Otgaar, Peters, & Merckelbach, Chapter 12 in this volume). It is vital to understand if these effects also occur in children because they have less experience with or knowledge about survival situations than adults. What our study and the studies reported above evidently show is that despite poorer recall rates for children than adults, both children's *and* adults' correct recall is superior when processing survival-related than moving-related information.

Critically, our study also revealed that false memories induced by suggestive pressure were more pronounced when information had to be memorized according to a survival scenario than a moving scenario. This parallels previous studies that have found spontaneous false memories to be elevated during survival processing (Howe & Derbish, 2010, and Chapter 5 in this volume; Otgaar & Smeets, 2010). Our study adds that *suggestion-based* false memories are also more likely to be elicited when survival information has to be remembered than when other information is presented. However, we found this effect to appear only for the Yield questions. The reason for this could be related to research showing that Yield questions are more related to memory processes whereas Shift questions are more linked to social processes (e.g., compliance, Mastroberardino & Marucci, in press). Furthermore, research shows that yielding to suggestive questions follows a negative linear developmental trend while susceptibility to shift scores sometimes undergo an inverted U-shaped developmental patterns because of developing factors such as Theory of Mind (e.g., Hünefeldt, Lucidi, Furia, & Rossi-Arnaud, 2008). It is then probably not surprising that our survival manipulation affected only the memory part of suggestibility. Our study also showed that overall, false memory rates were the highest in children, whereas true memory rates and accuracy were the highest in adults. This corroborates research showing that children are more vulnerable to the corrupting effects of suggestions than adults (e.g., Otgaar & Candel, 2011) because they have poorer monitoring skills than adults (see below).

One might contend that our effect is driven by the survival story being different on a number of dimensions. However, and as mentioned before, the survival and moving stories were comparable on a general level (i.e., going to another country) and differed only in the specific survival details. Of course, it is still possible that factors such as valence or arousal might be more predominant in the survival story (e.g., more emotionally negative, more arousing, or both) relative to the moving story, thereby leading to better memory performance of the survival group. However, in the survival memory paradigm, studies have shown that these factors (e.g., arousal) did not affect the survival recall advantage (e.g., Otgaar et al., 2011). Furthermore, studies show that the survival recall advantage remains even when scenarios are identical except for a survival component (e.g., Weinstein et al., 2008). Future studies should examine whether such findings are also relevant in an intentional memory paradigm.

The question remains what mechanisms could play a role in our observed effect that suggestion-based false memories are most likely to be formed when survival information is present. In our opinion, there are two explanations that might be related to this effect. First, recent research has shown that all types of false memories are caused by associative activation between related concepts in an individual's knowledge base (Howe et al., 2009; Otgaar, Howe, Peters, Smeets, & Moritz, submitted). This idea is derived from Associative-Activation Theory, which states that false memories arise because of increases in the connections in someone's knowledge base, which leads to more automaticity of associative activation. Although studies have shown that associative activation has a significant part in the development of spontaneous false memories (e.g., Howe & Derbish, 2010), it is likely that similar processes are at work in the current study.

That is, one might speculate that the presentation of a survival story resulted in participants' rapidly processing the content of this story—perhaps more rapidly than in the control condition. This spreading activation likely resulted in a better storage of the story with more correct elements of the story being activated, leading to the standard survival recall advantage. Moreover, it is likely that in the survival condition this associative activation has also triggered related, but not presented, concepts that could be reactivated when the suggestive questions are asked, leading to more false memories in the survival group than in the moving group (see also Toglia, Neuschatz, & Goodwin, 1999).

An alternative explanation is that the processing of survival-related information increases source-monitoring errors, resulting in an increase in false memories. A plethora of research has shown that some false memories are due to errors in source-monitoring abilities (Johnson, Hashtroudi, & Lindsay, 1993). Although it is possible that source-monitoring errors occur more often during survival processing, recent research has shown that source memory is not affected by the processing of survival-related information (Bröder, Krüger, & Schutte, 2011). It is more probable that source-monitoring errors have resulted in our finding that in general, children are more likely to produce suggestion-based false memories than adults. This finding has often been demonstrated and interpreted in terms of source monitoring (Ceci & Bruck, 1993). That is, research has shown that children have less well-developed source-monitoring abilities than adults and thereby are more at risk for suggestion-based false memories (Lindsay, Johnson, & Kwon, 1991).

It must be noted that in the current experiment, we did not find false memories to increase with age. That is, Associative-Activation Theory predicts that all types of false memories (suggestion-based and spontaneous) increase with age when relying on associative activation between related elements in a memory system (Otgaar et al., 2012). In our experiment, we found the usual suggestibility effects, with children being more at risk for suggestive questions and false memories than adults. It is likely that we did not find an age-related increase of suggestion-based false memories because in the current experiment, it is unclear to what extent the suggestive information was associatively related to the remembered information (stories). Furthermore, it is possible that children are more likely to accept "knowledge" from another source in the form of suggestive questions than adults if they do not receive contrary information. Indeed, studies show that children are highly likely to include extra knowledge in their developing knowledge base, thereby amplifying the susceptibility to false memories (Otgaar, Smeets, & Peters, 2012).

The finding that we found a decrease in suggestion-based false memories across age could also be due to the fact that in our design, adult experimenters carried out both the free recall test and the suggestive phase. It is possible that children were more likely to trust the experimenter's memory of the event than adult participants, thereby assenting more to the Yield questions. Furthermore, in the current study, we had more than twice as many suggestive questions than memory questions. This imbalance could potentially harm children's memory performance more than adults' performance.

Although one might question the adaptive nature of false memories, recent studies show that spontaneous false memories aid the solutions to complex problem-solving tasks (Howe et al., 2011). Furthermore, studies have also revealed that suggestion-based false memories can result in actual behavioral consequences (e.g., avoiding eating egg salad; Bernstein & Loftus, 2009). Such behavioral effects can easily be explained within a framework that takes into account that besides the negative effects of false memories (i.e., false accusations), false memories can also have more adaptive and positive consequences.

To recap, our study is the first to reveal that survival-related information amplifies the susceptibility to suggestion-based false memories in both children and adults. So, although evidence is increasing that our memory often works better within a survival-related context, it also leads to increased flexibility, thereby creating memory illusions. Whether these memory illusions are truly adaptive or are merely simple byproducts of our memory is completely open for future research endeavors.

ACKNOWLEDGMENTS

Dr. Otgaar's contributions were supported by a grant from the Netherlands Organization for Scientific Research (NWO; 016.135.052) and by a grant from the Edmund Hustinx Foundation.

REFERENCES

Aslan, A., & Bauml, K.-H. T. (2012). Adaptive memory: Young children show enhanced retention of fitness-related information. *Cognition, 122*, 118–122.

Bernstein, D. M., & Loftus, E. F. (2009). The consequence of false memories for food preferences and choices. *Perspectives on Psychological Science, 4*, 135–139.

Brainerd, C. J., Reyna, V. F., & Ceci, S. J. (2008). Developmental reversals in false memory: A review of data and theory. *Psychological Bulletin, 134*, 343–382.

Bröder, A., Krüger, N., & Schütte, S. (2011). The survival processing memory effect should generalise to source memory, but it doesn't. *Psychology, 2*, 896–901.

Burns, D. J., Burns, S. A., & Hwang, A. J. (2011). Adaptive memory: Determining the proximate mechanisms responsible for the memorial advantages of survival processing. *Journal of Experimental Psychology: Learning, Memory, and Cognition, 37*, 206–218.

Ceci, S. J., & Bruck, M. (1993). Suggestibility of the child witness: A historical review and synthesis. *Psychological Bulletin, 113*, 403–439.

Dewhurst, S. A., Thorley, C., Hammond, E. R., & Ormerod, T. C. (2011). Convergent, but not divergent, thinking predicts susceptibility to associative memory illusions. *Personality and Individual Differences, 51*, 73–76.

Endres, J. (1997). The suggestibility of the child witness: The role of individual differences and their assessment. *Journal of Credibility Assessment and Witness Psychology, 1*, 44–67.

Gould, S. J., & Lewontin, R.C. (1979). The spandrels of San Marco and the Panglossian paradigm: A critique on the adaptationist programme. *Proceedings of the Royal Society of London, 205*, 581–598.

Gudjonsson, G. H. (1984). A new scale of interrogative suggestibility. *Personality and Individual Differences, 5*, 303–314.

Howe, M. L. (2011). The adaptive nature of memory and its illusions. *Current Directions in Psychological Science, 20*, 312–315.

Howe, M. L., & Derbish, M. H. (2010). On the susceptibility of adaptive memory to false memory illusions. *Cognition, 115*, 252–267.

Howe, M. L., Garner, S. R., Charlesworth, M., & Knott, L. M. (2011). A brighter side to memory illusions: False memories prime children's and adults' insight-based problem solving. *Journal of Experimental Child Psychology, 108*, 383–393.

Howe, M. L., Wimmer, M. C., Gagnon, N., & Plumpton, S. (2009). An associative-activation theory of children's and adults' memory illusions. *Journal of Memory and Language, 60*, 229–251.

Hünefeldt, T., Lucidi, A., Furia, A., & Rossi-Arnaud, C. (2008). Age differences in the interrogative suggestibility of children's memory: Do shift scores peak around 5-6 years of age? *Personality and Individual Differences, 45*, 521–526.

Johnson, M. K., Hashtroudi, S., & Lindsay, D. S. (1993). Source monitoring. *Psychological Bulletin, 114*, 3–28.

Kang, S. H. K., McDermott, K. B., & Cohen, S. M. (2008). The mnemonic advantage of processing fitness-relevant information. *Memory and Cognition, 36*, 1151–1156.

Klein, S. B. (2012). A role for self-referential processing in tasks requiring participants to imagine survival on the savannah. *Journal of Experimental Psychology: Learning, Memory & Cognition, 38*, 1234–1242.

Klein, S. B., Cosmides, L., Tooby, J., & Chance, S. (2002). Decisions and the evolution of memory: Multiple systems, multiple functions. *Psychological Review, 109*, 306–329.

Kroneisen, M., & Erdfelder, E. (2011). On the plasticity of the survival processing effect. *Journal of Experimental Psychology: Learning, Memory, and Cognition, 37*, 1553–1562.

Lindsay, D. S., Johnson, M. K., & Kwon, P. (1991). Developmental changes in memory source monitoring. *Journal of Experimental Child Psychology, 52*, 297–318

Loftus, E. F. (2005). Planting misinformation in the human mind: A 30-year investigation of the malleability of memory. *Learning & Memory, 12*, 361–366.

Mastroberardino, S., & Marucci, F.S. (in press). Interrogative suggestibility: Was it just compliance or a genuine false memory? *Legal and Criminological Psychology*.

Nairne, J. S., & Pandeirada, J. N. S. (2010). Adaptive memory: Ancestral priorities and the mnemonic value of survival processing. *Cognitive Psychology, 61*, 1–22.

Nairne, J. S., Pandeirada, J. N. S., & Thompson, S. R. (2008). Adaptive memory: The comparative value of survival processing. *Psychological Science, 19*, 176–180.

Nairne, J. S., Thompson, S. R., & Pandeirada, J. N. S. (2007). Adaptive memory: Survival processing enhances retention. *Journal of Experimental Psychology: Learning, Memory, and Cognition, 17*, 239–243.

Nairne, J. S., VanArsdall, J. E., Pandeirada, J. N. S., & Blunt, J. R. (2012). Adaptive memory: Enhanced location memory after survival processing. *Journal of Experimental Psychology: Learning, Memory, and Cognition, 38*, 495–501.

Otgaar, H., & Candel, I. (2011). Children's false memories: Different false memory paradigms reveal different results. *Psychology, Crime & Law, 17*, 513–528.

Otgaar, H., & Smeets, T. (2010). Adaptive memory: Survival processing increases both true and false memory in adults and children. *Journal of Experimental Psychology: Learning, Memory, and Cognition, 36*, 1010–1016.

Otgaar, H., Howe, M. L., Peters, M., Smeets, T., & Moritz, S. (submitted). *The rise of memory illusions in childhood: An associative activation account*. Manuscript submitted for publication.

Otgaar, H., Smeets, T., Merckelbach, H., Jelicic, M., Verschuere, B., Galliot, A., & van Riel, L. (2011). Adaptive memory: Stereotype activation is not enough. *Memory & Cognition, 39*, 1033–1041.

Otgaar, H., Smeets, T., & Peters, M. (2012). Children's implanted false memories and additional script knowledge. *Applied Cognitive Psychology, 26*, 709–715.

Otgaar, H., Smeets, T., & Van Bergen, S. (2010). Picturing survival memories: Enhanced memory for fitness-relevant processing occurs for verbal and visual stimuli. *Memory & Cognition, 38*, 23–28.

Otgaar, H., Verschuere, B., Meijer, E. H., & van Oorsouw, K. (2012). The origin of children's implanted false memories: Memory traces or compliance. *Acta Psychologica, 139*, 397–403.

Raymaekers, L., Smeets, T., Otgaar, H., Peters, M., & Merckelbach, H. (this volume). Adaptive memory in survivors of childhood trauma.

Roediger, H. L., & McDermott, K. (1995). Creating false memories: Remembering words not presented in a list. *Journal of Experimental Psychology: Learning, Memory, and Cognition, 21*, 803–814.

Savine, A. C., Skullin, M. K., & Roediger, H. L., III (2011). Survival processing of faces. *Memory & Cognition, 39*, 1359–1373.

Schacter, D. L., Guerin, S. A., & St. Jacques, P. L. (2011). Memory distortion: An adaptive perspective. *Trends in Cognitive Sciences, 15*, 467–474.

Scullin, M. H., & Ceci, S. J. (2001). A suggestibility scale for children. *Personality and Individual Differences, 30*, 853–856.

Seamon, J. G., Bohn, J. M., Coddington, I. E., Inslee, E., Ebling, M. C., Grund, E. M., Haring, C. T., Jang, S., Kim. D., Liong, C., Paley, F. M., Pang, L. K., & Siddique, A. H. (2012). Can survival processing enhance story memory? Testing the generalizability of the adaptive memory framework. *Journal of Experimental Psychology: Learning, Memory, and Cognition, 38*, 1045–1056.

Smeets, T., Otgaar, H., Raymaekers, L., Peters, M. V., & Merckelbach, H. (2012). Survival processing in times of stress. *Psychonomic Bulletin & Review, 19*, 113–118.

Soderstrom, N. C., & McCabe, D. P. (2011). Are survival processing memory advantages based on ancestral priorities? *Psychonomic Bulletin and Review, 18*, 564–569.

Toglia, M. P., Neuschatz, J. S., & Goodwin, K. A. (1999). Recall accuracy and illusory memories: When more is less. *Memory, 7*, 233–256.

Weinstein, Y., Bugg, J. M., & Roediger, H. L., III (2008). Can the survival recall advantage be explained by basic memory processes? *Memory and Cognition, 36*, 913–919.

Wilson, S., Darling, S., & Dykes, J. (2010). Adaptive memory: Fitness relevant stimuli show a memory advantage in a game of pelmanism. *Psychonomic Bulletin & Review, 18*, 781–786.

APPENDIX

Example of the Survival Story Used in the Present Experiments (English Translation)

Sven, age 18, goes alone on a holiday to Hawaii. He leaves with a boat from Rotterdam to Hawaii. More then 150 people are on the boat. Halfway through the trip, the boat enters a dangerous storm that damages the boat heavily, and eventually the boat sinks. Sven is the only survivor and is stranded on an uninhabited island. Sven does not have any supplies to survive: no food, no drinks, no extra clothing. He decides to search for food and drink. He finds coconuts and other fruit to eat, but he knows that this food is not enough, so he decides to hunt. He makes a spear and a bow and arrow and captures several animals, such as fish and rabbits. However, there are also dangerous animals, like wild boars and tigers, on the island. To protect himself from those dangerous animals, he decides to make a large tree house with branches and leaves. He also makes a huge campfire on the beach in the hope that he will be discovered and saved. After 14 days he sees a large boat sailing to the island. The people on the boat luckily see his campfire and decide to save him.

Example of the Moving Story Used in the Present Experiments (English Translation)

Sven, age 18, moves to Hawaii. He is going to study biology in the capital city, Honolulu. He will stay for one year in Hawaii. He leaves from Rotterdam on a boat to Hawaii. More than 150 people are on the boat. When he arrives in Hawaii, he has to look for a new place. He finds a nice house on the beach. He is planning to purchase new furniture for his house. He buys a new couch, a table, and a large TV. The school in which he is going to study is very large. Approximately 1,000 people study at his new school. He unfortunately does not have any friends yet. After a couple of days he meets a boy named Scott at his school and they become good friends. Scott is from America and has also moved to Hawaii. He is going to study history in Hawaii. Together they are planning to do their best. After a year Sven returns to Holland while Scott stays in Hawaii. Sven has to move all his belongings back to Holland.

12

The Adaptive Value of Survival Processing in Childhood Trauma Victims

LINSEY RAYMAEKERS, HENRY OTGAAR, MAARTEN J. V. PETERS, AND TOM SMEETS

Prior studies have repeatedly shown that rating words for survival relevance can lead to superior subsequent recall in an incidental learning paradigm compared with several control conditions (e.g., moving scenario, pleasantness ratings, etc.). This finding, labeled the survival recall advantage, was first demonstrated by Nairne, Thompson, and Pandeirada in 2007. In their seminal study (Nairne et al., 2007), undergraduates were asked to rate 30 unrelated words for their relevance to a survival scenario (i.e., finding steady supplies of food and water and protecting oneself from predators when stranded on a foreign grassland) on a 5-point scale. This scenario was compared with two control conditions, namely a moving scenario (i.e., locating and purchasing a home and transporting one's belongings) or having participants rate the words for pleasantness (i.e., a standard deep processing control task). The authors found that, compared to both control conditions, processing words according to the survival scenario led to superior retention of the presented, unrelated words during a later surprise free recall task.

Numerous replications (e.g., Kang, McDermott, & Cohen, 2008; Kostic, McFarlan, & Cleary, 2012; Nairne & Pandeirada, 2008a; Nairne, Pandeirada, & Thompson, 2008; Otgaar et al., 2011; Smeets, Otgaar, Raymaekers, Peters, & Merckelbach, 2012) have underscored the apparent robustness of the survival recall advantage (but see Howe & Derbish, 2010; Klein, 2012; Otgaar & Smeets, 2010; Seamon et al., 2012). Moreover, this effect has also been shown to generalize to stimuli other than words, such as pictures (Nairne, VanArsdall, Pandeirada, & Blunt, 2012; Otgaar, Smeets, & van Bergen, 2010; but see Savine, Scullin, & Roediger, 2011). Finally, recent studies have shown that even children show superior memorability when information has to be processed in terms of its fitness value (Aslan & Bäuml, 2012; Otgaar & Smeets, 2010; Otgaar, Howe, Smeets, & Garner, submitted; see Otgaar, Howe, Smeets, Raymaekers, & van Beers, Chapter 12 in this volume). These studies demonstrate that the survival recall advantage is robust not only across a number of dimensions, but also across different samples of participants.

The robust survival recall advantage is commonly explained by referring to evolutionary psychological theories in which the human mind is conceptualized as a product of an evolutionary process reflecting the needs of our hunter-gatherer ancestors (e.g., Nairne & Pandeirada, 2008b). More specifically, according to this evolutionary interpretation, our memory system should be especially "crafted" to help us remember certain kinds of information better than others (e.g., Tooby & Cosmides, 2005). For example, Weinstein, Bugg, and Roediger (2008) reported findings consistent with an evolutionary account. These authors compared the evolutionary processing account (i.e., grasslands survival scenario) to the schematic processing account (i.e., city survival scenario) and found that the grasslands scenario produced better recall than did the modern, urban scenario that required a similar level of schematic processing (see also Nairne & Pandeirada, 2010).

Lately, however, several authors have questioned the evolutionary interpretation of the survival recall advantage. It has been suggested that the mnemonic advantages typically produced by survival processing are not the net result of evolutionary tuning *per se*, yet are caused by proximate mechanisms known to reliably propel memory performance (e.g., Butler, Kang, & Roediger, 2009; Howe & Otgaar, 2013; Kroneisen & Erdfelder, 2011; Soderstrom & McCabe, 2011). For example, Kroneisen and Erdfelder (2011) have argued that the survival recall advantage results from enhanced levels of richness and distinctiveness with which the information is processed (i.e., elaboration; see also Howe & Derbish, Chapter 5 in this volume). Others have suggested that the effect can, at least partly, be explained by the "fit" (i.e., congruity) between target items and processing (Butler et al., 2009; but see Nairne & Pandeirada, 2011). Furthermore, Burns, Burns, and Hwang (2011) found convincing evidence that item-specific and relational processing underlie the recall improvement resulting from survival processing.

Yet another potential explanation for the survival recall advantage involves self-referential processing. The idea is that when people relate to-be-remembered information to themselves, retention is improved. Indeed, previous research found that information related to the self is better encoded than other types of information (Rogers, Kuiper, & Kirker, 1999). In the Rogers et al. study, participants rated 40 adjectives on one of four tasks, each task representing a specific type of encoding. These tasks involved structural encoding (i.e., "Was the adjective presented in the same or in a bigger size type as the question?"), phonemic encoding (i.e., "Does the adjective rhyme with xxx"?), semantic encoding (i.e., "Does the adjective mean the same as xxx?"), or self-reference encoding (i.e., "Does the adjective describe you?"), with participants being instructed to answer each question with a simple "yes" or "no." On a subsequent surprise free recall task, the authors found superior recall for adjectives rated under the self-reference task relative to the other conditions. This finding indicates that the use of self during the encoding of words generates an elaborate, deep memory trace.

Given the superior effectiveness of self-referential processing over other mnemonic strategies (Rogers et al., 1999; see also Symons & Johnson, 1997, for a review) and the apparently robust survival recall advantage effect clearly holding a link with the self-referential processing framework, it is not farfetched to assume that self-referential processing causes the superior retention resulting from survival processing. Nairne et al. (2007, Study 4) investigated this issue by comparing survival processing with a self-referential processing task in which participants were instructed to rate how easily each presented word brought to mind an important personal experience (see Klein, Loftus, & Burton, 1989, for more information on the autobiographical self-referential processing). Results from Nairne et al. (2007, Experiment 4) showed a 16 percent recall advantage of survival processing over self-referent processing, suggesting that survival processing is superior in boosting recall relative to a standard self-referential processing task (see also Nairne et al., 2008).

It is important to note that even though Nairne et al. (2007) found the superior effect of survival processing over self-referential processing, it is not clear whether the authors *succeeded* in inducing self-referential processing to its fullest extent. Indeed, a recent paper by Klein (2012) shows that when using instructions explicitly requesting episodic memory retrieval, thereby manipulating self-referential processing, recall performance was statistically equivalent to that found with survival processing tasks. It was therefore concluded that the unawareness to ensure that task instructions require episodic retrieval is likely the cause of the preceding failures to find comparable recall with self-referential processing (for a theoretical consensus on self-referential processing typically involving episodic memory retrieval, see, for example, Conway & Pleydell-Pearce, 2000; Craik et al., 1999).

The purpose of the present chapter is twofold. First, to further attest to the generalizability of the survival recall advantage across different samples, we present two studies designed to investigate whether the survival recall advantage also occurs in people with a history of childhood trauma (childhood sexual abuse [CSA] and childhood parental loss [CPL]; Experiments 1 and 2). Second, our goal was to examine whether the use of an "abuse" scenario would target self-referential processing in a sample of people with a history of childhood trauma, which was expected to result in a recall advantage comparable to or even greater than the standard survival recall advantage (Experiment 1).

Though it is tempting to assume that *all* people would be relatively equally sensitive to the survival-related cues, one could speculate that certain people might be more "tuned" to preferentially process and retain fitness-relevant information than others. If that were the case, one could conclude that memory performance is shaped primarily by current selection pressures or those operating throughout development instead of ancestral priorities, a finding that would counter the evolutionary interpretation. Note, however, that the act of remembering would still be functional and goal-directed (Nairne, Pandeirada, Gregory, & VanArsdall, 2009; Nairne, Vasconcelos, & Pandeirada, 2011). Thus, examining survival processing among trauma-exposed individuals could shed new light on how previous "survival experiences" might mediate the survival recall advantage.

With respect to the way self-referential processing is addressed, one might argue that rating how easily a word brings to mind an important personal experience (e.g., Nairne et al., 2007) might not be as effective in boosting recall as, for example, having one retrieve an autobiographical memory in response to word presentation. Therefore, the current study elaborated on the consideration of self-referential processing by approaching it in a somewhat different manner. Namely, to ensure that participants rated the relevance of the words to personal survival, we developed an abuse scenario focusing on one's own personal abuse history. So, to test whether self-referential processing would lead to a comparable recall advantage (i.e., a recall advantage over control conditions that is comparable to the one resulting from survival processing) in people with a trauma history, an abuse scenario referring to one's own CSA history was created (Experiment 1). Here, participants reporting a history of CSA were asked to think back about the time they were abused, to focus on a single event, and to image the circumstances around it (i.e., who was with you, where did the event occur?). Controls received a trauma script in which they had to *imagine* being abused as a child.

As it is not unreasonable to assume that people who found themselves in a survival condition in the past would process information more efficiently in a survival setting than in a more neutral setting, these people (i.e., people with a history of CSA or CPL) were expected to show a larger survival recall advantage than people who have never before been in a survival situation (Experiments 1 and 2). To rule out the possibility that self-referential instead of survival processing specifically enhances processing, an abuse script was added to the procedure (Experiment 1; cf. supra). It was expected that by rating the relevance of the words according to an abuse scenario, self-referential processing in people with a history of CSA would be ensured, possibly leading to a recall advantage comparable to the standard survival-related recall advantage.

EXPERIMENT 1

Method

Participants

Participants with a history of CSA were recruited through advertisements in local newspapers as part of a larger research project. The advertisements invited people to participate in our research when they had a history of CSA, defined as the presence of "hand/mouth to genital" and "genital to genital" contact. To check whether these trauma victims had experienced their CSA history as traumatic, we asked them to rate the traumatic impact of their CSA on an 8-point Likert scale (1 = very low traumatic impact; 8 = very high traumatic impact). On average, people rated it as 7.30 (SD = 1.26), a high to very high traumatic impact. Controls were recruited by means of separate advertisements and flyers distributed in several public places. At time of enlistment, people were asked to answer some brief questions to check for the presence of a trauma history. Only people who never endured traumatic stress were invited to come to our lab and participate as a control subject. All participants received a small financial reimbursement for their participation. The study was approved by the Standing Ethical Committee of the Faculty of Psychology and Neuroscience, Maastricht University.

The final sample consisted of 78 participants: 39 (29 women) in the control group (M_{age} = 41.28 years, SD = 13.56) and 39 (34 women) in the CSA group (M_{age} = 44.31 years, SD = 9.74). The groups did not differ significantly in age or education level as measured on an 8-point scale (ps >.05). Moreover, there was no significant group difference in the proportion of female versus male participants, χ^2 (1) = 2.06, p = .13.

Design and Materials

Our study followed a 2 (group: control group vs. CSA group) × 3 (condition: survival script vs. abuse script vs. pleasantness) between-subjects design. Participants in the control and CSA groups were randomly assigned to one of the three conditions (for each condition, $n = 26$) and instructed to rate the 30 target words accordingly.

Survival processing paradigm. A Dutch translation of the stimulus materials employed in previous studies by Nairne et al. (e.g., Nairne et al., 2007) was used (see also Otgaar et al., 2011). The word list comprised 30 words from 30 unique categories (Van Overschelde, Rawson, & Dunlosky, 2004). To familiarize participants with the task, they received two practice words in advance. The instructions for each condition were as follows:

> *Survival scenario.* "In this task, we would like you to imagine that you were stranded in the grasslands of a foreign land, without any basic materials. Over the next few months, you'll need to find steady supplies of food and water and protect yourself from predators. We are going to show you some words, and we would like you to rate how relevant each word would be for you in this survival scenario. Some of the words may be relevant whereas other may not, it's up to you to decide."
>
> *Abuse scenario (CSA group).* "In this task, we would like you to think back on the time that you were sexually abused as a child. We would like you to think back on the experience—in case there were multiple experiences, try to focus on a single event—and the circumstances surrounding that experience (e.g., Where did it take place? Who was there with you?). We are now going to show you some words, and we would like you to rate how relevant each word would be for you in this abuse scenario. Some of the words may be relevant whereas others may not, it's up to you to decide."
>
> *Abuse scenario (controls).* "In this task, we would like you to imagine that you were sexually abused as a child. You are in no position to defend yourself. The next few months, you will have to be able to protect yourself from these kinds of abusive acts. We are now going to show you some words, and we would like you to rate how relevant each word would be for you in this abuse condition. Some of the words may be relevant whereas others may not, it's up to you to decide."
>
> *Pleasantness.* "In this task, we are going to show you some words, and we would like you to rate the pleasantness of each word. Some of the words may be pleasant whereas others may not, it's up to you to decide."

Procedure

After entering the laboratory, participants read and signed the informed consent. Participants with a trauma history were then asked to rate the impact of their CSA experiences. Hereafter, participants in both groups were randomly assigned to one of three conditions and were instructed to rate the 30 target words accordingly.

After receiving these instructions (see above), participants had to rate the relevance of the words for the imagined scenario on a 5-point scale (1 = "totally irrelevant"; 5 = "extremely relevant"). Words were sequentially presented on a computer screen in the same random order, each word lasting for 5 seconds, and participants were instructed to respond as fast as possible, preferably within the 5-second presentation interval. No information was provided about a later memory task. Subsequently, participants played Tetris for 2 minutes, serving as a distracter task. Finally, they received a surprise free recall task that lasted approximately 10 minutes in which they were asked to write down all the words they had encountered during the task.

Results and Discussion

Analyses of variance (ANOVA) were conducted with respect to correct recall. Figure 12.1 shows the mean proportion of correctly recalled words on the surprise recall task per group. There was no group-by-condition interaction, $F(2,72) = 1.83$, $p > .15$, $\eta_p^2 = 0.05$, nor a main effect of group, $F(1,74) = 2.81$, $p = .10$, $\eta_p^2 = 0.04$. In contrast, a significant main effect of condition was found, $F(2,74) = 5.55$, $p < .01$, $\eta_p^2 = 0.13$, with survival processing instructions yielding the best subsequent retention. That is, a Tukey honestly significant difference test showed that recall was significantly better in the survival condition compared to the abuse ($p < .01$) and pleasantness ($p = .05$) conditions. Also, rating words for pleasantness led to marginally better subsequent recall than rating words according to an abuse scenario ($p = .05$).

A second ANOVA was conducted with regard to extra-list intrusions. Figure 12.2 shows extra-list intrusions (i.e., false recall) retrieved on the surprise recall task per group (for the relevance of examining false recall in relation to the survival effect, see Howe, 2011). No interaction between group and condition, $F(2,72) = 0.50$, $p = .61$, $\eta_p^2 = 0.01$, or main effects of group ($F(1,72) = 1.33$, $p = .25$, $\eta_p^2 = 0.02$) or condition ($F(2,72) = 2.37$, $p = 0.10$, $\eta_p^2 = 0.06$) were found.

Table 12.1 presents the mean relevance ratings for the three incidental-learning conditions (survival vs. abuse vs. pleasantness) per group. When conducting analyses on the rating data, no main effect of group, $F(1,72) = 0.31$, $p > .50$, $\eta_p^2 = 0.09$, or a group-by-condition interaction effect, $F(2,72) = 0.88$, $p > .40$, $\eta_p^2 = 0.02$, emerged.

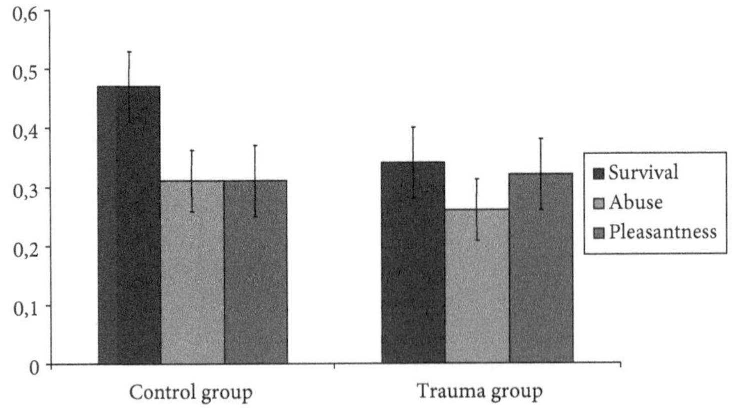

Figure 12.1 Experiment 1. Average proportion correct recall during the surprise free recall task sorted by condition and displayed separately for the control group (i.e., no history of childhood sexual abuse) and the trauma group (i.e., a history of childhood sexual abuse).

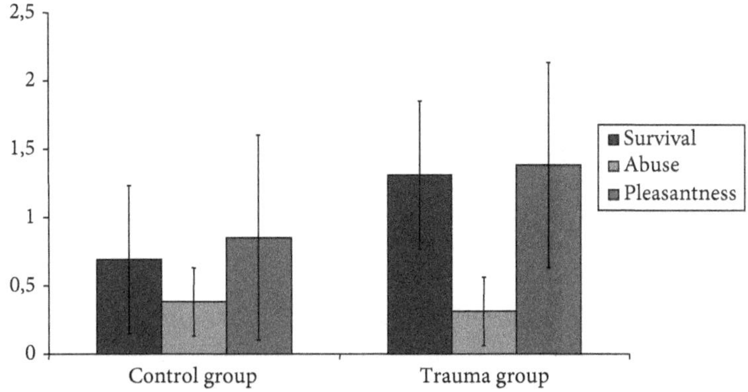

Figure 12.2 Experiment 1. Average false recall (i.e., extra-list intrusions) during the surprise recall task sorted by condition and displayed separately for the control group (i.e., no history of childhood sexual abuse) and the trauma group (i.e., a history of childhood sexual abuse).

Table 12.1. Mean Ratings for the Three Incidental-Learning Conditions Displayed per Group

	Control Group	**Trauma Group**
Survival condition	2.70 ± 0.48	2.82 ± 0.40
Abuse condition	1.94 ± 0.70	1.60 ± 0.73
Pleasantness condition	3.07 ± 0.81	3.04 ± 0.58

Data represent mean ± SEM.

There was, however, a significant effect of condition, $F(2,72) = 28.66$, $p <.001$, $\eta_p^2 = 0.44$, with participants rating the relevance of the words to the survival and pleasantness scenarios as more relevant than participants who rated the relevance of the words with respect to the abuse scenarios (see Table 12.1). To be more precise, a Tukey honestly significant difference test showed that significantly lower ratings were provided with respect to the abuse scenario than to the survival scenario ($p <.001$) and the pleasantness scenario ($p <.001$). Rating words according to the survival and pleasantness script yielded no significant difference in rating scores ($p = .23$). To investigate whether the rating data affected the survival recall advantage, we performed an ANCOVA on the correct recall data (for a similar procedure see Otgaar & Smeets, 2010), including mean rating data as a covariate. The covariate failed to reach significance ($p = .23$). Additionally, an ANCOVA was performed on recall of extra-list intrusions with rating included as a covariate. Like correct recall, this analysis showed that the ratings did not affect reporting extra-list items ($p > .10$).

Taken together, the results of Experiment 1 are generally consistent with initial findings reported by Nairne et al. (2007) in that survival processing led to higher recall rates than pleasantness and abuse processing. This effect was apparent in the control group as well as the trauma group (i.e., participants with a CSA history). To our knowledge, this is the first experiment illustrating the typical survival recall advantage in trauma victims. Moreover, in contrast to previous research by

Howe and Derbish (2010) and Otgaar and Smeets (2010), no differences in false recall between groups or scenarios were found. Also, the obtained differences in relevance ratings for the various scenarios did not significantly influence the pattern of our results, which is well in line with previous data showing that average ratings are unlikely to explain the survival processing advantages in recall (see, for example, Nairne et al., 2007). Finally, the findings of Experiment 1 seem to argue against the self-referential encoding effect in that the abuse scenario in the CSA group did not lead to recall rates that were comparable to those found following survival processing. So, although previous research (Rogers et al., 1999) has shown that information relating to the self is preferentially encoded and organized above all other types of information (i.e., the self-reference effect), we found no evidence for this effect in the current survival processing paradigm. Alternatively, one might argue that the inability to demonstrate a recall advantage in the abuse conditions is driven by their focusing on past experiences rather than on the future. Indeed, recent findings have suggested that the future-directed temporal orientation of the survival instructions (i.e., a planning component) is the key component mediating the high levels of recall found with survival processing (Klein, Robertson, & Delton, 2011). In the present study, the survival scenario differs from the abuse scenarios in that it endorses a form of priming about the future (i.e., rating words useful to future survival; "over the next few months"), whereas the abuse scenarios are more focused on the past (i.e., rating words relevant to survival while thinking back at the abusive past [CSA group], or rating words relevant to survival while imaging that you were an abuse victim [control group]). All in all, in showing no memory advantages for the abuse scenarios, our data accord well with the idea that future episodic memory may play an important role in the adaptive memory effect (see also Klein, Robertson, & Delton, 2010).

One might argue that the present study was hampered by low statistical power causing a failure to detect an interaction between our main variables of interest, namely group and condition. Also, in calling attention to the generalizability of these results, a second experiment was conducted to investigate survival processing in a different trauma population, namely undergraduates with a CPL history. If the findings of Experiment 1 (i.e., a typical survival recall advantage in a trauma history group rather than an amplified advantage) held up, participants with a CPL history were expected to show a survival recall advantage comparable to that of a trauma-free control group.

EXPERIMENT 2

Method

Participants

Participants were 77 undergraduate students (M_{age} = 21.07, *SD* = 2.56; 25 men) enrolled at Maastricht University. All participants signed an informed consent and participated in exchange for course credit. The sample consisted of 37 (16 men) participants with a CPL history (M_{age} = 21.80 years, *SD* = 2.77) and 40 (9 men) controls (M_{age} = 20.11 years, *SD* = 1.94). CPL was defined as loss of a parent through permanent separation or death before the age of 12. The study was approved by the Standing Ethical Committee of the Faculty of Psychology and Neuroscience, Maastricht University.

Design and Materials

Our study followed a 2 (group: CPL group vs. control group) × 2 (condition: survival vs. pleasantness) between-subjects design. Participants in the control and CPL groups were randomly assigned to one of the two conditions and instructed to rate the 30 target words accordingly (survival condition: $n = 39$; pleasantness condition: $n = 38$).

Survival processing paradigm. A Dutch translation of the stimulus materials employed in previous studies by Nairne et al. (e.g., Nairne et al., 2007) was used (see also Otgaar et al., 2010). The word list comprised 30 words from 30 unique categories (van Overschelde et al., 2004). To familiarize participants with the task, they received two practice words in advance.

Procedure

After entering the laboratory, participants read and signed the informed consent. Hereafter, participants in both groups were randomly assigned to one of two conditions (i.e., survival vs. pleasantness) and were instructed to rate the 30 target words accordingly. The instructions were identical to the survival and pleasantness instructions used in Experiment 1. Words were sequentially presented on a computer screen in the same random order, each word lasting for 5 seconds, and participants were instructed to respond within the 5-second presentation interval. No information was provided about a later memory task. Subsequently, participants played Tetris for 2 minutes, which served as a distracter task. Finally, they received a surprise free recall task that lasted approximately 10 minutes in which they were asked to write down all the words they had encountered during the rating task.

Results and Discussion

Figure 12.3 shows the mean proportion of correct recall per group.[1] ANOVA showed that the interaction effect between group and condition fell short of significance, $F(1,73) = 0.01$, $p = .94$, $\eta_p^2 < 0.01$. A significant main effect of condition was found, $F(1,73) = 7.61$, $p < .01$, $\eta_p^2 = 0.10$, with survival processing ($M = 0.47$, $SD = 0.10$) yielding superior subsequent retention relative to processing words according to a pleasantness condition ($M = 0.41$, $SD = 0.10$). Also, a main effect of group was found, $F(1,73) = 4.62$, $p < .05$, $\eta_p^2 = 0.06$. People reporting a history of CPL recalled significantly fewer words than controls ($M = 0.41$, $SD = 0.11$ vs. $M = 0.46$, $SD = 0.10$).

Figure 12.4 shows the mean proportion of false recall (i.e., extra-list intrusions) retrieved on the surprise recall task per group. ANOVA yielded no group-by-condition interaction effect, $F(1,73) = 0.16$, $p = .69$, $\eta_p^2 < 0.01$. A significant main effect of condition was found, $F(1,73) = 4.82$, $p < .05$, $\eta_p^2 = 0.06$, with survival processing resulting in enhanced recall of extra-list intrusions relative to rating words for pleasantness. No significant main effect of group was found $F(1,73) = 0.35$, $p = .56$, $\eta_p^2 = 0.06$.

Table 12.2 presents the mean ratings for the two incidental-learning conditions (survival vs. pleasantness) per group (CPL group vs. control group). When conducting analyses on the rating data, a significant effect of condition emerged,

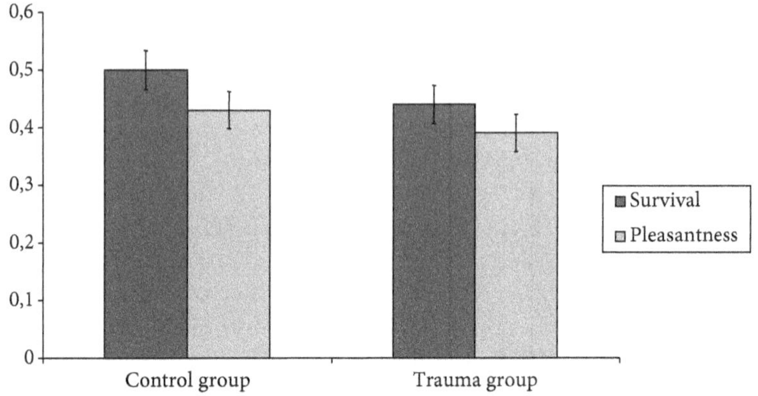

Figure 12.3 Experiment 2. Average proportion correct recall during the surprise free recall task sorted by condition and displayed separately for the control group (i.e., no history of childhood parental loss) and the trauma group (i.e., a history of childhood parental loss).

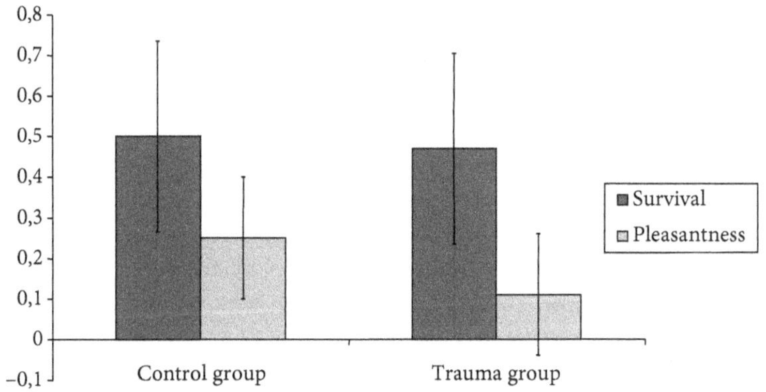

Figure 12.4 Experiment 2. Average false recall (i.e., extra-list intrusions) during the surprise recall task sorted by condition and displayed separately for the control group (i.e., no childhood parental loss) and the trauma group (i.e., a history of childhood parental loss).

Table 12.2. Mean Ratings for the Two Incidental-Learning Conditions Displayed per Group

	Control Group	**Trauma Group**
Survival condition	2.66 ± 0.46	2.51 ± 0.46
Pleasantness condition	2.83 ± 0.67	3.01 ± 0.55

Data represent mean ± SEM.

$F(1,75) = 6.99$, $p = .01$, $\eta_p^2 = 0.09$, with pleasantness ratings ($M = 2.91$, $SD = 0.62$) being higher than ratings provided in relation to the survival scenario ($M = 2.58$, $SD = 0.46$). No main effect of group was found, $F(1,75) = 0.01$, $p > .90$, $\eta_p^2 < 0.01$, and also the group-by-condition interaction remained nonsignificant, $F(1,75) = 1.72$, $p > .15$, $\eta_p^2 = 0.02$. It would be counterintuitive to assume that the lower ratings in the survival processing group led to enhanced recall of the words relative to the pleasantness scenario. Nevertheless, as per Experiment 1, we performed an ANCOVA on the correct recall data, including mean rating data as a covariate. Similar to in Experiment 1, the covariate failed to reach significance ($p = .85$).

Taken together, the findings from Experiment 2 are well in line with those of Experiment 1 in that the standard survival recall advantage appeared in controls as well as people with a trauma history. Again, we found no interaction between group and condition, suggesting that people with a trauma history are equally, not more, prone to the mnemonic benefits of survival processing. Thus, the idea that people with a trauma history are more disposed to show a superior survival recall advantage compared with nontraumatized controls was not supported. Analyses concerning recall of extra-list intrusions showed that survival processing resulted in increased recall of extra-list intrusions relative to the pleasantness control condition. These findings are in line with previous data (see for example Otgaar & Smeets, 2010). Note that some authors argue that such memory inaccuracies (i.e., an increased false memory propensity) may also reflect adaptive significance (for examples and explanations, see Howe & Derbish, 2010; see also Toglia, Neuschatz, & Goodwin, 1999). Moreover, the relevance rating of Experiment 2, much like those in Experiment 1 and previous studies, were shown not to have influenced the survival processing advantages in subsequent recall (see for example Nairne et al., 2007).

GENERAL DISCUSSION

The main findings of Experiments 1 and 2 can be summarized as follows. The current data are generally consistent with results initially reported by Nairne et al. (2007). That is, survival processing led to higher recall rates during a surprise free recall task than other processing conditions such as pleasantness (Experiments 1 and 2) and abuse-related processing (Experiment 1). This effect was apparent in the control as well as the trauma (i.e., victims with a history of CSA and CPL) groups. In other words, the current study failed to show a heightened survival recall advantage for people with a history of childhood trauma. Moreover, although previous research (Rogers et al., 1999) has shown that information relating to the self is preferentially encoded and organized above all other types of information (i.e., the self-reference effect), we found no evidence that this effect is relevant to the typical survival recall advantage. Specifically, abuse-relevant processing instructions (i.e., a script developed to induce self-referential processing) did not lead to superior recall during the surprise free recall task in people with a history of CSA.

Although many studies seem to indicate that the survival recall advantage results from ancestral priorities (e.g., Nairne & Pandeirada, 2008b, 2010; Weinstein et al., 2008), the memory functionality principle does not exclude the possibility that memory might be shaped primarily by current or developmental selection pressures (e.g., Nairne et al., 2009). This idea had led to the speculation that people with a history of childhood trauma (i.e., "trauma victims") may be even more susceptible to the survival processing effect (i.e., showing enhanced survival recall advantage) than nontraumatized controls, a hypothesis that was not supported by the present findings. The current findings seem to underscore previous research failing to illustrate the existence of special memory mechanisms that are unique to experiencing traumatic events (see, for an overview, McNally, 2003; Shobe & Kihlstrom, 1997). In line with this, Howe et al. (2004) demonstrated that basic memory processes do not tend to differ as a function of child maltreatment (see also Eisen, Qin, Goodman, & Davis, 2002; Goodman, Bottoms, Rudy, Davis, & Schwartz-Kenny, 2001). Specifically, the development of true and false memories in maltreated and non-maltreated children was examined using a standardized Deese-Roediger-McDermott (DRM) paradigm (Deese, 1959; Roediger & McDermott, 1995). Three age groups were studied: children aged 5 to 7, children aged 8 to 9, and children aged 10 to 12. It was shown that both true and false memories tended to increase with age in both maltreated and non-maltreated children. Moreover, both maltreated and non-maltreated children exhibited false memories, illustrating that maltreated children were just as susceptible to false memory effects as their non-maltreated counterparts. Also, the authors reported that there were remarkable similarities in false recall and recognition performance for both maltreated versus non-maltreated children; this appears to be consistent with research that has looked at other basic memory processes in these groups of children (e.g., Eisen et al., 2002). In a more recent paper by Howe and colleagues (2011), DRM performance of maltreated and non-maltreated children aged 6 to 12 years old was compared with respect to the ability to inhibit true and false memories for neutral and emotional material (Howe, Toth, & Cichetti, 2011). Again, the authors found that both groups of children, regardless of age, could inhibit the output of true and false emotional information and, although to a lesser degree, the output of true and false neutral material. The current data add to the body of literature illustrating an absence of memory differences between maltreated and non-maltreated children and add an emotional component to this line of research.

In line with previous findings reported by Nairne et al. (2007, 2008) and in contrast to findings reported by Klein (2012), Experiment 1 showed that the self-referential aspect of the survival task is not a likely candidate to explain the mnemonic advantage of survival processing. One could argue that the current experiment failed to induce self-referential processing. On the other hand, as participants rated the words as significantly less relevant in relation to the abuse scenario than relevance ratings in the other scripts, one could however also argue that deep, thoughtful encoding (i.e., self-referential processing) did occur, although this latter explanation seems rather doubtful.

An alternative explanation for the failure to demonstrate a recall advantage in the abuse conditions focuses on the future-directed temporal orientation of the survival instructions versus the past-directed temporal orientation of the abuse instructions. Specifically, in the current study, the survival scenario differs from the abuse scenarios in that it primes participants about the future, whereas the abuse scenarios do not. As recent findings by Klein et al. (2010, 2011) have suggested that the future-oriented nature of the survival instructions forms the key component mediating the high levels of recall found with survival processing, one would not expect to see memory differences in the abuse scenarios. Although this explanation sounds appealing, it seems to argue against the idea that imagining the future and remembering the past entail the same neural processes (e.g., Schacter, 2012).

As to the limitations of the present paper, it should be mentioned that it was assumed that a history of CSA or CPL matches a standard survival condition in some aspects. We, however, did not include a measure questioning this issue. Furthermore, we asked people with a history of CSA to rate words according to an abuse scenario. It was expected that this experimental setup would induce self-referential processing in people with such a trauma history. Given that our participants showed the lowest recall and relevance ratings in the abuse condition compared to the survival and the pleasantness conditions, it could be argued that our self-referential induction was unsuccessful. Another limitation is concerned with the potential high level of variance accompanying the abuse scenario in which trauma victims had to think back about their abusive past. That is, a CSA history tends to differ between victims in terms of nature, relation to the perpetrator, level of memorability, etc. Such differences might have compromised the uniformity of the control script, constituting another factor in which the survival scenario might be considered to diverge from the abuse scenario. With respect to the abuse scenario for control subjects, the potential differential ability to imagine such fictitious abuse events might have also contributed to differences between this script and the standard survival script. Lastly, Experiment 1 involved a nonrandom sample of people with a CSA history who responded to advertisements and might not have been representative of *all* individuals with a CSA history.

Future research might focus on using words that are more relevant to an abuse scenario to more thoroughly examine the effect of self-referential processing in people with a history of CSA. Furthermore, studying survival processing in people who have found themselves in survival conditions other than those related to traumatic stress would be interesting. It can be suggested that abuse survivors represent a different population than, for example, earthquake survivors, war veterans, or better yet, jungle survivors.

In sum, past research showed the robust effect of survival processing in several undergraduate samples (e.g., Kang et al., 2008; Nairne & Pandeirada, 2008a, 2008b; Nairne et al., 2007, 2008; Otgaar et al., 2010). Recently, this effect was also demonstrated in children (Aslan & Bäuml, 2012; Otgaar & Smeets, 2010; Otgaar et al., Chapter 11 in this volume). To our knowledge, the present study is the first to illustrate the survival recall effect in childhood trauma victims. Also, the present study was the first to examine the influence of rating words according to an abuse scenario, a new (i.e., not used in previous studies) scenario that could also be viewed as a form of survival and potentially inducing self-referential processing in people with a history of abuse. To conclude, the present data illustrate the adaptive role of the memory system in people with a history of CSA and CPL as well as matched community and undergraduate samples consisting of control subjects without a trauma history. Given that similar survival recall advantage rates emerged in different samples (namely children, undergraduates, a community sample, and different trauma samples), it can be concluded that people are relatively equally susceptible to the survival recall advantage effect. Thus, it appears that the survival effect can be considered such a robust phenomenon that even drastic, negative life events are not of significant influence on the processes underlying the survival effect. A different way of looking at the present data is by concluding that current or developmental selection pressures do not seem to contribute to the survival recall advantage, a conclusion that matches the evolution account. Whatever the case may be, we hope that future studies will unravel the exact principles that guide the robustness of the survival recall advantage.

ACKNOWLEDGMENTS

Linsey Raymaekers was supported by a grant from The Netherlands Organization for Scientific Research (NWO 400 06 024). Henry Otgaar's contribution was supported by a grant from The Netherlands Organization for Scientific Research (016.135.052).

We thank Sophie van der Zee, Christine Bouw, Latifa Abidi, and Natasja Verleysen for their assistance in data collection.

NOTE

1. Two outliers (deviating more than 2 *SDs* from the mean) were eliminated from analyses and replaced by the mean scores.

REFERENCES

Aslan, A., & Bäuml, K.-H. T. (2012). Adaptive memory: Young children show enhanced retention of fitness-related information. *Cognition*, *122*, 118–122.

Burns, D. J., Burns, S. A., & Hwang, A. J. (2011). Adaptive memory: Determining the proximate mechanisms responsible for the memorial advantages of survival processing. *Journal of Experimental Psychology: Learning, Memory, and Cognition*, *37*, 206–218.

Butler, A. C., Kang, S. H. K., & Roediger, H. L. (2009). Congruity effects between materials and processing tasks in the survival processing paradigm. *Journal of Experimental Psychology: Learning, Memory, and Cognition*, *35*, 1477–1486.

Conway, M. A., & Pleydell-Pearce, C. W. (2000). The construction of autobiographical memories in the self-memory system. *Psychological Review*, *107*, 261–288.

Craik, F. I. M., Moroz, T. M., Moscovitch, M., Stuss, D. T., Winocur, G., Tulving, E., & Kapur, S. (1999). In search of the self: A positron emission tomography study. *Psychological Science*, *10*, 26–34.

Deese, J. (1959). On the prediction of occurrence of particular verbal intrusions in immediate recall. *Journal of Experimental Psychology*, *58*, 17–22.

Eisen, M. L., Qin, J., Goodman, G. S., & Davis, S. L. (2002). Memory and suggestibility in maltreated children: Age, stress arousal, dissociation, and psychopathology. *Journal of Experimental Child Psychology*, *83*, 167–212.

Goodman, G. S., Bottoms, B. L., Rudy, L., Davis, S. L., & Schwartz-Kenney, B. M. (2001). Effects of past abuse experiences on children's eyewitness memory. *Law and Human Behavior*, *25*, 269–298.

Howe, M. L. (2011). The adaptive nature of memory and its illusions. *Current Directions in Psychological Science*, *20*, 312–315.

Howe, M. L., Cicchetti, D., Toth, S. L., & Cerrito, B. M. (2004). True and false memories in maltreated children. *Child Development*, *75*, 1402–1417.

Howe, M. L., & Derbish, M. H. (2010). On the susceptibility of adaptive memory to false memory illusions. *Cognition*, *115*, 252–267.

Howe, M. L., & Otgaar, H. (2013). Proximate mechanisms and the development of adaptive memory. *Current Directions in Psychological Science*, *22*, 16-22.

Howe, M. L., Toth, S. L., & Cicchetti, D. (2011). Can maltreated children inhibit true and false memories for emotional information? *Child Development*, *82*, 967–981.

Kang, S. H. K., McDermott, K. B., & Cohen, S. M. (2008). The mnemonic advantage of processing fitness-relevant information. *Memory & Cognition*, *36*, 1151–1156.

Klein, S. B. (2012). A role for self-referential processing in tasks requiring participants to imagine survival on the savannah. *Journal of Experimental Psychology: Learning, Memory, and Cognition, 37*, 206–218.

Klein, S. B., Loftus, J., & Burton, H. (1989). Two self-reference effects: The importance of distinguishing between self-descriptiveness judgements and autobiographical retrieval in self-referent encoding. *Journal of Personality and Social Psychology, 56*, 853–865.

Klein, S., Robertson, T. E., & Delton, A.W. (2011). The future-orientation of memory: Planning as a key component mediating the high levels of recall found with survival processing. *Memory, 19*, 121–139.

Klein, S. B., Robertson, T. E., & Delton, A. W. (2010). Facing the future: Memory as an evolved system for planning future acts. *Memory & Cognition, 38*, 13–22.

Kostic, B., McFarlan, C.C., & Cleary, A.M. (2012). Extensions of the survival advantage in memory: Examining the role of ancestral context and implied social isolation. *Journal of Experimental Psychology: Learning, Memory, and Cognition, 38*, 1091–1098.

Kroneisen, M., & Erdfelder, E. (2011). On the plasticity of the survival processing effect. *Journal of Experimental Psychology: Learning, Memory, and Cognition, 37*, 1553–1562.

McNally, R. J. (2003). *Remembering trauma*. Cambridge, MA: Belknap Press/Harvard University Press.

Nairne, J. S., & Pandeirada, J. N. S. (2008a). Adaptive memory: Is survival processing special? *Journal of Memory and Language, 59*, 377–385.

Nairne, J. S., & Pandeirada, J. N. S. (2008b). Adaptive memory: Remembering with a stone-age brain. *Current Directions in Psychological Science, 17*, 239–243.

Nairne, J. S., & Pandeirada, J. N. S. (2010). Adaptive memory: Ancestral priorities and the mnemonic value of survival processing. *Cognitive Psychology, 61*, 1–22.

Nairne, J. S., & Pandeirada, J. N. S. (2011). Congruity effects in the survival processing paradigm. *Journal of Experimental Psychology: Learning, Memory, and Cognition, 37*, 539–549.

Nairne, J. S., Pandeirada, J. N. S., Gregory, K. J., & VanArsdall, J. E. (2009). Adaptive memory: Fitness relevance and the hunter-gatherer mind. *Psychological Science, 20*, 740–746.

Nairne, J. S., Pandeirada, J. N. S., & Thompson, S. R. (2008). Adaptive memory: The comparative value of survival processing. *Psychological Science, 19*, 176–180.

Nairne, J. S., Thompson, S. R., & Pandeirada, J. N. S. (2007). Adaptive memory: Survival processing enhances retention. *Journal of Experimental Psychology: Learning, Memory, and Cognition, 33*, 263–273.

Nairne, J. S., VanArsdall, J. E., Pandeirada, J. N., & Blunt, J. R. (2012). Adaptive memory: Enhanced location memory after survival processing. *Journal of Experimental Psychology: Learning, Memory, & Cognition, 38*, 495–501.

Nairne, J. S., Vasconcelos, M., & Pandeirada, J. N. S. (2011). Adaptive memory and learning. In N. M. Seel (Ed.), *Encyclopedia of the sciences of learning* (pp. 118–121). New York: Springer.

Otgaar, H., Howe, M., Smeets, T., & Garner, S.R. (submitted). Developmental trends in adaptive memory. Manuscript submitted for publication.

Otgaar, H., &, Smeets, T. (2010). Adaptive memory: Survival processing increases both true and false memory in adults and children. *Journal of Experimental Psychology: Learning, Memory, and Cognition, 36*, 1010–1016.

Otgaar, H., Smeets, T., Merckelbach, H., Jelicic, M., Verschuere, B., Galliot, A., & van Riel, L. (2011). Adaptive memory: stereotype activation is not enough. *Memory and Cognition, 39*, 1033–1041.

Otgaar, H., Smeets, T., & van Bergen, S. (2010). Picturing survival memories: Enhanced memory after fitness-relevant processing occurs for verbal and visual stimuli. *Memory & Cognition, 38*, 23–28.

Roediger, H. L., & McDermott, K. B. (1995). Creating false memories: Remembering words not presented in lists. *Journal of Experimental Psychology: Learning, Memory and Cognition, 24*, 803–814.

Rogers, T. B., Kuiper, N. A, & Kirker, W. S. (1999). Self-reference and the encoding of personal information. In F. Baumeister (Ed.), *The self in social psychology* (pp. 139–149). Philadelphia: Psychology Press.

Savine, A. C., Skullin, M. K., & Roediger, H. L., III (2011). Survival processing of faces. *Memory & Cognition, 39*, 1359–1373.

Schacter, D.L. (2012). Adaptive constructive processes and the future of memory. *American Psychologist, 67*, 603–613.

Seamon, J. G., Bohn, J. M., Coddington, I. E., Ebling, M. C., Grund, E. M., Haring, C. T.,… Siddique, A. H. (2012). Can survival processing enhance story memory? Testing the generalizability of the adaptive memory framework. *Journal of Experimental Psychology: Learning, Memory, and Cognition, 38*, 1045–1056.

Shobe, K. K., & Kihlstrom, J. F. (1997). Is traumatic memory special? *Current Directions in Psychological Science, 6*, 70–74.

Smeets, T., Otgaar, H., Raymaekers, L., Peters, M. J. V., & Merckelbach, H. (2012). Adaptive memory in times of stress. *Psychonomic Bulletin & Review, 19*, 113–118.

Soderstrom, N. C., & McCabe, D. P. (2011). Are survival processing memory advantages based on ancestral priorities? *Psychonomic Bulletin and Review, 18*, 564–569.

Symons, C. S., & Johnson, B. T. (1997). The self-reference effect in memory: A meta-analysis. *Psychological Bulletin, 121*, 371–394.

Toglia, M. P., Neuschatz, J. S., & Goodwin, K. A. (1999). Recall accuracy and illusionary memories: When more is less. *Memory, 7*, 233–256.

Tooby, J., & Cosmides, L. (2005). Conceptual foundations of evolutionary psychology. In D. M. Buss (Ed.), *The handbook of evolutionary psychology* (pp. 5–67). Hoboken, NJ: Wiley.

Van Overschelde, J. P., Rawson, K. A., & Dunlosky, J. (2004). Category norms: An updated and expanded version of the Battig and Montague (1969) norms. *Journal of Memory and Language, 50*, 289–335.

Weinstein, Y., Bugg, J. M., & Roediger, H. L. III (2008). Can the survival recall advantage be explained by basic memory processes? *Memory and Cognition, 36*, 913–919.

13

Recognizing Others: Adaptive Changes to Person Recognition Throughout the Lifespan

CORRINA MAGUINNESS AND FIONA N. NEWELL

INTRODUCTION

Humans are social and we have evolved in close proximity with one another, surrounded by different identities almost everywhere we turn. The accurate perception of others is a fundamental aspect of social cognition, allowing us to detect the intention, attention, and identity of an individual (among other attributes) (Bruce & Young, 1986). Indeed, humans have developed an expertise for face perception, which arguably exceeds the perception of other stimuli: we can remember thousands of faces as "unique." When we consider the influence that this skill has on our behavior it is not difficult to understand its evolutionary significance and how it is likely to have assisted in negotiating our survival, mediating, among other things, approach-avoidance behavior (Engell, Haxby, & Todorov, 2007; Todorov, 2008), mate selection (Little, Jones, & DeBruine, 2011; Rhodes, 2006; Rhodes, Simmons, & Peters, 2005), the recognition of familiar, unrelated others (Bruce, Henderson, Newman, & Burton, 2001; Burton, Wilson, Cowen, & Bruce, 1999; Rossion, Schiltz, & Crommelinck, 2003), and the recognition of kin (DeBruine et al., 2009; Maloney & Dal Martello, 2006). Yet for such a skill to be truly adaptive we must consider that recognizing others is not purely dependent on our memory for faces alone but that other social information, particularly information in the voice, can be used to identify others. Like the face, the voice conveys information that can act as an identity signature and affect recognition and thus can be thought of as an "auditory face" (Belin, Fecteau, & Bédard, 2004).

In this chapter we explore the selection pressures that were likely to have mediated the development of our ability to recognize others. We consider how face recognition reflects a unique cognitive process with genetic underpinnings, highlighting face recognition as a phylogenetic adaptation (i.e., within-species adaptation). Although recent research has highlighted the influence of genes on face recognition abilities, this chapter discusses evidence that face recognition abilities are rapidly acquired and can be influenced by our surrounding environment. This experience-dependent malleability suggests that face recognition is also a product of ontogenetic adaptation (i.e., individual development across the lifespan), which has direct consequences on our ability to discriminate and remember faces. Although most research on face perception has been based on the recognition of faces from static images, here we also discuss research that highlights our ability to recognize faces under naturalistic conditions, such as the face in motion. Finally, we review the neural processes underlying face recognition and how they may share similar neural underpinnings to voice recognition, with the argument that such common processing of face and voice information most likely evolved to support person recognition throughout the lifespan.

FACE RECOGNITION: IT'S IN OUR GENES

Genetic Fitness: Recognizing Faces

Faces are arguably the most dominant visual stimuli in our surroundings and we, in turn, spend a vast amount of our time perceiving faces in order to gather relevant social information. Even when we are not specifically attending to faces they capture our attention (Langton, Law, Burton, & Schweinberger, 2008). Our fascination with faces is not surprising when we consider the array of social signals that the face conveys, including the identity, intent, and attractiveness of an individual. Our perceptual system has evolved with the capacity to efficiently extract this information to assist in interacting with our social world (Leopold & Rhodes, 2010). Indeed, the emergence of such a skill is likely to have facilitated our survival and to have developed in line with one of the most dominant selection pressures, that of sexual selection. Our ability to effectively deduce differences across individual faces has assisted in determining the genetic fitness of mate choices. Indeed, it is argued that across cultures we have evolved a common percept of what facial characteristics are deemed attractive (Little et al., 2011). These biological standards of beauty include sexual dimorphism (femininity cues in female faces or masculinity cues in male faces), bilateral symmetry (how similar each half of the face is to the other), and averageness (facial features that are determined as the mathematical average for a given population) (for reviews see Little et al., 2011; Rhodes, 2006) and are suggested to be expressions of genetic fitness cues including fertility (Johnston, 2000; Johnston, Hagel, Franklin, Fink, & Grammer, 2001; Law Smith et al., 2006) and immunocompetence (Lie, Rhodes, & Simmons, 2008). Figure 13.1 illustrates how averageness can enhance perceived attractiveness.

For example, it has been suggested that visual features indicating femininity in a female face (short, narrow jaw and fuller lips) serve as a cue to fertility throughout the lifespan (Johnston, 2000; Johnston & Franklin, 1993; Law Smith et al., 2006; Perrett et al., 1998). In addition, we rate as both healthier and more attractive the faces of individuals associated with "good quality" immune genes, compared to those who are not (Roberts, Little, Gosling, Perrett, et al., 2005). These genes, which are thought to be important in mate selection, are called major histocompatibility complex (MHC) genes and are associated with immune defense. However, to avoid inbreeding and to enhance the immunocompetency of offspring, it has been

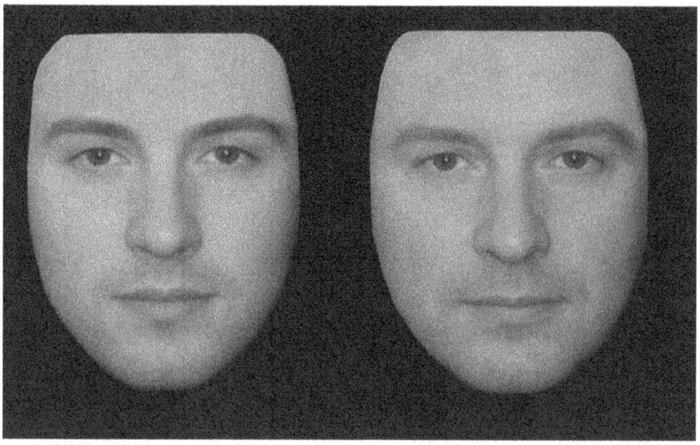

Figure 13.1 Example of average face images, created using a composite of 100 individual male faces. Each image reflects an average face for males in their mid-20s (left) and mid-40s (right). Typically, both images will be deemed more attractive than their individual counterparts.

suggested that we should prefer individuals who exhibit an MHC combination different from our own. For example, often mating patterns are observed to be non-random and reflect either "assortative" (similar genotypes/phenotypes in mating partners) or "disassortative" (dissimilar genotypes/phenotypes in mating partners) choices and can be understood under the genetic compatibility hypothesis (Neff & Pitcher, 2005). However, faces of individuals who possess dissimilar immune genes are not necessarily those that are rated as being highly attractively (Roberts, Little, Gosling, Jones, et al., 2005; Roberts, Little, Gosling, Perrett, et al., 2005), suggesting that disassortative mating is not the only influence on preferences for faces.

However, other perceptual studies demonstrate that we do use facial cues of kinship (familial resemblance) to effectively avoid inbreeding, which may have a negative effect on genetic fitness. For example, when our own face is morphed with an opposite-sex face we tend to find the resulting face less attractive than if it is morphed with a same-sex face. However, participants who share no genetic relatedness to the face images do not show this pattern for the same- and opposite-sex face images (DeBruine, Jones, Little, & Perrett, 2008; DeBruine, 2004b). Moreover, this is particularly true for opposite-sex siblings: for example, women who have brothers find the faces of self-resembling unfamiliar male faces less attractive than women who do not have brothers (DeBruine et al., 2011). However, if there is no identity change in face images we find the faces of male and female "siblings" similarly attractive (Bülthoff & Newell, 2004), suggesting that features that are attractive are not sex-specific (Casey, Mernagh, & Newell, 2009).

Faces that are deemed attractive induce activity in areas of the brain associated with reward (O'Doherty et al., 2003). Attractive faces have also been suggested to be better remembered than their less attractive counterparts (Marzi & Viggiano, 2010; Tsukiura & Cabeza, 2011), but this finding is not always found (Light, Hollander, & Kayra-Stuart, 1981; Wickham & Morris, 2003). Thus, face recognition serves a largely adaptive function, as we can accurately deduce cues from the face that may assist in increasing genetic transmission by producing healthy offspring. Evidence suggests that the perceived attractiveness of a face shares a relationship with sexual behavior, increasing mating success (Rhodes et al., 2005). Although these traits are preferred across cultures, it is important to note that they vary in line with our familiarity to faces. For example, average faces are deemed more attractive when they reflect the average of our own group prototype rather than another group (Potter & Corneille, 2008), and increasing exposure to a face also increases its perceived attractiveness (Peskin & Newell, 2004).

As discussed, face recognition appears to affect mate selection, and in turn face recognition also endows us with the ability to identify our own kin, and we are adept at determining the genetic relatedness of individual faces (DeBruine et al., 2009; Maloney & Dal Martello, 2006). DeBruine (2004a) reported that we tend to find the faces of infants who resemble ourselves as more appealing than those who do not, even when we are unaware that a similarity may exist. Here, DeBruine (2004a) transformed participants' faces to reflect neotenous face proportions and then morphed this face with that of an unfamiliar infant morph, such that the resulting infant face-morph did not appear older through the inclusion of self-related facial information. There is evidence to suggest that this effect may be more evident in male than female participants, as men may rely heavily on facial likeness as a verification of paternity (Platek, Burch, Panyavin, Wasserman, & Gallup, 2002). Other research suggests that the faces of newborn infants also draw attention (Brosch, Sander, & Scherer, 2007). Together these findings propose a largely adaptive function, when we consider that protecting infants is fundamental to reproductive success (Brosch et al., 2007).

Genetic Underpinnings: Recognizing Faces

The abilities discussed above, including kin recognition, are largely harnessed by our capacity to detect often subtle differences between faces, and this is particularly true when determining the unique identity of an individual face. Although the process appears effortless, recognizing and remembering faces as individual is a challenge for the perceptual system. Each face we encounter shares the same basic visual-feature organization as the next, yet we can accurately discriminate between these faces. Furthermore we can recognize the face of a familiar person even over large changes in visual image appearance, such as viewpoint, expression, or illumination (Bruce, 1982; Bruce et al., 2001; Burton et al., 1999) (Fig. 13.2). Recent findings suggest that our capacity for remembering faces is robustly separable from other general cognitive abilities, including measures of immediate and delayed recall (Wilhelm, Herzmann, Kunina, & Sommer, 2007). In addition, face recognition performance appears to follow a unique cognitive trajectory, continuing to improve until just beyond the age of 30 years. In contrast, other cognitive skills associated with person recognition, for example name recognition, cease to show age improvements almost a decade earlier (Germine, Duchaine, & Nakayama, 2011).

Recent research has suggested that the ability to recognize other faces varies across individuals. This variation is inherently dissociable from individual differences in other general cognitive abilities (Herzmann, Kunina, Sommer, & Wilhelm, 2010; Wilhelm et al., 2010). Differences in face recognition performance are so evident across individuals that they have been shown to be *quantitatively* measureable (Russell, Chatterjee, & Nakayama, 2012). For example, recent evidence has suggested

Figure 13.2 Although face-images 1 and 2 share more visual similarity than faces 1 and 3, the perceptual system must become tuned to recognizing the face of the same individual over large changes in visual appearance and therefore recognize faces 1 and 3 as the same facial identity (images from Langner et al., 2010).

the existence of "super-recognizers:" individuals with extraordinary face recognition ability that greatly surpasses average face recognition performance (Russell, Duchaine, & Nakayama, 2009). In fact, the degree of enhanced performance observed in these individuals is quantitatively similar to the degree of deficit expressed in face recognition performance in developmental prosopagnosics (also referred to as *congenital* prosopagnosia) (Russell et al., 2009). Developmental prosopagnosics are individuals who have a specific impairment in face recognition performance that is not attributable to brain trauma or insult (see further discussion below).

Several studies have suggested that rather like the highly heritable and neurobiological correlates that have been observed for general intelligence (for review see Toga & Thompson, 2005), the degree of expertise expressed in face perception may also be largely genetically determined. This genetic component suggests that in humans, face recognition ability may have largely evolved as a phylogenetic adaptation (within-species adaptation) (Johnson, 2011). The most compelling evidence that genetics features as a contributor in face recognition ability comes from the results observed through twin studies (Wilmer et al., 2010; Zhu et al., 2010). What is important about these studies in the current context is that the role of genetics can be studied while also controlling for environmental influences, as twins who are raised together largely share similar environmental factors, the same as nonidentical twins or other siblings. Wilmer et al. (2010) observed that monozygotic (MZ) twins (i.e., developed from one fertilized oocyte) show highly correlated face recognition performance. In fact, the correlation between their performance was more than double the correlation observed for dizygotic (DZ) twins (i.e., developed from different fertilized oocytes) raised in the same environment (correlation 0.70 vs. 0.29, respectively). As the MZ and DZ twin sets both shared similar environmental factors, the observed differences can be assumed to largely reflect genetic underpinnings (Wilmer et al., 2010). However, despite these recent findings that the ability to perceive faces is influenced by genetics, the associated adaptive function of these individual differences is, for the moment, unclear.

Zhu et al. (2010) provided similar evidence for face recognition ability being highly correlated for MZ and less so for DZ twins raised together. Intriguingly, they found no evidence for such a relationship for object recognition abilities across twin sets. They did, however, demonstrate a similar pattern of correlational strength for the "face inversion effect" across twin sets. Image inversion has been shown to disrupt face recognition, yet it is thought to elicit less disruption in the recognition of other objects and is suggested to reflect a method of processing, holistic or configural processing, that is largely specific to face perception (Maurer, Le Grand, & Mondloch, 2002; Yin, 1969; Young, Hellawell, & Hay, 1987). Thus, the effect Zhu et al. (2010) observed for face recognition performance suggests that it reflects the perceptual processes that are recruited specifically for face recognition and not generic low-level processing differences across individuals.

However, it is also important to note that not all of the variance in face recognition performance in the aforementioned twin studies was explained by genetic similarity. The statistical models used in such twin studies (Rabe-Hesketh, Skrondal, & Gjessing, 2008) take into consideration an error measurement that includes the contribution of nonfamilial influences (i.e., the environmental influence unique to each twin). Thus, although the ability to recognize faces may reflect a *unique* and *highly heritable* perceptual ability, environmental influences were also reported to be an important contributor in face recognition performance (a more in-depth discussion of the role of environmental factors in face recognition ability follows later in the chapter). What perhaps is intriguing, however, is that the effect of genetics on face recognition performance was more likely to explain the variance in face recognition performance for the older age group of twins (13–19 years) than for the younger age group (7–12 years) (Zhu et al., 2010). Thus, although environmental factors have a role to play in face recognition ability, the older we become, the more likely it is that variability in facial recognition performance across individuals may relate to familial influence. For example, Burton et al. (1999) demonstrated that even police officers who were practiced in forensic identification (average of 13.5 years of service) showed similar performance in recognizing unfamiliar faces as untrained control participants, suggesting that individual differences in face recognition performance may not reflect "learned" abilities. As such, the greater the genetic similarity between twins, as is the case for MZ twins, the more likely they will share a similar ability to recognize other individuals' faces.

In addition to twin studies, our understanding of the genetic basis of face recognition has also been informed by the study of *developmental* prosopagnosia. Prosopagnosia ("face blindness") refers to a condition first documented by Bodamer in 1947 (see Ellis & Florence, 1990, for translated paper) in which patients demonstrated a distinct deficit in recognizing individual faces, whereas other visual abilities remained intact. Beyn and Knyazeva (1962) described this condition as reflecting a "failure in recognition at its most highly specific and individual level" (p. 158). While the first cases documented largely reflected acquired prosopagnosia, which results from insult to, and dysfunction of, the cortical face processing network (Barton, 2008; Busigny et al., 2010; Rossion et al., 2003; Schiltz et al., 2006), later studies demonstrated the existence of developmental prosopagnosia. Developmental prosopagnosics have atypical face processing, including impairment in recognizing the identity of individual faces, and such deficits emerge in the absence of insult or atypical neurodevelopment (Lee, Duchaine, Wilson, & Nakayama, 2010). Developmental prosopagnosia has been shown to be largely heritable, with multiple cases observed in the same family (Duchaine, Germine, & Nakayama, 2007; Grueter et al., 2007; Lee et al., 2010; Schmalzl, Palermo, & Coltheart, 2008). Such findings may implicate the existence of an evolved neural substrate, largely located in the ventral stream of the cortex, for the recognition of faces (Young et al., 2008), with genetics arguably playing a role in both typical and atypical development.

FACE RECOGNITION: DEVELOPMENTAL AND ENVIRONMENTAL INFLUENCE

That face recognition evolved as a phylogenetic adaptation is also evident from the infant literature, which suggests that newborns exhibit a preference for looking at "face-like" stimuli (Fantz, 1963; Valenza, Simion, Cassia, & Umiltà, 1996) and for visually tracking stimuli that schematically represent a face versus those that do not (Goren, Sarty, & Wu, 1975; Johnson, Dziurawiec, Ellis, & Morton, 1991). Such evidence suggests that the newborn perceptual system may be biased to attend to and process "face-like" stimuli (Johnson, 2005) (Fig. 13.3 shows an example of stimuli). There is also evidence to suggest that even with such an immature visual system, infants show a preference for their mother's face over that of a stranger (Pascalis, de Schonen, Morton, Deruelle, & Fabre-Grenet, 1995); however, this may be mediated by single–feature processing, such as hairstyle (Pascalis et al., 1995).

Although such precocious face perception abilities may reflect phylogenetic adaptation, it is also important to consider that faces are arguably the most dominant stimulus in the infant's environment (Sinha, Balas, & Ostrovsky, 2007). As such, with the multitude of faces in the infant's surrounding world, face information may be rapidly acquired, in turn biasing the perceptual preferences of infants. Indeed, for such a skill to be *truly* adaptive, it must have evolved to accommodate and become tuned to the specific pressures of the surrounding environment. Evidence also suggests that face recognition is malleable to change, showing adaptive changes in line with ontogenetic individual developmental experience (Johnson, 2011). These adaptive changes are evident across the lifespan, being largely mediated by our degree of exposure to different face types. Perhaps two of the best examples of such adaptation are highlighted by the *other-race effect* (for a review, see Meissner & Brigham,

Figure 13.3 Infants prefer to look at visual stimuli that schematically represent a face (left) versus those that do not (right). (Stimuli adapted from Valenza et al., 1996.)

2001) and the *own-age bias* (for a review, see Rhodes & Anastasi, 2011). These influences are discussed in detail in the following two sections.

Face Recognition: The Effect of Target Race

As noted, we are experts at recognizing and remembering faces, and we can remember thousands of faces as unique. Yet, as impressive as this ability is, our memory for faces is also susceptible to a significant degree of error when the face we are to recognize belongs to an ethnic or other-race group with which we are unfamiliar. We perceive individual faces of other-race faces as sharing a large degree of similarity (Vizioli, Rousselet, & Caldara, 2010) and as a consequence exhibit increased "false alarm" rates in memory for other-race faces in comparison with faces of the race by which we are familiar (Meissner & Brigham, 2001). This phenomenon is known in the literature as the "other-race effect" or "own-race bias" (Meissner & Brigham, 2001), and it emerges during the early stages of development through the process of perceptual narrowing (Kelly et al., 2007). For example, 3-month-old infants can discriminate between both own-race and other-race faces. However, by the age of 9 months the infant's perceptual system has "narrowed," and the ability to discriminate between two faces is restricted to own-race faces (Kelly et al., 2007). This own-race bias is mediated by exposure to same-race faces in the infant's environment. This exposure has direct consequences on our understanding of how we represent faces in memory (Valentine, 1991), by forming representations based on a prototype that is most beneficial for interacting within our surrounding social environment. Chiroro, Tredoux, Radaelli, and Meissner (2008) demonstrated how finely tuned this ability is. They observed that white and black South Africans demonstrate an own-race bias *only* for the faces of the ethnic groups with which they have perceptual experience (i.e., South African faces). When presented with faces of white and black Americans, they failed to observe enhanced recognition for "own-race" faces (Chiroro et al., 2008). What this suggests is that the perceptual system is highly sensitive to the degree of variation even within what may be classed as "same-race" faces. This mechanism is likely to be largely adaptive when we consider that our perceptual system becomes tuned to detecting and remembering the faces of those we are surrounded by and in turn more likely to encounter in the future. As such, the false alarm rates for other-race faces can be equated to false memories observed in other memory tasks, emphasizing that due to ontogenetic adaptation, our memory for recognizing faces has most likely evolved to use past information to effectively predict future encounters (Nairne & Pandeirada, 2008a).

Face Recognition: The Effect of Aging and Target Age of Face

As noted, face recognition performance appears to peak at just beyond the age of 30 years (Germine et al., 2011). Subsequently, performance appears to decline as a function of increasing age, with decline evident at around 50 years of age (Bowles et al., 2009). This decrease in face recognition performance appears to be specific to the recognition of faces in comparison to the recognition of other within-category objects (Boutet & Faubert, 2006). What is important to note is that although such changes are evident, they are dissociable from changes in general cognitive ability that occur with the aging process. Throughout the lifespan, even until old age, face recognition appears to represent a unique cognitive ability (Hildebrandt, Wilhelm, Schmiedek, Herzmann, & Sommer, 2011). These deficits in face recognition performance have been suggested to be mediated by a number of factors, including poorer sensory acuity typically associated with the aging process (Barnes, De L'Aune, & Schuchard, 2011; Norton, McBain, & Chen, 2009; Owsley, Sekuler, & Boldt, 1981), changes in how faces are encoded (Chaby, Narme, & George, 2011; Firestone, Turk-Browne, & Ryan, 2007; Gao et al., 2009; Grady et al., 1995; Murray, Halberstadt, & Ruffman, 2010), and marked differences in the underlying neural activity associated with processing faces (Grady, 2002; Haxby et al., 1994; Lee, Grady, Habak, Wilson, & Moscovitch, 2011), including less lateralized activation in face processing areas (Lee et al., 2011).

Perhaps one of the most striking behavioral effects that demonstrate this age-related decline is the higher level of "false alarm" memory responses that older adults report during face recognition tasks compared to younger adults (Bartlett & Leslie, 1986; Bartlett, Leslie, Tubbs, & Fulton, 1989; Rhodes, Castel, & Jacoby, 2008). For example, older adults show a bias for stating that they find a novel face familiar even when they have not encountered the face before. Although such findings may be explained by the underlying changes in sensory acuity and neuronal activity mentioned previously, it is also important to consider that older adults have been exposed to a larger number of faces across the lifespan than younger adults. This larger "database" of faces in memory (or what Valentine [1991] referred to as "face-space") may elicit more false alarm responses. By increasing the number of typical faces in face memory, newly encountered faces may be likely to elicit stronger signals of resemblance to another face within this large face-space, thus leading to false recognition. This may be one key reason why, as we get older, our ability to uniquely identify a face may appear to decline.

However, most studies that have reported a decline in face recognition performance in older compared to younger adults have largely used images of younger adult faces as stimuli. Studies that have used face images of older adults as stimuli have reported that age-related decline in face recognition is less evident for the recognition of older adult faces relative to younger adult faces (Firestone, Turk-Browne & Ryan 2007; Fulton & Bartlett, 1991; Lamont, Stewart-Williams, & Podd, 2005). This effect is reported as the "own-age bias" (for review, see Rhodes & Anastasi, 2011) and is similar to the other-race effect in that it is thought to emerge due to the extensive exposure that an individual has with faces of his or her own age group. This suggests that throughout the lifespan, even until old age, face recognition is malleable to effects such as the degree of exposure to particular faces in our surrounding environment.

Facing Pressure: How Social Information Affects Face Recognition

As illustrated by the own-age bias and other-race effect, face recognition appears to be tailored to facilitate optimal interaction with our social environment. Recent evidence has suggested that "high-level" social information can in turn affect the "low-level" perceptual processing of faces. For example, Bliss-Moreau, Barrett, and Wright (2008) demonstrated that information that assists in determining whether an individual may be helpful or threatening can directly affect how a face is perceived. They observed that unfamiliar faces, displaying a neutral expression, were more likely to be perceived as "negative" or "positive" when they had been respectively paired with statements with negative or positive valence versus neutral statements. This effect was particularly evident when the statements were *social in nature* (i.e., the behavior directly referenced another individual). Furthermore, faces paired with such negative social information have been shown to dominate visual awareness over faces paired with neutral or positive statements (Anderson, Siegel, Bliss-Moreau, & Barrett, 2011). In addition, it appears that such effects may also be evident for our memory for faces. For example, Chiappe et al. (2004) and Mealey, Daood, and Krage (1996) observed that the faces of "cheaters" (faces accompanied with cheating-related information at the learning stage) were subsequently better remembered than faces paired with "trustworthy"/"cooperative" or neutral information. However, others have suggested that the faces of "cheaters" and "cooperative" individuals may be similarly remembered, particularly when we are primed to process the presence of such behaviors (Felisberti & Pavey, 2010). Such findings suggest that without having direct experience with individual people, we use information gathered by others to assist in determining who may be helpful or threatening (Anderson et al., 2011) and that this in turn directly affects our memory and perception for faces.

It may be argued that face perception when influenced by social information may reflect processes related to survival, and as such our memory system may be biased toward remembering fitness-relevant information (Nairne & Pandeirada, 2008a; Nairne, Thompson, & Pandeirada, 2007). Indeed it has been documented that *survival processing* (i.e., processing material, such as words, within the framework of a survival scenario) in comparison to a control scenario enhances recall (Nairne et al., 2007). However, research findings using the classic survival processing paradigm have failed to demonstrate similar survival processing enhancement on our memory for faces (Savine, Scullin, & Roediger, 2011). For example, when Savine et al. (2011) had participants rate faces for perceived helpfulness under primed "survival" (e.g., lost in a desert and in danger of attack from predators) versus nonsurvival scenarios, they failed to observe a relative memory enhancement for faces viewed in the "survival" scenarios. Thus, for the moment it remains unclear whether our memory for faces may be modulated by general survival information (Savine et al., 2011). However, as discussed, many reports suggest that more specific survival-relevant social information can directly affect our perception of and memory for faces (Anderson et al., 2011). Intriguingly, recent evidence suggests that fitness-related information in faces that may affect mate choice (sexual dimorphism in male faces) enhances recall of additionally presented non-face information, such as objects (Allan, Jones, DeBruine, & Smith, 2012).

FACE RECOGNITION: UNDERLYING NEURAL PROCESSES

Many researchers have argued for domain-specificity in face processing, and that the cognitive processes involved in face recognition appear to be distinct and specific for processing faces in comparison with other stimuli (Duchaine, Dingle, Butterworth, & Nakayama, 2004; McKone, Kanwisher, & Duchaine, 2007). This remains controversial, however, because others have suggested that face processing reflects a particular type of perceptual expertise in discriminating between similar visual stimuli, as discussed below. Evidence for domain-specificity largely originated from neuropsychological investigations that reported selective deficits to face processing following damage to particular regions of the temporal lobe of the brain, which are involved in the processing of faces. Subsequent functional magnetic resonance imaging (fMRI) revealed a face module, known as the fusiform face area (FFA), that is located in visual areas of the ventral stream and is selectively activated when viewing images of faces (Kanwisher, McDermott, & Chun, 1997). In particular, this area is activated when viewing face images over images of other non-face objects or scrambled face images (suggesting that the area is not simply responsive to visual qualities of the image [e.g., luminance]). This area is active not only when we are perceiving faces but also when we are imagining them (O'Craven & Kanwisher, 2000), and electrical stimulation of this region specifically distorts the perception of faces over other stimuli (Parvizi et al., 2012). Moreover, the FFA is thought to be part of a larger neural network involved in the processing of faces, including the occipital face area (OFA) and areas in the superior temporal sulcus (STS) (Haxby, Hoffman, & Gobbini, 2000).

Although some have argued that this face-sensitive activation may reflect "domain-specificity" (Kanwisher et al., 1997) for the processing of faces, it is important to note that the FFA has been suggested to also reflect an area recruited for within-category visual discrimination of other non-face stimuli, which requires a similar level of visual "expertise" for individual discrimination (Gauthier, Skudlarski, Gore, & Anderson, 2000). This "expertise hypothesis" (Gauthier & Nelson, 2001; Gauthier et al., 2000) suggests that the observed cortical specialization for faces is a product of perceptual exposure and training in discriminating between similar visual stimuli. For example, it has been shown that individuals who are experts (or who are trained to become experts) in within-category discrimination of non-face stimuli, such as cars or birds, show similar underlying neural activity when perceiving faces or other "expertise" stimuli (Gauthier et al., 2000; Rossion, Gauthier, Goffaux, Tarr, & Crommelinck, 2002). In addition, highly familiar non-face stimuli that require holistic processing due to their multipart visual organization (Bilalic, Langner, Ulrich, & Grodd, 2011) elicit activation in face selective areas. However, this finding remains controversial, and a number of studies have reported greater neural activation to faces in comparison with objects of expertise in the FFA (Grill-Spector, Knouf, & Kanwisher, 2004; McKone et al., 2007). It is also important to note that expertise in face recognition and the involvement of the FFA in face processing appear to develop implicitly across all neurotypical individuals. On the other hand, the involvement of the FFA in discriminating between other "similar" stimuli often involves explicit perceptual discrimination training. Moreover, others have argued that normal face recognition does not rely on the development of normal object recognition, as might be assumed from the expertise hypothesis (Germine, Cashdollar, Düzel, & Duchaine, 2011).

However, a bridging of these two viewpoints may be realized through the "interactive specialization" approach (Johnson, 2011). Interactive specialization considers both the predisposition for cortical activation to, for instance, face stimuli (intrinsic factors) and the level of exposure to faces (extrinsic factors), which modulates the degree of localization and specialization of this cortical activity. Thus, under this model areas of the brain may start out as having broadly tuned functions, and with development they gradually become more refined for more specialized functions (Johnson, 2000, 2001), with direct consequences on localization and specialization of neural activation in the brain. Several developmental studies have reported that children show differential neural activation to face stimuli in comparison to adults. For example, activation in face-specific areas of the brain of children when viewing face stimuli is typically not observed, as the areas are relatively immature in terms of both specialization of function and volume (Scherf, Behrmann, Humphreys, & Luna, 2007). When activation is observed in face-selective areas, it is often much less localized than it is in the adult brain (Passarotti et al., 2003). Passarotti et al. (2003) stated that the neural systems involved in face processing continue to experience changes in terms of their functional specialization into late childhood. This account of increasing specialization in brain function is also in line with behavioral evidence supporting the idea that face recognition ability continues to improve into adulthood (Germine et al., 2011).

BEYOND THE FACE: PERSON RECOGNITION IS MULTISENSORY

As demonstrated throughout this chapter, much research has been dedicated to understanding both the perceptual and neurological processes that underlie human face recognition. Although this has brought us far in understanding how the brain processes facial identity, the contribution of such studies to our understanding of person recognition in the real world is limited to information in the visual domain and more specifically, to the use of static images of faces as stimuli. Such investigations belie the true nature of the processes involved in the recognition of individuals we encounter every day, as faces move and speak and are thus inherently dynamic, multisensory stimuli.

Dynamic Faces

In the real world we rarely experience a face in static form and most often see the face in motion, across a large range of viewpoints and expression changes. It would seem odd if we encountered someone who did not move his or her face. Indeed, static face images extracted from moving sequences are viewed as less appealing than their dynamic counterparts, a finding referred to as the "frozen face" effect (Post, Haberman, Iwaki, & Whitney, 2012). In addition, it has been reported that most regions of interest involved in the core and extended face processing system are more successfully located through the use of dynamic (vs. static) face images in functional localizers (Fox, Iaria, & Barton, 2009).

The bias in using static images in studies of face perception has been suggested to reflect an underlying assumption about face perception itself: that recognition is based on our ability to recognize the face through invariant characteristics while ignoring changes that occur through movement (Bruce & Young, 2012). Yet evidence from an emerging literature suggests that our ability to recognize a face may depend on our ability to process these dynamic changes and in turn create a more robust representation of the face in memory. For example, the way someone moves his forehead when in thought may be peculiar to that individual, and thus we may recognize the person not in spite of this movement but perhaps in light of it. Thus, spatiotemporal information associated with an individual's face may act as a unique signature to facilitate the recognition of that person (Bülthoff, Cunningham, & Wallraven, 2011; Knappmeyer, Thornton, & Bülthoff, 2003).

Although the face can move in numerous ways, movements can be largely classed as either *rigid* or *non-rigid* motion (O'Toole, Roark, & Abdi, 2002). Rigid motion refers to whole-head movements (e.g., rotation, nodding) whereas non-rigid motion refers to movements made through deformations of the face (i.e., the internal feature movement observed when making expressive movements or when speaking) (Knappmeyer et al., 2003; O'Toole et al., 2002). Although the role for both types of information in learning new faces remains somewhat unclear (Christie & Bruce, 1998; Lander & Bruce, 2003; Lander & Chuang, 2005; Pike, Kemp, Nicola, & Phillips, 1997), what is apparent is the beneficial role that non-rigid motion plays in the recognition of familiar others, particularly under degraded visual conditions. Indeed, recovering identity information from motion may be necessary for recognition under challenging viewing environments when facial form information may be compromised (e.g., in dim lighting or at a distance). Under these conditions motion may help in determining if a face is familiar or unfamiliar. For example, Knight and Johnston (1997) demonstrated that when visual images of famous faces were degraded by presenting the images in their negative photograph form, participants were more accurate at identifying the target face when it was shown as a video clip rather than a static image. Furthermore, they observed this dynamic enhancement to be larger for upright in comparison with inverted faces. Therefore, the enhancement is not solely due to low-level motion signals but relates to a "motion encoding system," which is refined to upright faces (Hill & Johnston, 2001). Lander, Christie, and Bruce (1999) similarly demonstrated that dynamic information can assist in the recognition of famous faces seen under suboptimal viewing conditions, including negatives, "threshold" (one-bit-per-pixel black and white images) or inverted images, although the facilitation effect on inverted faces was small in comparison with other viewing conditions.

What is also evident is that preserving *characteristic motion* cues affects performance when learning new faces (Wallis, 1998; Wallis & Bülthoff, 2001) and recognizing familiar others (Lander et al., 1999). That is, motion presented in forward form is more beneficial for the perceptual system than disrupted motion (e.g., reversed or scrambled). This can be understood in terms of the *temporal-association hypothesis* (Wallis & Bülthoff, 1999). Under this hypothesis the visual system learns to associate multiple views of, for example, a face when they are not only physically similar but also *temporally correlated* (Wallis & Bülthoff, 2001). Thus, if multiple views of a face are seen in close temporal proximity, it is most likely that they all correspond to the same face and are thus integrated into a robust single representation of the face in memory. Wallis and Bülthoff (2001) illustrated this with a rather extreme example in which they observed that participants learned to associate a view of a completely different identity with the same face when all images were presented in spatiotemporal coherence. They presented four views of a face in smooth rotation and switched the identity of an intermediate view and observed that participants were likely to regard all face-views as belonging to the one person, having a direct effect on their memory of that face. Yet when the view sequence was "jumbled," disrupting spatiotemporal association, this effect disappeared. Spatiotemporal association has been suggested to be more relevant for the recognition of faces over objects and may in part reflect the nature in which we encounter faces in movement in the natural environment (Liu, 2007).

Audiovisual Person Recognition

Although the voice, like the face, is uniquely associated with a person's identity, most research has focused on how we process vocal information for the purpose of perceiving speech, and it is only in recent years that attention has turned to how we process paralinguistic information in speech to support speaker recognition. Yet in evolutionary terms speech is only a recent addition, and our ancestors have been deducing information from vocalizations to support recognition for millions of years (Belin et al., 2004). The fundamental nature of voice recognition is apparent not just from birth, with newborns showing a preference for their mother's voice (DeCasper & Fifer, 1980), but there is evidence to suggest that this ability to discriminate voices may in fact emerge *in utero* (Fifer & Moon, 1994).

The voice, like the face, is a unique identity marker; it conveys similar identity information, including age, gender, ethnicity, and familiarity, as the face and has thus been referred to as an "auditory face" (Belin et al., 2004). Indeed, the perceptual system is tasked with a similar problem in decoding identity in either the voice or the face. Like faces, voices share the same basic perceptual features across individuals (acoustic parameters such as pitch and timbre), and thus the perceptual system must represent a unique "vocal signature" for each human voice in memory by perceiving subtle differences in these acoustic parameters (Belin, Bestelmeyer, Latinus, & Watson, 2011; Belin et al., 2004). Although our ability to recognize voices is not as efficient as our ability to recognize faces (Hanley, Smith, & Had, 1998; Stevenage et al., 2012; Stevenage, Howland, & Tippelt, 2011), we can deduce identity information from the voice alone (Kreiman & Papcun, 1991; Papcun, Kreiman, & Davis, 1989; Schweinberger, Herholz, & Sommer, 1997).

Indeed, the neocortex has developed a network that supports voice recognition that is structured along the same principles as the network that supports face recognition. As the FFA has been implicated in the recognition of facial identity (Grill-Spector et al., 2004; Kanwisher et al., 1997; Parvizi et al., 2012), recent findings have revealed voice selective regions, specifically temporal voice areas (TVA), which are located in middle and anterior regions of the superior temporal sulcus (mSTS and aSTS, respectively) (Belin, Zatorre, Lafaille, Ahad, & Pike, 2000; Belin et al., 2004; Belin, Zatorre, & Ahad, 2002). These areas are sensitive to, among other things, identity information in speech. Similar to the prosopagnosic literature, insult to these areas, particularly when localized to the right hemisphere, can result in deficits in voice recognition known as phonagnosia (Van Lancker & Canter, 1982), and recent evidence also suggests the existence of developmental phonagnosia (Garrido et al., 2009). In contrast, damage to left hemisphere areas often result in deficits in speech perception (aphasia) but with intact voice recognition (Van Lancker & Canter, 1982).

Integrating the Face and Voice for Person Recognition

It appears that the brain has evolved to efficiently process face and voice information in a similar fashion. Recent evidence suggests that vocal and facial information is integrated and shared across modalities at earlier stages of processing (Ghazanfar, 2011; Ghazanfar & Schroeder, 2006) than were previously assumed (Burton, Bruce, & Johnston, 1990). Pioneering work by von Kriegstein and colleagues reported intriguing findings that the FFA is active during familiar voice recognition, even when there is no visual input (von Kriegstein, Kleinschmidt, Sterzer, & Giraud, 2005). This adds to the growing evidence that activation in unisensory areas of the brain can be modulated by cross-modal input (Alais, Newell, & Mamassian, 2010; Driver & Noesselt, 2008; Driver & Spence, 2000; Ghazanfar & Schroeder, 2006) and suggests that the ventral temporal lobe may represent such person-identity information *independent* of the input modality. Indeed, it has been reported that the even the haptic and the visual modalities may encode face information in a similar way, for the purpose of supporting face recognition (Casey & Newell, 2007).

In addition, von Kriegstein et al. (2005) observed that when the task of the participant was to focus on the identity of the voice, activation between the FFA and TVA was apparent. As mentioned, such areas are involved in the processing of identity information from the voice (Belin et al., 2011). Such direct sharing of information is evolutionarily advantageous as it allows for robust person recognition, through cross-modal compensation, under both suboptimal viewing and/or listening conditions. This may be evident with decline in the sensory modalities, as occurs with the aging process (Maguinness, Setti, Burke, Kenny, & Newell, 2011) or in a dimly lit or noisy room, where it may be difficult to see a person's face or to hear the person's voice.

Research reported by von Kriegstein and Giraud (2006) has provided further evidence for the existence of an audiovisual person recognition network. They demonstrated in an fMRI study that the learning of audiovisual face and voice pairings increased the functional coupling between face and voice areas in the cortex. Furthermore, this functional coupling was observed only for natural audiovisual pairings that offer some level of cross-modal redundancy (e.g., a face and a voice vs. a phone and a ring tone). Blank, Anwander, and von Kriegstein (2011) conducted a follow-up study using diffusion tensor imaging (DTI). DTI offers the possibility of assessing the connectivity of brain areas through probabilistic tractography. Blank et al. (2011) demonstrated direct structural connections between the FFA and the voice identity-sensitive areas in aSTS and mSTS. Such direct sharing of information between regions may be used to optimize person recognition. Thus, at least for the naturalistic coupling of face and voice stimuli, the multisensory effects may result from direct interactions between voice and face recognition areas in the brain, as opposed to top-down modulation from supramodal areas, as was once thought.

In line with the findings reported by von Kriegstein and colleagues, Schweinberger, Kloth, and Robertson (2011) revealed evidence for face and voice interactions occurring at strikingly early stages of processing. They demonstrated through electroencephalogram (EEG) recordings that event-related potential (ERP) responses occur in the frontocentral areas at around 50 to 80 milliseconds (ms) following audiovisual presentation. These ERP responses occurred earlier and were more pronounced (super-additive) for audiovisual presentations compared to visual or auditory unimodal presentations. Moreover, they also noted that audiovisual presentations induced a larger N170 response than visual-only presentations of faces (Schweinberger et al., 2011). The N170 response is associated with, and consistently evoked by, the visual presentation of a face and is thus thought to be a key marker of processing that is specific to faces. Schweinberger et al. (2011) argued that such early audiovisual processing is mandatory; however, it may not reflect audiovisual integration for the purpose of recognition. For example, detecting audiovisual incongruencies in person recognition emerges at a later stage in the processing stream: mismatches in face and voice identity elicit differences in ERP responses at around 250 to 600 ms (Schweinberger et al., 2011). However, it is important to note that this is most likely the optimal way for person perception to have evolved, when we consider that it is unlikely that we would ever encounter a face and a voice articulating speech in spatial and temporal alignment that did not belong to the same identity. As such, these incongruency effects tell us a lot about the optimal nature of the processing ability.

The neurological effects observed above have also been shown to have direct behavioral consequences. For example, Schweinberger, Robertson, and Kaufmann (2007) presented individuals with dynamic face and voice pairings, preserving the cross-modal spatial and temporal alignment of the faces and voices, while manipulating the congruency of identity in these face and voice pairings (such that a voice could be presented with a noncorresponding facial identity, but the face would appear to be articulating with the voice in synchrony). They noted behavioral benefits and costs for identifying familiar voices when they were respectively accompanied with either a congruent or incongruent facial identity. Yet, importantly, this was particularly true for voices accompanied with time-synchronized dynamic facial information; static facial information either reduced or removed the effects of cross-modal modulation, demonstrating that spatial and temporal alignment is crucial for audiovisual integration in person recognition. Indeed, dynamic information assists in recognizing individuals not just in audiovisual conditions but also across modalities, and we are efficient at matching a voice to a moving face even when the utterance content is completely different (Kamachi, Hill, Lander, & Vatikiotis-Bateson, 2003; Lander, Hill, Kamachi, & Vatikiotis-Bateson, 2007; Munhall & Buchan, 2004).

Research from von Kriegstein and Giraud (2006) suggests how audiovisual information is also beneficial in creating a robust percept of a person in memory. They observed increased accuracy for recognizing voices that had been originally paired in the learning phase with their associative faces. Similarly, O'Mahony and Newell (2012) observed that participants who were trained on face and voice pairings were faster in detecting the familiarity of a person when they were presented with congruent face and voice information at the testing stage. Intriguingly, they did not observe the same recognition enhancement for congruent face and name pairings, suggesting that it is the naturalistic coupling of face and voice information that enhances person recognition and not the pairing of the face with additional arbitrary information (i.e., the name of the person).

IN SUMMARY

Person recognition underpins many other cognitive tasks, including our ability to socially interact with the world. The fundamental nature of this skill is highlighted by the fact that person recognition appears to be a unique cognitive ability, which may be genetically determined. However, person recognition is adapted to and tailored toward the pressures of our surrounding environment and is malleable to change throughout development, even into old age. This interaction of genetics and environment has ultimately resulted in our high level of expertise in recognizing others, which in turn is likely to have mediated our survival throughout our evolutionary history.

ACKNOWLEDGMENTS

Face stimuli in Figure 13.2 are taken from the Radboud Faces Database (RaFD) (Langner et al., 2010) and are available for download, with permission, from http://www.socsci.ru.nl:8180/RaFD2/RaFD?p=main

REFERENCES

Alais, D., Newell, F. N., & Mamassian, P. (2010). Multisensory processing in review: from physiology to behaviour. *Seeing and Perceiving*, *23*(1), 3–38. doi:10.1163/187847510X488603

Allan, K., Jones, B. C., DeBruine, L. M., & Smith, D. S. (2012). Evidence of adaptation for mate choice within women's memory. *Evolution and Human Behavior*, *33*(3), 193–199. doi:10.1016/j.evolhumbehav.2011.09.002

Anderson, E., Siegel, E. H., Bliss-Moreau, E., & Barrett, L. F. (2011). The visual impact of gossip. *Science*, *332*(6036), 1446–1448. doi:10.1126/science.1201574

Barnes, C. S., De L'Aune, W., & Schuchard, R. A. (2011). A test of face discrimination ability in aging and vision loss. *Optometry and Vision Science*, *88*(2), 188–199. doi:10.1097/OPX.0b013e318205a17c

Bartlett, J. C., & Leslie, J. E. (1986). Aging and memory for faces versus single views of faces. *Memory & Cognition*, *14*(5), 371–381. doi:10.3758/BF03197012

Bartlett, J. C., Leslie, J. E., Tubbs, A., & Fulton, A. (1989). Aging and memory for pictures of faces. *Psychology and Aging*, *4*(3), 276–283. doi:10.1037//0882-7974.4.3.276

Barton, J. J. S. (2008). Structure and function in acquired prosopagnosia: lessons from a series of 10 patients with brain damage. *Journal of Neuropsychology*, *2*(1), 197–225. doi:10.1348/174866407X214172

Belin, P., Bestelmeyer, P. E. G., Latinus, M., & Watson, R. (2011). Understanding voice perception. *British Journal of Psychology*, *102*(4), 711–725. doi:10.1111/j.2044-8295.2011.02041.x

Belin, P., Fecteau, S., & Bédard, C. (2004). Thinking the voice: neural correlates of voice perception. *Trends in Cognitive Sciences*, *8*(3), 129–135. doi:10.1016/j.tics.2004.01.008

Belin, P., Zatorre, R. J., & Ahad, P. (2002). Human temporal-lobe response to vocal sounds. *Cognitive Brain Research*, *13*(1), 17–26. doi:10.1016/S0926-6410(01)00084-2

Belin, P., Zatorre, R. J., Lafaille, P., Ahad, P., & Pike, B. (2000). Voice-selective areas in human auditory cortex. *Nature*, *403*(6767), 309–312. doi:10.1038/35002078

Beyn, E. S., & Knyazeva, G. R. (1962). The problem of prosopagnosia. *Journal of Neurology, Neurosurgery, and Psychiatry, 25*, 154–158. doi:10.1136/jnnp.25.2.154

Bilalic, M., Langner, R., Ulrich, R., & Grodd, W. (2011). Many faces of expertise: fusiform face area in chess experts and novices. *Journal of Neuroscience, 31*(28), 10206–10214. doi:10.1523/JNEUROSCI.5727-10.2011

Blank, H., Anwander, A., & von Kriegstein, K. (2011). Direct structural connections between voice- and face-recognition areas. *Journal of Neuroscience, 31*(36), 12906–12915. doi:10.1523/JNEUROSCI.2091-11.2011

Bliss-Moreau, E., Barrett, L. F., & Wright, C. I. (2008). Individual differences in learning the affective value of others under minimal conditions. *Emotion, 8*(4), 479–493. doi:10.1037/1528-3542.8.4.479

Boutet, I., & Faubert, J. (2006). Recognition of faces and complex objects in younger and older adults. *Memory & Cognition, 34*(4), 854–864. doi:10.3758/BF03193432

Bowles, D. C., McKone, E., Dawel, A., Duchaine, B., Palermo, R., Schmalzl, L., Rivolta, D., et al. (2009). Diagnosing prosopagnosia: effects of ageing, sex, and participant-stimulus ethnic match on the Cambridge Face Memory Test and Cambridge Face Perception Test. *Cognitive Neuropsychology, 26*(5), 423–455. doi:10.1080/02643290903343149

Brosch, T., Sander, D., & Scherer, K. R. (2007). That baby caught my eye... attention capture by infant faces. *Emotion, 7*(3), 685–689. doi:10.1037/1528-3542.7.3.685

Bruce, V. (1982). Changing faces: visual and non-visual coding processes in face recognition. *British Journal of Psychology, 73*(1), 105–116. doi:10.1111/j.2044-8295.1982.tb01795.x

Bruce, V., Henderson, Z., Newman, C., & Burton, A. M. (2001). Matching identities of familiar and unfamiliar faces caught on CCTV images. *Journal of Experimental Psychology: Applied, 7*, 207–218. doi:10.1037//1076-898X.7.3.207

Bruce, V., & Young, A. (1986). Understanding face recognition. *British Journal of Psychology, 77*(3), 305–327. doi:10.1111/j.2044-8295.1986.tb02199.x

Bruce, V., & Young, A. (2012). *Face Perception.* Sussex: Psychology Press.

Burton, A. M., Bruce, V., & Johnston, R. A. (1990). Understanding face recognition with an interactive model. *British Journal of Psychology, 81*, 361–380. doi:10.1111/j.2044-8295.1990.tb02367.x

Burton, A. M., Wilson, S., Cowan, M., & Bruce, V. (1999). Face recognition in poor-quality video: evidence from security surveillance. *Psychological Science, 10*(3), 243–248. doi:10.1111/1467-9280.00144

Bülthoff, H. H., Cunningham, D. W., & Wallraven, C. (2011). Dynamic aspects of face processing in humans. In S. Z. Li & A. K. Jain (Eds.), *Handbook of face recognition* (pp. 575–596). London: Springer London. doi:10.1007/978-0-85729-932-1

Bülthoff, I., & Newell, F. N. (2004). Categorical perception of sex occurs in familiar but not unfamiliar faces. *Visual Cognition, 11*(7), 823–855. doi:10.1080/13506280444000012

Busigny, T., Joubert, S., Felician, O., Ceccaldi, M., & Rossion, B. (2010). Holistic perception of the individual face is specific and necessary : Evidence from an extensive case study of acquired prosopagnosia. *Neuropsychologia, 48*(14), 4057–4092. doi:10.1016/j.neuropsychologia.2010.09.017

Casey, S. J., Mernagh, M., & Newell, F. N. (2009). Are attractive facial characteristics peculiar to the sex of a face? *Quarterly Journal of Experimental Psychology, 62*(5), 833–843. doi:10.1080/17470210802592030

Casey, S. J., & Newell, F. N. (2007). Are representations of unfamiliar faces independent of encoding modality? *Neuropsychologia, 45*(3), 506–513. doi:10.1016/j.neuropsychologia.2006.02.011

Chaby, L., Narme, P., & George, N. (2011). Older adults' configural processing of faces: role of second-order information. *Psychology and Aging*, *26*(1), 71–79. doi:10.1037/a0020873

Chiappe, D., Brown, A., Dow, B., Koontz, J., Rodriguez, M., & McCulloch, K. (2004). Cheaters are looked at longer and remembered better than cooperators in social exchange situations. *Evolutionary Psychology*, *2*(2), 108–120.

Chiroro, P. M., Tredoux, C. G., Radaelli, S., & Meissner, C. A. (2008). Recognizing faces across continents: the effect of within-race variations on the own-race bias in face recognition. *Psychonomic Bulletin & Review*, *15*(6), 1089–1092. doi:10.3758/PBR.15.6.1089

Christie, F., & Bruce, V. (1998). The role of dynamic information in the recognition of unfamiliar faces. *Memory & Cognition*, *26*(4), 780–90. doi:10.3758/BF03211397

DeBruine, L. M. (2004a). Resemblance to self increases the appeal of child faces to both men and women. *Evolution and Human Behavior*, *25*(3), 142–154. doi:10.1016/j.evolhumbehav.2004.03.003

DeBruine, L. M. (2004b). Facial resemblance increases the attractiveness of same-sex faces more than other-sex faces. *Proceedings of the Royal Society: Biological Sciences*, *271*(1552), 2085–90. doi:10.1098/rspb.2004.2824

DeBruine, L. M., Jones, B. C., Little, A. C., & Perrett, D. I. (2008). Social perception of facial resemblance in humans. *Archives of Sexual Behavior*, *37*(1), 64–77. doi:10.1007/s10508-007-9266-0

DeBruine, L. M., Jones, B. C., Watkins, C. D., Roberts, S. C., Little, A. C., Smith, F. G., & Quist, M. C. (2011). Opposite-sex siblings decrease attraction, but not prosocial attributions, to self-resembling opposite-sex faces. *Proceedings of the National Academy of Sciences of the United States of America*, *108*(28), 11710–11714. doi:10.1073/pnas.1105919108

DeBruine, L. M., Smith, F. G., Jones, B. C., Roberts, S. C., Petrie, M., & Spector, T. D. (2009). Kin recognition signals in adult faces. *Vision Research*, *49*, 38–43. doi:10.1016/j.visres.2008.09.025

DeCasper, A. J., & Fifer, W. P. (1980). Of human bonding: newborns prefer their mothers' voices. *Science*, *208*(4448), 1174–1176. doi:10.1126/science.7375928

Driver, J., & Noesselt, T. (2008). Multisensory interplay reveals crossmodal influences on "sensory-specific" brain regions, neural responses, and judgments. *Neuron*, *57*(1), 11–23. doi:10.1016/j.neuron.2007.12.013

Driver, J., & Spence, C. (2000). Multisensory perception: beyond modularity and convergence. *Current Biology*, *10*(20), R731–R735. doi:10.1016/S0960-9822(00)00740-5

Duchaine, B. C., Dingle, K., Butterworth, E., & Nakayama, K. (2004). Normal greeble learning in a severe case of developmental prosopagnosia. *Neuron*, *43*(4), 469–473. doi:10.1016/j.neuron.2004.08.006

Duchaine, B., Germine, L., & Nakayama. (2007). Family resemblance: ten family members with prosopagnosia and within-class object agnosia. *Cognitive Neuropsychology*, *24*(4), 419–430. doi:10.1080/02643290701380491

Ellis, H. D., & Florence, M. (1990). Bodamer's (1947) paper on prosopagnosia. *Cognitive Neuropsychology*, *7*(2), 81–105. doi:10.1080/02643299008253437

Engell, A. D., Haxby, J. V., & Todorov, A. (2007). Implicit trustworthiness decisions: automatic coding of face properties in human amygdala. *Journal of Cognitive Neuroscience*, *19*(9), 1508–1519. doi:10.1162/jocn.2007.19.9.1508

Fantz, R. L. (1963). Pattern vision in newborn infants. *Science*, *140*(3564), 296–297. doi:10.1126/science.140.3564.296

Felisberti, F. M., & Pavey, L. (2010). Contextual modulation of biases in face recognition. *PLoS ONE*, *5*(9), e12939. doi:10.1371/journal.pone.0012939

Fifer, W. P., & Moon, C. M. (1994). The role of mother's voice in the organization of brain function in the newborn. *Acta Paediatrica Supplementum*, *83*(s397), 86–93. Retrieved from http://dx.doi.org/10.1111/j.1651-2227.1994.tb13270.x

Firestone, A., Turk-Browne, N. B., & Ryan, J. D. (2007). Age-related deficits in face recognition are related to underlying changes in scanning behavior. *Neuropsychology, Development, and Cognition*, *14*(6), 594–607. doi:10.1080/13825580600899717

Fox, C. J., Iaria, G., & Barton, J. J. S. (2009). Defining the face processing network: optimization of the functional localizer in fMRI. *Human Brain Mapping*, *30*(5), 1637–1651. doi:10.1002/hbm.20630

Fulton, A., & Bartlett, J. C. (1991). Young and old faces in young and old heads: the factor of age in face recognition. *Psychology and Aging*, *6*(4), 623–630. doi:10.1037//0882-7974.6.4.623

Gao, L., Xu, J., Zhang, B., Zhao, L., Harel, A., & Bentin, S. (2009). Aging effects on early-stage face perception: an ERP study. *Psychophysiology*, *46*(5), 970–983. doi:10.1111/j.1469-8986.2009.00853.x

Garrido, L., Eisner, F., McGettigan, C., Stewart, L., Sauter, D., Hanley, J. R., Schweinberger, S. R., et al. (2009). Developmental phonagnosia: a selective deficit of vocal identity recognition. *Neuropsychologia*, *47*(1), 123–131. doi:10.1016/j.neuropsychologia.2008.08.003

Gauthier, I., & Nelson, C. A. (2001). The development of face expertise. *Current Opinion in Neurobiology*, *11*(2), 219–224. doi:10.1016/S0959-4388(00)00200-2

Gauthier, I., Skudlarski, P., Gore, J. C., & Anderson, A. W. (2000). Expertise for cars and birds recruits brain areas involved in face recognition. *Nature Neuroscience*, *3*(2), 191–197. doi:10.1038/72140

Germine, L., Cashdollar, N., Düzel, E., & Duchaine, B. (2011). A new selective developmental deficit: impaired object recognition with normal face recognition. *Cortex*, *47*(5), 598–607. doi:10.1016/j.cortex.2010.04.009

Germine, L. T., Duchaine, B., & Nakayama, K. (2011). Where cognitive development and aging meet: face learning ability peaks after age 30. *Cognition*, *118*(2), 201–210. doi:10.1016/j.cognition.2010.11.002

Ghazanfar, A. A. (2011). Unity of the senses for primate vocal communication. In M. M. Murray & M. T. Wallace (Eds.), *The neural bases of multisensory processes* (pp. 653–666). Boca Raton, FL: CRC Press. doi:10.1201/b11092-41

Ghazanfar, A. A., & Schroeder, C. E. (2006). Is neocortex essentially multisensory? *Trends in Cognitive Sciences*, *10*(6), 278–285. doi:10.1016/j.tics.2006.04.008

Goren, C. C., Sarty, M., & Wu, P. Y. K. (1975). Visual following and pattern discrimination of face-like stimuli by newborn infants. *Pediatrics*, *56*(4), 544–549. Retrieved from http://pediatrics.aappublications.org/cgi/content/abstract/56/4/544

Grady, C. L. (2002). Age-related differences in face processing: a meta-analysis of three functional neuroimaging experiments. *Canadian Journal of Experimental Psychology*, *56*(3), 208–220. doi:10.1037/h0087398

Grady, C. L., McIntosh, A. R., Horwitz, B., Maisog, J. M., Ungerleider, L. G., Mentis, M. J., Pietrini, P., et al. (1995). Age-related reductions in human recognition memory due to impaired encoding. *Science*, *269*(5221), 218–221. doi:10.1126/science.7618082

Grill-Spector, K., Knouf, N., & Kanwisher, N. (2004). The fusiform face area subserves face perception, not generic within-category identification. *Nature Neuroscience*, *7*(5), 555–562. doi:10.1038/nn1224

Grueter, M., Grueter, T., Bell, V., Horst, J., Laskowski, W., Sperling, K., Halligan, P. W., et al. (2007). Hereditary prosopagnosia: The first case series. *Cortex*, *43*(6), 734–749. doi:10.1016/S0010-9452(08)70502-1

Hanley, J. R., Smith, S. T., & Had, J. (1998). I recognise you but I can't place you :An investigation of familiar-only experiences during tests of voice and face recognition. *Quarterly Journal of Experimental Psychology*, *51A*(1), 179–195.

Haxby, J., Hoffman, E. A., & Gobbini, M. I. (2000). The distributed human neural system for face perception. *Trends in Cognitive Sciences*, *4*(6), 223–233. doi:10.1016/S1364-6613(00)01482-0

Haxby, J. V., Maisog, J. M., Horwitz, B., Ungerleider, L. G., Mentis, M. J., Salerno, J. A., Pietrini, P., et al. (1994). Age-related processing changes in cortical blood flow activation of faces and location during visual processing of faces and location. *Journal of Neuroscience*, *14*(3), 1450–1462.

Herzmann, G., Kunina, O., Sommer, W., & Wilhelm, O. (2010). Individual differences in face cognition: brain-behavior relationships. *Journal of Cognitive Neuroscience*, *22*(3), 571–589. doi:10.1162/jocn.2009.21249

Hildebrandt, A., Wilhelm, O., Schmiedek, F., Herzmann, G., & Sommer, W. (2011). On the specificity of face cognition compared with general cognitive functioning across adult age. *Psychology and Aging*, *26*(3), 701–715. doi:10.1037/a0023056

Hill, H., & Johnston, A. (2001). Categorizing sex and identity from the biological motion of faces. *Current Biology*, *11*(11), 880–885. doi:10.1016/S0960-9822(01)00243-3

Johnson, M. H. (2000). Functional brain development in infants: elements of an interactive specialization framework. *Child Development*, *71*(1), 75–81. doi:10.1111/1467-8624.00120

Johnson, M. H. (2001). Functional brain development in humans. *Nature Reviews Neuroscience*, *2*(7), 475–483. doi:10.1038/35081509

Johnson, M. H. (2005). Subcortical face processing. *Nature Reviews Neuroscience*, *6*(10), 766–774. doi:10.1038/nrn1766

Johnson, M. H. (2011). Face processing as a brain adaptation at multiple timescales. *Quarterly Journal of Experimental Psychology*, *64*(10), 1873–1888. doi:10.1080/17470218.2011.590596

Johnson, M. H., Dziurawiec, S., Ellis, H., & Morton, J. (1991). Newborns' preferential tracking of face-like stimuli and its subsequent decline. *Cognition*, *40*(1-2), 1–19. doi:10.1016/0010-0277(91)90045-6

Johnston, V. S. (2000). Female facial beauty: the fertility hypothesis. *Pragmatics & Cognition*, *8*(1), 107–122. doi:10.1075/pc.8.1.06joh

Johnston, V. S., & Franklin, M. (1993). Is beauty in the eye of the beholder? *Ethology and Sociobiology*, *14*(3), 183–199. doi:10.1016/0162–3095(93)90005–3

Johnston, V. S., Hagel, R., Franklin, M., Fink, B., & Grammer, K. (2001). Male facial attractiveness: evidence for hormone-mediated adaptive design. *Evolution and Human Behavior*, *22*(4), 251–267. doi:10.1016/S1090-5138(01)00066-6

Kamachi, M., Hill, H., Lander, K., & Vatikiotis-Bateson, E. (2003). Putting the face to the voice: matching identity across modality. *Current Biology*, *13*(19), 1709–1714. doi:10.1016/j.cub.2003.09.005

Kanwisher, N., McDermott, J., & Chun, M. M. (1997). The fusiform face area: a module in human extrastriate cortex specialized for face perception. *Journal of Neuroscience*, *17*(11), 4302–4311.

Kelly, D. J., Quinn, P. C., Slater, A. M., Lee, K., Liezhong, G., & Pascalis, O. (2007). The other-race effect develops during infancy evidence of perceptual narrowing. *Psychological Science*, *18*(12), 1084–1089. doi:10.1111/j.1467-9280.2007.02029.x.

Knappmeyer, B., Thornton, I. M., & Bülthoff, H. H. (2003). The use of facial motion and facial form during the processing of identity. *Vision Research*, *43*(18), 1921–1936. doi:10.1016/S0042-6989(03)00236-0

Knight, B., & Johnston, A. (1997). The role of movement in face recognition. *Visual Cognition*, *4*(3), 265–273. doi:10.1080/713756764

Kreiman, J., & Papcun, G. (1991). Comparing discrimination and recognition of unfamiliar voices. *Speech Communication*, *10*(3), 265–275. doi:10.1016/0167-6393(91)90016-M

Lamont, A. C., Stewart-Williams, S., & Podd, J. (2005). Face recognition and aging: effects of target age and memory load. *Memory & Cognition*, *33*(6), 1017–1024. doi:10.3758/BF03193209

Lander, K., & Bruce, V. (2003). The role of motion in learning new faces. *Visual Cognition*, *10*(8), 897–912. doi:10.1080/13506280344000149

Lander, K., Christie, F., & Bruce, V. (1999). The role of movement in the recognition of famous faces. *Memory & Cognition*, *27*(6), 974–985. doi:10.3758/BF03201228

Lander, K., & Chuang, L. (2005). Why are moving faces easier to recognize? *Visual Cognition*, *12*(3), 429–442. doi:10.1080/13506280444000382

Lander, K., Hill, H., Kamachi, M., & Vatikiotis-Bateson, E. (2007). It's not what you say but the way you say it: matching faces and voices. *Journal of Experimental Psychology: Human Perception and Performance*, *33*(4), 905–914. doi:10.1037/0096-1523.33.4.905

Langner, O., Dotsch, R., Bijlstra, G., Wigboldus, D. H. J., Hawk, S. T., & van Knippenberg, A. (2010). Presentation and validation of the Radboud Faces Database. *Cognition & Emotion*, *24*(8), 1377–1388. doi:10.1080/02699930903485076

Langton, S. R. H., Law, A. S., Burton, A. M., & Schweinberger, S. R. (2008). Attention capture by faces. *Cognition*, *107*(1), 330–342. doi:10.1016/j.cognition.2007.07.012

Law Smith, M. J., Perrett, D. I., Jones, B. C., Cornwell, R. E., Moore, F. R., Feinberg, D. R., Boothroyd, L. G., et al. (2006). Facial appearance is a cue to oestrogen levels in women. *Proceedings of the Royal Society B Biological Sciences*, *273*(1583), 135–140. doi:10.1098/rspb.2005.3296

Lee, Y., Duchaine, B., Wilson, H. R., & Nakayama, K. (2010). Three cases of developmental prosopagnosia from one family: detailed neuropsychological and psychophysical investigation of face processing. *Cortex*, *46*(8), 949–964. doi:10.1016/j.cortex.2009.07.012

Lee, Y., Grady, C. L., Habak, C., Wilson, H. R., & Moscovitch, M. (2011). Face processing changes in normal aging revealed by fMRI adaptation. *Journal of Cognitive Neuroscience*, *23*(11), 3433–3447. doi:10.1162/jocn_a_00026

Leopold, D. A., & Rhodes, G. (2010). A comparative view of face perception. *Journal of Comparative Psychology*, *124*(3), 233–251. doi:10.1037/a0019460.

Lie, H. C., Rhodes, G., & Simmons, L. W. (2008). Genetic diversity revealed in human faces. *Evolution*, *62*(10), 2473–2486. doi:10.1111/j.1558-5646.2008.00478.x

Light, L. L., Hollander, S., & Kayra-Stuart, F. (1981). Why attractive people are harder to remember. *Personality and Social Psychology Bulletin*, *7*(2), 269–276. doi:10.1177/014616728172014

Little, A. C., Jones, B. C., & DeBruine, L. M. (2011). Facial attractiveness: evolutionary based research. *Philosophical Transactions of the Royal Society of London. Series B, Biological Sciences*, *366*(1571), 1638–1659. doi:10.1098/rstb.2010.0404

Liu, T. (2007). Learning sequence of views of three-dimensional objects: the effect of temporal coherence on object memory. *Perception*, *36*(9), 1320–1333. doi:10.1068/p5778

Maguinness, C., Setti, A., Burke, K., Kenny, R. A., & Newell, F. N. (2011). The effect of combined sensory and semantic components on audio-visual speech perception in older adults. *Frontiers in Aging Neuroscience*, *3*:19. doi: 10.3389/fnagi.2011.00019

Maloney, L. T., & Dal Martello, M. F. (2006). Kin recognition and the perceived facial similarity of children. *Journal of Vision*, *6*(10), 1047–1056. doi:10.1167/6.10.4

Marzi, T., & Viggiano, M. P. (2010). When memory meets beauty: insights from event-related potentials. *Biological Psychology*, *84*(2), 192–205. doi:10.1016/j.biopsycho.2010.01.013

Maurer, D., Le Grand, R., & Mondloch, C. J. (2002). The many faces of configural processing. *Trends in Cognitive Sciences*, *6*(6), 255–260. doi:10.1016/S1364-6613(02)01903-4

McKone, E., Kanwisher, N., & Duchaine, B. C. (2007). Can generic expertise explain special processing for faces? *Trends in Cognitive Sciences*, *11*(1), 8–15. doi:10.1016/j.tics.2006.11.002

Mealey, L., Daood, C., & Krage, M. (1996). Enhanced memory for faces of cheaters. *Ethology and Sociobiology*, *17*(2), 119–128. doi:10.1016/0162-3095(95)00131-X

Meissner, C. A., & Brigham, J. C. (2001). Thirty years of investigating the own-race bias in memory for faces: a meta-analytic review. *Psychology, Public Policy, and Law*, *7*(1), 3–35. doi:10.1037//1076-8971.7.1.3

Munhall, K. G., & Buchan, J. N. (2004). Something in the way she moves. *Trends in Cognitive Sciences*, *8*(2), 51–53. doi:10.1016/j.tics.2003.12.006

Murray, J. E., Halberstadt, J., & Ruffman, T. (2010). The face of aging: sensitivity to facial feature relations changes with age. *Psychology and Aging*, *25*(4), 846–850. doi:10.1037/a0019864

Nairne, J. S., Thompson, S. R., & Pandeirada, J. N. S. (2007). Adaptive memory: survival processing enhances retention. *Journal of Experimental Psychology. Learning Memory and Cognition*, *33*(2), 263–273.

Nairne, J. S., & Pandeirada, J. N. S. (2008a). Adaptive memory: remembering with a stone age brain. *Current Directions in Psychological Science*, *17*(4), 239–244. doi:10.1111/j.1467-8721.2008.00582.x

Neff, B. D., & Pitcher, T. E. (2005). Genetic quality and sexual selection: an integrated framework for good genes and compatible genes. *Molecular Ecology*, *14*(1), 19–38. doi:10.1111/j.1365-294X.2004.02395.x

Norton, D., McBain, R., & Chen, Y. (2009). Reduced ability to detect facial configuration in middle-aged and elderly individuals: associations with spatiotemporal visual processing. *Journal of Gerontology: Psychological Sciences*, *64B*(3), 328–334. doi:10.1093/geronb/gbp008

O'Craven, K. M., & Kanwisher, N. (2000). Mental imagery of faces and places activates corresponding stimulus-specific brain regions. *Journal of Cognitive Neuroscience*, *12*(6), 1013–1023. doi:10.1162/08989290051137549

O'Doherty, J., Winston, J., Critchley, H., Perrett, D., Burt, D. M., & Dolan, R. J. (2003). Beauty in the smile: the role of medial orbitofrontal cortex in facial attractiveness. *Neuropsychologia*, *41*(2), 147–155. doi:10.1016/S0028-3932(02)00145-8

O'Mahony, C., & Newell, F. N. (2012). Integration of faces and voices, but not faces and names, in person recognition. *British Journal of Psychology*, *103*(1), 73–82. doi:10.1111/j.2044-8295.2011.02044.x

O'Toole, A. J., Roark, D. A., & Abdi, H. (2002). Recognizing moving faces: a psychological and neural synthesis. *Trends in Cognitive Sciences*, *6*(6), 261–266. doi:10.1016/S1364-6613(02)01908-3

Owsley, C., Sekuler, R., & Boldt, C. (1981). Aging and low-contrast vision: face perception. *Investigative Ophthalmology & Visual Science, 21*(2), 362–365. Retrieved from http://www.iovs.org/content/21/2/362.short

Papcun, G., Kreiman, J., & Davis, A. (1989). Long-term memory for unfamiliar voices. *Journal of the Acoustical Society of America, 85*(2), 913–925. doi:10.1121/1.397564

Parvizi, J., Jacques, C., Foster, B. L., Withoft, N., Rangarajan, V., Weiner, K. S., & Grill-Spector, K. (2012). Electrical stimulation of human fusiform face-selective regions distorts face perception. *Journal of Neuroscience, 32*(43), 14915–14920. doi:10.1523/JNEUROSCI.2609-12.2012

Pascalis, O., de Schonen, S. D. E., Morton, J., Deruelle, C., & Fabre-Grenet, M. (1995). Mother's face recognition by neonates: a replication and an extension. *Infant Behavior and Development, 18*(1), 79–85. doi:10.1016/0163-6383(95)90009-8

Passarotti, A. M., Paul, B. M., Bussiere, J. R., Buxton, R. B., Wong, E. C., & Stiles, J. (2003). The development of face and location processing: an fMRI study. *Developmental Science, 6*(1), 100–117. NIH Public Access. doi:10.1111/1467-7687.00259

Perrett, D. I., Lee, K. J., Penton-Voak, I., Rowland, D., Yoshikawa, S., Burt, D. M., Henzi, S. P., et al. (1998). Effects of sexual dimorphism on facial attractiveness. *Nature, 394*(6696), 884–887. Nature Publishing Group. doi:10.1038/29772

Peskin, M., & Newell, F. N. (2004). Familiarity breeds attraction: effects of exposure on the attractiveness of typical and distinctive faces. *Perception, 33*(2), 147–157. doi:10.1068/p5028

Pike, G. E., Kemp, R. I., Nicola, A. T., & Phillips, K. C. (1997). Recognizing moving faces : the relative contribution of motion and perspective view information. *Visual Cognition, 4*(4), 409–437. doi:10.1080/713756769

Platek, S. M., Burch, R. L., Panyavin, I. S., Wasserman, B. H., & Gallup, G. G. (2002). Reactions to children's faces: resemblance affects males more than females. *Evolution and Human Behaviour, 23*(3), 159–166. doi:10.1016/S1090-5138(01)00094-0

Post, R. B., Haberman, J., Iwaki, L., & Whitney, D. (2012). The frozen face effect: why static photographs may not do you justice. *Frontiers in Psychology, 3*(22). doi:10.3389/fpsyg.2012.00022

Potter, T., & Corneille, O. (2008). Locating attractiveness in the face space: faces are more attractive when closer to their group prototype. *Psychonomic Bulletin & Review, 15*(3), 615–622. doi:10.3758/PBR.15.3.615

Rabe-Hesketh, S., Skrondal, A., & Gjessing, H. K. (2008). Biometrical modeling of twin and family data using standard mixed model software. *Biometrics, 64*(1), 280–288. doi:10.1111/j.1541-0420.2007.00803.x

Rhodes, G. (2006). The evolutionary psychology of facial beauty. *Annual Review of Psychology, 57*, 199–226. doi:10.1146/annurev.psych.57.102904.190208

Rhodes, G., Simmons, L. W., & Peters, M. (2005). Attractiveness and sexual behavior: Does attractiveness enhance mating success? *Evolution and Human Behavior, 26*(2), 186–201. doi:10.1016/j.evolhumbehav.2004.08.014

Rhodes, M. G., & Anastasi, J. S. (2011). The own-age bias in face recognition: a meta-analytic and theoretical review. *Psychological Bulletin, 138*(1), 146–174. doi:10.1037/0882-7974.23.2.239

Rhodes, M. G., Castel, A. D., & Jacoby, L. L. (2008). Associative recognition of face pairs by younger and older adults: the role of familiarity-based processing. *Psychology and Aging, 23*(2), 239–249. doi:10.1037/0882-7974.23.2.239

Roberts, S. C., Little, A. C., Gosling, L. M., Jones, B. C., Perrett, D. I., Carter, V., & Petrie, M. (2005). MHC-assortative facial preferences in humans. *Biology Letters, 1*(4), 400–403. doi:10.1098/rsbl.2005.0343

Roberts, S. C., Little, A. C., Gosling, L. M., Perrett, D. I., Carter, V., Jones, B. C., Penton-Voak, I., et al. (2005). MHC-heterozygosity and human facial attractiveness. *Evolution and Human Behavior*, *26*(3), 213–226. doi:10.1016/j.evolhumbehav.2004.09.002

Rossion, B., Caldara, R., Seghier, M., Schuller, A., Lazeyras, F., & Mayer, E. (2003). A network of occipito-temporal face-sensitive areas besides the right middle fusiform gyrus is necessary for normal face processing. *Brain*, *126*(Pt 11), 2381–2395. doi:10.1093/brain/awg241

Rossion, B., Gauthier, I., Goffaux, V., Tarr, M. J., & Crommelinck, M. (2002). Expertise training with novel objects leads to left-lateralized facelike electrophysiological responses. *Psychological Science*, *13*(3), 250–257.. doi:10.1111/1467-9280.00446

Rossion, B., Schiltz, C., & Crommelinck, M. (2003). The functionally defined right occipital and fusiform "face areas" discriminate novel from visually familiar faces. *NeuroImage*, *19*(3), 877–883. doi:10.1016/S1053-8119(03)00105-8

Russell, R., Chatterjee, G., & Nakayama, K. (2012). Developmental prosopagnosia and super-recognition: no special role for surface reflectance processing. *Neuropsychologia*, *50*(2), 334–340. doi:10.1016/j.neuropsychologia.2011.12.004

Russell, R., Duchaine, B., & Nakayama, K. (2009). Super-recognizers: people with extraordinary face recognition ability. *Psychonomic Bulletin & Review*, *16*(2), 252–257. doi:10.3758/PBR.16.2.252

Savine, A. C., Scullin, M. K., & Roediger, H. L. (2011). Survival processing of faces. *Memory & Cognition*, *39*(8), 1359–1373. doi:10.3758/s13421-011-0121-0

Scherf, K. S., Behrmann, M., Humphreys, K., & Luna, B. (2007). Visual category-selectivity for faces, places and objects emerges along different developmental trajectories. *Developmental Science*, *10*(4), F15–F30. doi:10.1111/j.1467-7687.2007.00595.x

Schiltz, C., Sorger, B., Caldara, R., Ahmed, F., Mayer, E., Goebel, R., & Rossion, B. (2006). Impaired face discrimination in acquired prosopagnosia is associated with abnormal response to individual faces in the right middle fusiform gyrus. *Cerebral Cortex*, *16*(4), 574–586. doi:10.1093/cercor/bhj005

Schmalzl, L., Palermo, R., & Coltheart, M. (2008). Cognitive heterogeneity in genetically based prosopagnosia: a family study. *Journal of Neuropsychology*, *2*(1), 99–117. doi:10.1348/174866407X256554

Schweinberger, S. R., Herholz, A., & Sommer, W. (1997). Recognizing famous voices: influence of stimulus duration and different types of retrieval cues. *Journal of Speech, Language and Hearing Research*, *40*(2), 453–463. Retrieved from http://jslhr.highwire.org/cgi/content/abstract/40/2/453

Schweinberger, S. R., Kloth, N., & Robertson, D. M. C. (2011). Hearing facial identities: brain correlates of face—voice integration in person identification. *Cortex*, *47*(9), 1026–1037. doi:10.1016/j.cortex.2010.11.011

Schweinberger, S. R., Robertson, D., & Kaufmann, J. M. (2007). Hearing facial identities. *Quarterly Journal of Experimental Psychology*, *60*(10), 1446–1456. doi:10.1080/17470210601063589

Schweinberger, S. R., Walther, C., Zäske, R., & Kovács, G. (2011). Neural correlates of adaptation to voice identity. *British Journal of Psychology*, *102*(4), 748–764. doi:10.1111/j.2044-8295.2011.02048.x

Sinha, P., Balas, B., & Ostrovsky, Y. (2007). Discovering faces in infancy. Paper presented to the Vision Sciences Society, May 11–16, Florida.

Stevenage, S. V., Howland, A., & Tippelt, A. (2011). Interference in eyewitness and earwitness recognition. *Applied Cognitive Psychology*, *25*(1), 112–118. doi:10.1002/acp.1649

Stevenage, S. V., Neil, G. J., Barlow, J., Dyson, A., Eaton-Brown, C., & Parsons, B. (2012). The effect of distraction on face and voice recognition. *Psychological Research*. doi:10.1007/s00426-012-0450-z

Todorov, A. (2008). Evaluating faces on trustworthiness. An extension of systems for recognition of emotions signaling approach/avoidance behaviors. *Annals of the New York Academy of Sciences, 1124*, 208–224. doi:10.1196/annals.1440.012

Toga, A. W., & Thompson, P. M. (2005). Genetics of brain structure and intelligence. *Annual Review of Neuroscience, 28*, 1–23. doi:10.1146/annurev.neuro.28.061604.135655

Tsukiura, T., & Cabeza, R. (2011). Remembering beauty: roles of orbitofrontal and hippocampal regions in successful memory encoding of attractive faces. *Neuroimage, 54*(1), 653–660. doi:10.1016/j.neuroimage.2010.07.046.

Valentine, T. (1991). A unified account of the effects of distinctiveness, inversion, and race in face recognition. *Quarterly Journal of Experimental Psychology, 43A*, 161–204. doi:10.1080/14640749108400966

Valentine, T., & Ferrara, A. (1991). Typicality in categorization, recognition and identification: evidence from face recognition. *British Journal of Psychology, 82*, 87–102. doi:10.1111/j.2044-8295.1991.tb02384.x

Valenza, E., Simion, F., Cassia, V. M., & Umiltà, C. (1996). Face preference at birth. *Journal of Experimental Psychology. Human Perception and Performance, 22*(4), 892–903. doi:10.1037/0096-1523.22.4.892

Van Lancker, D. R., & Canter, G. J. (1982). Impairment of voice and face recognition in patients with hemispheric damage. *Brain and Cognition, 1*(2), 185–195. doi:10.1016/0278-2626(82)90016-1

Vizioli, L., Rousselet, G. A., & Caldara, R. (2010). Neural repetition suppression to identity is abolished by other-race faces. *Proceedings of the National Academy of Sciences of the United States of America, 107*(46), 20081–20086. doi:10.1073/pnas.1005751107

von Kriegstein, K., & Giraud, A.-L. (2006). Implicit multisensory associations influence voice recognition. *PLoS Biology, 4*(10), e326. doi:10.1371/journal.pbio.0040326

von Kriegstein, K., Kleinschmidt, A., Sterzer, P., & Giraud, A.-L. (2005). Interaction of face and voice areas during speaker recognition. *Journal of Cognitive Neuroscience, 17*(3), 367–376. doi:10.1162/0898929053279577

Wallis, G. (1998). Temporal order in human object recognition learning. *Journal of Biological Systems, 6*(3), 299–313. doi:10.1142/S0218339098000200

Wallis, G., & Bülthoff, H. (1999). Learning to recognize objects. *Trends in Cognitive Sciences, 3*(1), 22–31. doi:10.1016/S1364-6613(98)01261-3

Wallis, G., & Bülthoff, H. H. (2001). Effects of temporal association on recognition memory. *Proceedings of the National Academy of Sciences USA, 98*(8), 4800–4804. doi:10.1073/pnas.071028598

Wickham, L., & Morris, P. (2003). Attractiveness, distinctiveness, and recognition of faces: Attractive faces can be typical or distinctive but are not better recognized. *American Journal of Psychology, 116*(3), 455–468. doi:10.2307/1423503

Wilhelm, O., Herzmann, G., Kunina, O., Danthiir, V., Schacht, A., & Sommer, W. (2010). Individual differences in perceiving and recognizing faces: One element of social cognition. *Journal of Personality and Social Psychology, 99*(3), 530–548. doi:10.1037/a0019972

Wilhelm, O., Herzmann, G., Kunina, O., & Sommer, W. (2007). Face cognition: a set of distinct mental abilities. *Nature Proceedings*. Retrieved from http://hdl.handle.net/10101/npre.2007.1385.1

Wilmer, J. B., Germine, L., Chabris, C. F., Chatterjee, G., Williams, M., Loken, E., Nakayama, K., et al. (2010). Human face recognition ability is specific and highly heritable. *Proceedings of the National Academy of Sciences USA*, *107*(11), 5238–5241. doi:10.1073/pnas.0913053107

Yin, R. (1969). Looking at upside-down faces. *Journal of Experimental Psychology*, *81*(1), 141–145. doi:10.1037/h0027474

Young, A. W., DeHaan, E. H. F., & Bauer, R. M. (2008). Face perception: a very special issue. *Journal of Neuropsychology*, *2*(1), 1–14. doi:10.1348/174866407X269848

Young, A. W., Hellawell, D., & Hay, D. C. (1987). Configural information in face perception. *Perception*, *16*(6), 747–759. doi:10.1068/p160747

Zhu, Q., Song, Y., Hu, S., Li, X., Tian, M., Zhen, Z., Dong, Q., et al. (2010). Heritability of the specific cognitive ability of face perception. *Current Biology*, *20*(2), 137–142. doi:10.1016/j.cub.2009.11.067

14

Perceptual Versus Conceptual Memory Processes in a Chimpanzee (*Pan troglodytes*)

JENNIFER VONK AND KELLY W. MOSTELLER

In humans, it has been established that conscious retrieval of previously viewed stimuli is enhanced by deeper levels of processing, with deeper levels referring to meaningful processing of the semantic content of the stimuli (Craik & Lockhart, 1972). Various strategies for organizing information in memory have proven to be beneficial. Categorizing stimuli into meaningful clusters or units based on semantic relatedness can increase the number of items that can be recalled (Calfee & Peterson, 1968; Paris, 1978). Paris (1978) demonstrated that younger children (second graders) did not spontaneously organize list items categorically, whereas older children (sixth graders) did. This finding indicates that the use of semantic information to actively construct organization in memory develops over time with greater knowledge, working memory, and metacognitive capacity. An interesting question therefore is whether nonhumans, in the absence of formal training and semantic labels, can also benefit from the opportunity to organize information categorically. Given that such strategies benefit human memory, it is possible that organizing information categorically also accrues benefits to nonhumans with regard to recalling, for example, edible versus nonedible foods.

When humans view stimuli that are not intrinsically meaningful, their memory for patterns is aided by the addition of a label for the pattern (Calfee & Peterson, 1968). Such results suggest that semantic labels facilitate memory for complex stimuli. Thus, the evolution of language in humans has undoubtedly enhanced the human capacity for long-term and associative memory. Many categories are linked through the use of language rather than perceptually shared features. For instance, humans can form categories for constructs as abstract as "positive qualities in colleagues" and "fundamental human rights." Nonhumans have been shown to represent abstract concepts as well, but these categories are usually recognizable by shared perceptual features, such as animals versus non-animals (Vonk & MacDonald, 2002, 2004; Vonk, Jett, & Mosteller, 2012), or identity (Vonk, 2003). Some experiments suggest that species such as pigeons may use observable features such as axis of orientation rather than abstract categories to discriminate images of different types of animals (Cook, Wright, & Drachman, 2012). Comparisons between human and nonhuman memory processes are therefore limited by an inability to conclusively determine how complex stimuli are processed and to utilize recall as well as recognition tests (but see Basile & Hampton, 2011). Researchers have shown several components of episodic-like memory in animals such as gorillas (Schwartz, Hoffman, & Evans, 2005) and corvids (Clayton, Yu, & Dickinson, 2001) despite reliance on recognition paradigms, but the jury is still out on whether this evidence constitutes palinscopic memory in which memory is oriented toward the past (Tulving & Lepage, 2000). To qualify as a component of episodic memory, palinscopic memory must involve reflecting upon specific past events rather than simply recalling facts. Some have referred to this aspect of memory as "mental time travel" (Tulving, 1999). The human proficiency with symbolic systems, such as language, greatly aids researchers' ability to assess the effects of various stimulus manipulations on recall versus recognition. Thus, our understanding of human memory systems is in some ways far advanced in comparison to our understanding of nonhuman memory.

Although nonhumans do not share the human capacity for language, several species have shown precursors to language-like abilities, being able to learn arbitrary symbols to represent real-life objects (Richards, Wolz, & Herman, 1984; Savage-Rumbaugh, Sevcik, & Hopkins, 1988) and even being able to use these symbols to communicate with humans (Savage-Rumbaugh, Sevcik, Brakke, & Rumbaugh, 1990; Sevcik & Savage-Rumbaugh, 1994). Non-primate species such as dolphins and sea lions have been shown to follow complex commands that require an appreciation for the syntactical structure of the command—for example, "take the ball from outside and put it through the blue hoop" (Gisiner & Schusterman, 1992; Herman, Kuczaj, & Holder, 1993). Pepperberg (2005) has argued, in contrast to the idea of evolutionary precursors to human-like language, that "certain linguistically relevant cognitive abilities common to humans and parrots are significant for the evolution of complex communication, and that necessary neural substrates for behavioral precursors to language can evolve in any reasonably complex vertebrate brain, given the right socioecological selection pressures" (p. 239). It is possible, then, that these nonhuman species share the capacity to organize information based on natural categories either through the perception of shared features or association through labeling. It is also possible that the memory abilities of "language-" or symbol-trained animals may be expected to differ from that of non–symbol-trained animals (see also Beran, 2011). Examining the effects of categorization on memory retrieval can help elucidate conscious versus automatic memory strategies and the manner in which organization aids memory.

The use of categorical information when constructing memories can be revealed by memory errors as well as successes. In free recall tests, humans have been shown to falsely remember words that are semantically related to words on a study list (Deese, 1959; Roediger & McDermott, 1995). For example, when presented with words like "sleep," "bed," "dream," "stars" and "night," subjects often falsely recall words such as "pillow" and "moon," even though "pillow" and "moon" were not on the list. Subjects also tend to falsely "recognize" items that are associated with list items on recognition tests, making more errors or false alarms to words semantically related to those on the test list (Roediger & McDermott, 1995). Swartz, Chen, and Terrace (1991) developed a list memory paradigm for use with nonhuman primates in which subjects view a list of items (typically arbitrary pictures such as a house, cat, beach ball, etc.) on a touch-sensitive monitor and, at the end of the list, select, by touching, only those items that were just presented as part of the list. Swartz and colleagues have shown that orangutans organize their responses spatially by touching items in a left–right response pattern across the test screen (Swartz, Himmanen, & Shumaker, 2007); however, this finding does not necessarily imply organization at the representational level during retrieval. A logical extension of this work is to include categorically related images within these lists to examine whether participants organize their recall by clustering categorically related images in memory, evidenced by touching the images from particular categories consecutively. Alternatively, when given limited opportunities to selected "studied" items, one could ascertain whether nonhumans select items from categorical groupings.

There is an abundance of evidence that a variety of species are capable of forming categories at various degrees of abstraction (Roberts & Mazmanian, 1988; Vonk, Jett, & Mosteller, 2012; Vonk, Jett, Mosteller, & Galvan, 2013; Vonk & MacDonald, 2002, 2004; Zentall, Wasserman, Lazareva, Thompson, & Ratterman, 2008), with evidence that language is not necessary for the formation of even abstract categories (Fize, Cauchoix, & Fabre-Thorpe, 2011; Vonk & MacDonald, 2002, 2004; Vonk et al., 2012). What is lacking is evidence that information is organized in memory according to these abstract categories. Humans readily use semantic categories to organize list items into meaningful clusters, reducing lists to meaningful units that can be more easily recalled (Cofer, Bruce, & Reicher, 1966; Cohen, 1963; Guerin & Miller, 2008). Humans also use mental imagery to aid in recall and generally recall more items when items are processed at a deeper (i.e. more conceptual) level (Craik & Lockhart, 1972; Lockhart & Craik, 1990), suggesting an advantage for conceptual encoding at the retrieval stage. Surprisingly, researchers have not yet tackled the interesting question of whether nonhumans might also benefit from encoding list items at a deeper, conceptual level. Basile and Hampton (2012) recently used depth of processing to investigate perceptual priming in rhesus macaques. However, they primarily used the depth of processing manipulation (categorization) to ensure that monkeys processed primes rather than to directly compare the effects on memory to conditions of shallow processing. In their study, monkeys did not recall primes more accurately than unprimed images despite the deeper processing.

However, many studies have revealed parallel memory processes in humans and nonhumans, such as the finding that both rats (Reed, Chih-Ta, Aggleton, & Rawlins, 1991; Reed & Richards, 1996) and monkeys (Parker, Wilding, & Akerman, 1998) share with humans the tendency to better remember more distinct items on a list, known as the von Restorff (1933), "oddball," or "isolation" effect. These findings are consistent with research demonstrating that primates find it relatively easy to learn non–match-to-sample identity tasks in which they are to select the nonmatching comparison image, also known as an oddity task (Davenport & Menzel, 1960; Davis, Leary, Stevens, & Thompson, 1967). The nonmatching item is more distinctive given that the target and the other comparison are exactly the same. Thus, some primates may be drawn to items that are dissimilar to other items presented simultaneously. In fact, researchers have found that some species of primate acquire oddity more quickly than they learn identity matching (Overman, McLain, Ormsby, & Brooks, 1983; Platt, Brannon, Briese, & French, 1996). However, these studies manipulated the *perceptual* distinctiveness of items within sets by having odd items differ on the basis of some visual feature such as shape or color. In contrast, and more relevant to the current work, recently Beran (2011) examined the isolation effect by presenting chimpanzees with four-item lists in which the isolate item was categorically, rather than perceptually, distinct from the rest of the homogenous sets. He found that two of four chimpanzees tested were more likely to select the isolate, from a different category, when choosing between that item and a distractor, than to select an item from the same list position that was categorically homogenous with other items in its list. This finding suggests that at least some chimpanzees spontaneously organize their memory with regard to semantic categories.

Although Beran (2011) examined categorical distinctiveness in memory, prior studies have not investigated the extent to which nonhumans prioritize perceptual similarity or category membership when they encode or retrieve visual stimuli. That is, researchers have not explicitly contrasted the contribution of conceptual rather than perceptual processes in nonhuman memory encoding or retrieval. Theorists have postulated that perceptual and conceptual processes involve two separate encoding processes (Intraub, 1984; Loftus & Ginn, 1984; Potter, 1976), with conceptual processes potentially operating in the absence of the visual stimulus, whereas perceptual processes require the presence of the stimulus. Here, we do not adopt that view, as we are referring to processes that occur not just during encoding but also during retrieval of the items from memory. That is, it is presumed that items may be encoded into and maintained in storage at only a perceptual level even when the stimulus is no longer present. In addition, the processes need not be independent, as many categories are formed based on perceptual features, yet more abstract categories can be inferred beyond the relationship of particular salient perceptual characteristics between the exemplars. The question of the extent to which these processes work in similar ways in humans and nonhumans is only beginning to provoke careful empirical investigation (Fize et al., 2011).

Brodbeck (1997) demonstrated perceptual priming in pigeons in an elegant paradigm modeled after the experimental literature on priming in humans (Schacter, 1994; Tulving & Schacter, 1990). However, despite the ease with which Brodbeck's study could be adapted to study the possibility of conceptual priming in nonhuman primates, researchers have been slow to adopt such methodologies. Recently Basile and Hampton (2012) failed to replicate his findings with rhesus macaques, which could explain the lack of published findings on this topic. Basile and Hampton alluded to the possibility that different mechanisms may be at work in pigeon versus monkey classification, suggesting that pigeons may process stimuli at a local level whereas monkeys may process stimuli more globally—suggesting greater parallels to human processing at least with regard to this aspect. Some studies in the area of cognitive neuroscience indicate significant overlap in brain activation between humans and monkeys working on tasks such as go/no-go procedures (Erickson & Desimone, 1999; Rainer, Rao, & Miller, 1999) and paired-associate tasks (Naya, Sakai, & Miyashita, 1996; Sakai & Miyashita, 1991), in which monkeys are primed to make a response to an associate of a prime, demonstrating prospective memory in some cases. These procedures demonstrate short-term and working memory processes similar to those used in lexical decision tasks with humans (Brunel & Lavigne,

2009). These types of studies, along with the isolation effect studies emphasizing semantic distinctiveness (as in Beran's 2011 study), have the potential to indicate whether nonhumans' representations of pictorial stimuli are semantic or conceptual as well as perceptual in nature. Evidence for semantic representations in nonhumans would indicate that language is not necessary for the formation of a semantic code and would suggest that concepts are acquired in the absence of verbal or symbolic labels, particularly if evidence for such organization is found in a non–symbol- or language-trained animal. Such evidence would also suggest that representations may be maintained and organized in memory in the absence of a linguistic code to aid in the organization process. Further, effects of conceptual information on memory may indicate conscious, rather than merely automatic, retrieval processes.

In the study presented here, we presented an adult chimpanzee with short lists of pictures in a modified serial list paradigm and, at test, presented him with both the studied items and distractor items that were either perceptually (e.g., yellow canary, yellow canoe) or conceptually (yellow canary, owl) similar to studied list items. Although it is certainly the case that items can be matched on various features such as body parts, and shape as well as color, in this study, perceptually matched items were matched primarily on the basis of color based on the results of an unpublished study, which found greater attention to color versus shape matches (Vonk, unpublished data). A greater tendency to select perceptual versus conceptual distractors would support the hypothesis that at least some nonhumans attend more to the perceptual features of stimuli than to their semantic referents when encoding or retrieving the items. It was predicted that if language is critical for categorizing items and for clustering in memory for later recall, nonhumans may be more likely to mistake perceptually similar items for studied items. The chimpanzee should predominantly select the three unrelated target items at test, but his errors could be either predominantly perceptual matches or conceptual matches. If the chimpanzee erred more often by selecting conceptual matches, this finding could support the idea that nonhumans without language training use semantic categories to organize information in memory and aid memory retrieval. For these to be truly memory errors (i.e., qualify as "false" memories), selection of the perceptual and conceptual matches should occur only when the target item was not selected. Alternatively, the chimpanzee could select a particular target and one or more of its "matches" if he was organizing the items into such categorical groupings on the basis of perceptual or conceptual similarity at retrieval. Thus, our procedure allowed for a flexible manner in which to investigate any potential organization of the chimpanzee's categorical representations. A preference for selecting items that conceptually matched the targets would be evidence for semantic or categorical organization at either the level of encoding or retrieval (but would not distinguish between the two).

In a second experiment, we manipulated category membership within our lists, with one item being distinctive by virtue of it not belonging to the same category as the other list items. We also manipulated the level of category (concrete or intermediate and the distance of the odd item from the other items in the set in terms of its categorical relatedness) to determine whether items perceptually rather than categorically distinct were more likely to be better remembered at test. As such, this experiment more directly attempted to assess the connection between memory and categorical organization.

GENERAL METHOD

Participant

A single, captive, adult male chimpanzee, named Joe, approximately 17 years of age, participated in these experiments. Joe was raised in the entertainment industry as a juvenile but has been residing alone at the Mobile Zoo in Wilmer, Alabama, for many years. He was tested in his indoor enclosure where zoo visitors could not observe the testing sessions, although he had free access to his outdoor enclosure during testing sessions and could come and go freely. Joe was not on a restricted diet during testing but was given additional treats as food rewards during testing. He had free access to water at all times. Joe was experimentally naïve when training in this experiment began, although he was simultaneously trained on a two-choice procedure to be used for a variety of other studies (Vonk, 2013; Vonk et al., 2013; Vonk & Vedder, 2013). Testing complied with the institutional animal care and use review board, and the zoo was compliant with U.S. Department of Agriculture regulations.

Materials

A durable Panasonic Toughbook CF-18 laptop computer and a 21-inch capacitative touchscreen monitor welded and encased within a hydraulic lift cart represented the experimental apparatus. Experiments were programmed using RealBasic 2006 for Windows. Stimuli consisted of two-dimensional photographs approximately 300 × 200 mm downloaded from various sources and edited using Adobe Photoshop CS2. Nine images were used in each of three sets for Experiment 1 for a total of 27 images. Figure 14.3 later in the chapter lists the items and rate of selection of each item by set. Nine images were used in each of 12 sets in Experiment 2 for a total of 108 images.

An incorrect response was followed by an unpleasant buzzer tone and a brief time-out with a blank black screen. Correct responses were followed by a pleasant tone and a blank white screen and paired with food reinforcement, which consisted of dried fruits and special treats such as honey-roasted peanuts, banana pellets, dried banana chips, yogurt-covered raisins, wafer cookies, etc. Food was presented by hand.

Procedure

The computer cart was pushed up against an open space between the steel bars separating the human experimenter from the chimpanzee in the inside enclosure, allowing the subject access to the touchscreen monitor. During testing the experimenter was centered behind the computer cart observing the animal's responses on the laptop monitor, which was positioned directly behind the touchscreen monitor. The experimenter could not provide any cues because she could not see the touchscreen or the chimpanzee's face or hands during the trial. She was always standing directly behind the touchscreen with her eyes on the laptop screen, positioned behind the touchscreen while the chimpanzee was responding. The experimenter stepped forward and presented Joe with a food reward, as described above, immediately following a correct response, which was signaled by the tone and white screen at a consistent location. Trials continued automatically until the end of a session. Joe completed as many sessions in a day as he was motivated to perform, intermixed with sessions of two-choice discrimination concept formation tasks, or match-to-sample tasks, which involved different stimuli. Typically he was given three to five sessions of memory tests in a single day, two or three times a week. The experiment took place over a period of more than 2 years as the chimpanzee was sometimes tested only on other experiments for weeks at a time.

Training

Joe was first trained on the task of touching a single image on the screen. If he touched anywhere else on the screen other than on the image itself within a set period of time (which was gradually reduced to 2 seconds), a buzzer sounded and the trial was considered incorrect. If he selected the image on the screen a pleasant tone sounded and he was reinforced with a small food reward, and the next image was presented at another random location on the screen. Initially he was presented with only the study trials, with each session consisting of 27 trials. Three images appeared in each of the nine locations on the screen once each across all study trials. Only a single item appeared on a screen during each trial. The images appeared in random locations and orders. Eventually, when Joe was consistently able to select the image within 2 seconds without touching any other part of the screen for all 27 trials, the test screen was introduced. On the test screen, the three studied target images appeared in random locations on the screen. Six distractors filled the other locations on the screen. Thus, all nine items appeared simultaneously on the screen during the test. Joe was given three test trials within each session in which he was rewarded if he touched one of the studied items and not rewarded if he touched a nonstudied item. Correct responses were paired with the pleasant tone and food reinforcement while incorrect responses were paired with unpleasant buzzer sounds and no reinforcement. Images disappeared when selected. After three selections the session ended. Figure 14.1 depicts a representation of the training procedure.

The length of the target/study list was limited to three items because we needed all target items and their related distractors to appear simultaneously on the test screen at the end of the study trials. Given the maximum size of the durable touchscreen at 21 inches, it was determined that no more than nine images should be displayed at a reasonable size on the screen at a given time.

In training, Joe was presented with homogenous sets of images, with the distracter images coming from two other homogenous sets, until he met a criterion of three correct choices on the test screen for three consecutive sessions—that is, sets contained items belonging to a single theme: Spiderman, Sponge Bob, Shrek, motherboards, and sports cars. These training items were chosen because they did not include categories that were being used in categorization tests that Joe was working on simultaneously (Vonk et al., 2013). However, by the time Joe reached the testing phase, he had completed the primary categorization task. Test items were then chosen to include targets and conceptual distracters that Joe would have encountered previously, either in reality or in images that he had categorized previously, such as foods, buildings, other animals such as birds, and flowers, which he could see from his outdoor enclosure. Although the categories of some items overlapped with previously learned discriminations, all of the images used for this experiment were new to Joe. Following criterion, he was presented with a transfer set in which the studied items came from a different set. In this manner we could ensure that Joe had learned the general

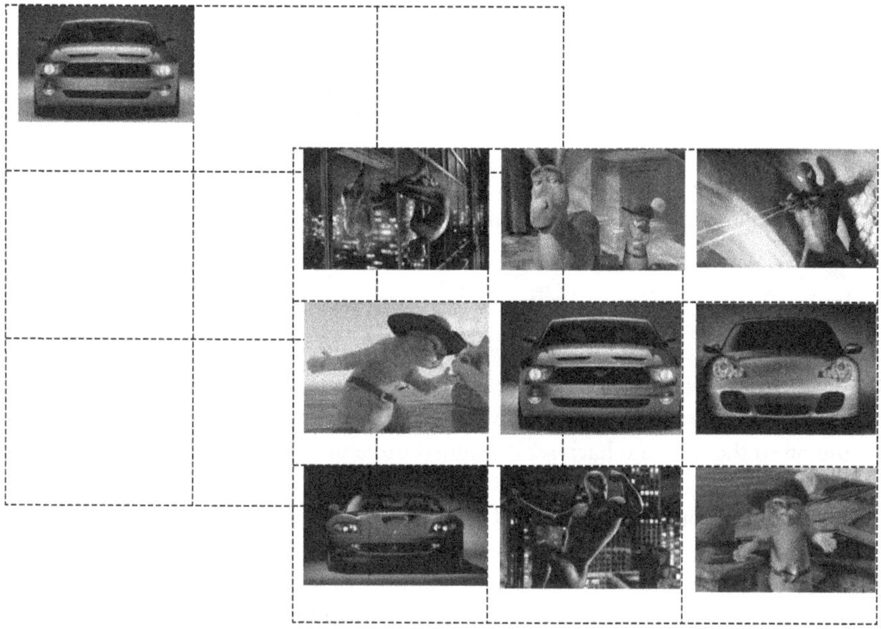

Figure 14.1 A schematic representation of the basic experimental procedure used during training. Top left depicts a possible study trial in which a single image appears in one of the nine locations on the screen. Bottom right depicts the test screen in which all three of the targets (in this case sports cars) appear along with distractors from two other training categories (Shrek and Spiderman). The subject's task is to select all three of the cars on three test trials. An image disappears when selected.

procedure and did not just select the specific images that belonged to the first set of items. Several transfer sessions with novel sets of images were required before transfer was obtained, and Joe proceeded to testing. Once Joe mastered the basic procedure he was given reinforcement for approximately every third response to the study items rather than rewarding him for every correct response. He was reinforced with small items such as raisins and dried cranberries on study trials and with a larger treat (e.g., large chunks of dried fruit, such as mango or papaya) for every correct response on the test screen. Because responses during study trials were simply orienting responses to ensure that Joe saw the test items, it was not considered necessary to reward each correct response. However, during testing Joe was required to make a discrimination between previously studied and unstudied items. Thus, to maintain his understanding of the task, during the test, he was rewarded for each correct choice.

EXPERIMENT 1: CONCEPTUAL VERSUS PERCEPTUAL DISTRACTORS

To determine whether the chimpanzee organized items in memory categorically or perceptually, a serial list paradigm was adopted in which target items, which appeared during the study phase, were matched with items that (a) belonged to the same categories but shared few perceptual features (conceptual matches) or (b) did not belong to the same categories but shared perceptual features, such as color (perceptual matches). Use of Visual Similarity Duplicate Image Finder (version 4.2.0.1), available at MindGems.com, confirmed that the conceptual matches did not share more than 50 percent similarity with the targets. All of the items were presented at test to determine whether, when errors were made, they were more often from the categorical/conceptual match or perceptual match category. In addition, because the chimpanzee could select three items, we could determine whether he chose items related to target items as a coherent set based on category membership.

Method

During testing, Joe completed 55 sessions of each of three sets of test stimuli. He completed all 55 sessions of Test 1 before beginning Test 2 and all 55 sessions of Test 2 before beginning Test 3. He was not required to reach a criterion level of responding before moving on to the next set. Each set contained three unrelated targets. Each target was matched to an image that was from the same category but did not appear perceptually similar (conceptual matches), and an image that shared perceptual features such as color but did not belong to the same category (perceptual matches) to create the six distracters to appear on the test screen. The number of sets that could be presented was limited due to the difficulty in finding images that could meet the stringent criterion for belonging to the same category without sharing perceptual features with the target, or sharing perceptual features without possibly belonging to the same category (at any level of category membership). Although there was some concern that the distractors would also become familiar over the course of 55 sessions with the same set of images, once Joe learned the basic procedure it was still the case that they did not appear within the study session of 27 trials, while the target items did. Thus, the target items were always those that had just been presented in the trials immediately prior to testing.

The goal of the training procedure was to teach Joe to select the items that were presented during the study trials immediately preceding testing. Training items were chosen to appear perceptually similar to create "easily discriminable" sets of three items that were distinctive from the distracters. The images used in training were not necessarily intrinsically meaningful to chimpanzees (e.g., sports cars, Spiderman, Shrek characters) but displayed clear color patterns and shapes that did not overlap between categories. For instance, sports cars are not found in a chimpanzee's natural world but are real objects nonetheless. Shrek characters were cartoon drawings. Spiderman images depicted a human in disguise. We chose these images to avoid presenting items from natural categories that Joe had been previously trained to select or not to select (Vonk et al., 2013). Data from training were not analyzed with regard to perceptual versus conceptual memory processes and were used simply for the purpose of training Joe on the goal of the procedure (i.e., to select on the test screen only those items that had just been presented during the 27 "study" trials). Ideally, after learning the goal of the experimental procedure, test choices would reflect explicit use of memory rather than an overall sense of familiarity with the images.

Analyses

We analyzed whether Joe's choices of the targets, conceptual matches, and perceptual matches were above chance for each set using one-sample *t*-tests. We also compared his choice of conceptual versus perceptual distracters overall and when he did not choose a given target. In the latter case, we examined how often he chose items that had been deemed conceptual or perceptual matches for the chosen target (match-to-target chosen) and therefore were errors that should not be categorized as "memory confusions" *per se*. In this way, we could examine in particular his memory failures: Was he more likely to choose images that conceptually or perceptually matched the targets he failed to select, or distractor items that were paired with the other targets he had in fact selected? Chance levels were determined as a function of the number of options available to Joe at each choice and could not be set to a standard level as the number of available options changed with each selection. Alpha was set to .05 for all tests. We also examined Joe's pattern of responding to examine whether there were particular images he preferred to select, regardless of whether they were targets or conceptual or perceptual matches. In comparing his choices across these different types of stimuli, we used Cochran's *Q* test to verify if his choices were evenly distributed among the stimuli.

Results and Discussion

Training

Joe required 107 sessions to reach criterion with the first set of training images (Spiderman). Joe did not show immediate transfer to the first set of transfer stimuli (sports cars) and received over 175 sessions with this set of stimuli before demonstrating transfer on a third set of stimuli (Sponge Bob). Transfer was determined when he chose all three target images at testing on first presentation of a new list.

Testing

Across all three test sets, targets were chosen significantly above chance, one-sample *t*-tests ($t(2) = 6.35, p = .02$), indicating that Joe did indeed understand the goal of the procedure. Conceptual distracters were chosen at levels not quite significantly below chance, ($t(2) = 3.49, p = .07$). Perceptual distracters were chosen at levels that did not significantly differ from chance. Chance was set at .33, given that this was the average probability across all the possibilities for choosing an item of a given type, given all the other options on a trial, based on prior choices. However, there was no significant difference in Joe's likelihood to choose perceptual over conceptual distractors (28 vs. 25 percent), paired *t*-test, $t(2) = .66, p = .58$. The percentage of choices of each type of image across the three sets of images is depicted in Figure 14.2. Clearly, Joe was no more likely overall to choose images that were categorically related to targets or that appeared perceptually similar to targets. Perhaps this was due to the fact that he chose targets above chance and thus made fewer errors of any kind.

However, our overall analysis may not have been sensitive enough to detect true confusions of memory. Therefore, in addition, for each test, to more accurately examine the issue of memory "confusions" we considered Joe's choices in relation to his choice of targets. If he failed to choose a given target, we examined whether he chose its conceptual or perceptual match in that session, and we also examined whether he chose a match for a target he had chosen on that trial, which would not be considered a memory "confusion." That is, we considered the items that were deemed

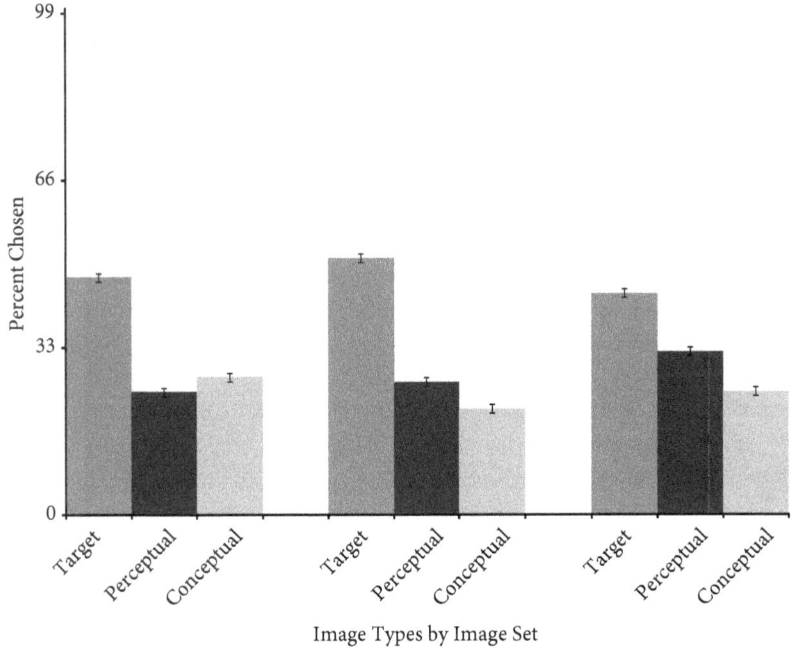

Figure 14.2 Percentage of choices of studied items (target), conceptual matches, and perceptual matches across test sets in Experiment 1. Chance was determined to be 33 percent.

as conceptual or perceptual matches for a target as its matches and labeled choices of those items as "match for target chosen." We subtracted the number of choices of targets from the overall number of opportunities to make choices (three choices per session across 55 sessions = 165) to calculate the *N* from which to calculate percentage of choices of each of the possible options (conceptual confusion, perceptual confusion, or match-to-target chosen). The percentage of Joe's choices of each of these types of stimuli across each test set appear in Figure 14.3. Friedman's two-way analysis of ranks indicates that the distributions of choices of each of these types of stimuli are essentially the same ($p = .097$). However, one-sample *t*-tests revealed that only the choices of match-to-target-chosen differed significantly from chance, $t(2) = 6.84$, $p = .02$. Chance was calculated at .25, given that, once one target was chosen, there remained an opportunity to make two such choices out of eight options. Of the remaining eight items, two were the other two targets, two were matches for the target that was chosen, and the remaining four items were perceptual and conceptual matches for the remaining targets (allowing for two each possible conceptual and perceptual confusions out of eight choices). Thus, Joe was most likely to choose stimuli that matched targets he also chose on the same trial, suggesting that these choices did not reflect memory confusions but might in fact reflect an organization of memory around those particular items.

It is possible that the procedure of rewarding Joe immediately for a correct choice influenced his responses on the following trials, such that reinforcing him for choosing canary may have encouraged him to select owl on the next choice. However, we elected to reinforce his response immediately, rather than following all three choices,

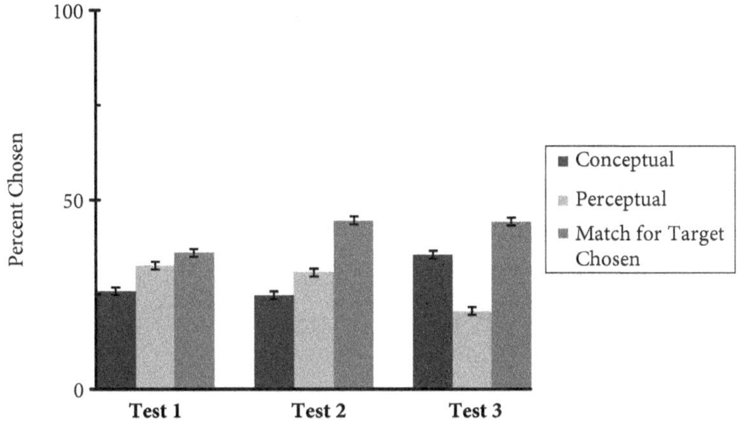

Figure 14.3 Percentage of choices for conceptual or perceptual matches for studied items not chosen (memory confusions) AND percentage of choices for matches (conceptual and perceptual) for studied items (targets) that WERE selected on a given trial, such that those choices should NOT be deemed memory confusions, as a function of test set in Experiment 1. Chance was determined to be 25 percent.

to avoid treating his choice of a target item as incorrect. Given Joe's prior testing history (Vonk et al., 2013), in which he was typically reinforced for every correct response, we wished to continue with a similar practice on test trials in which there was a choice to be made so as not to confuse him regarding the procedure. This strategy differs from that during the study trials, in which no choice had to be made (there was only one image on the screen and Joe simply had to make an orienting response).

Joe's choices were determined primarily by particular stimuli within each set. Figure 14.4 depicts the number of choices of individual items as a function of whether the items were targets, conceptual distractors, or perceptual distractors, across test sets. When ignoring the category of target, conceptual distractor, or perceptual distractor across sets, several of the images were chosen at rates significantly different from chance, according to binomial tests, with chance being determined as .126 given that there were three options to choose first among nine options on each test screen ($p = .11$), then among eight options ($p = .125$), then among seven options ($p = .14$). The average of these three probabilities is .126. Rates of choosing the stimuli at various probabilities against chance are indicated on Figure 14.4. In addition, the distributions of choice of each stimulus were not equivalent according to Cochran's Q test, $p < .001$. When examining Figure 14.4, one can see preferences for particular stimuli within each set that were not necessarily determined by whether the image was a target or a conceptual or perceptual match to a target. This finding occurred even though Joe tended to select targets at higher rates overall. One can see some idiosyncratic preferences; however, a general theme is that Joe selected images of foods and animals and other more familiar objects, as well as brightly colored images such as flowers at higher rates. Thus, it is possible that Joe's selections were driven more by personal meaningfulness rather than connections between the list items, which is an interesting finding in its own right and further

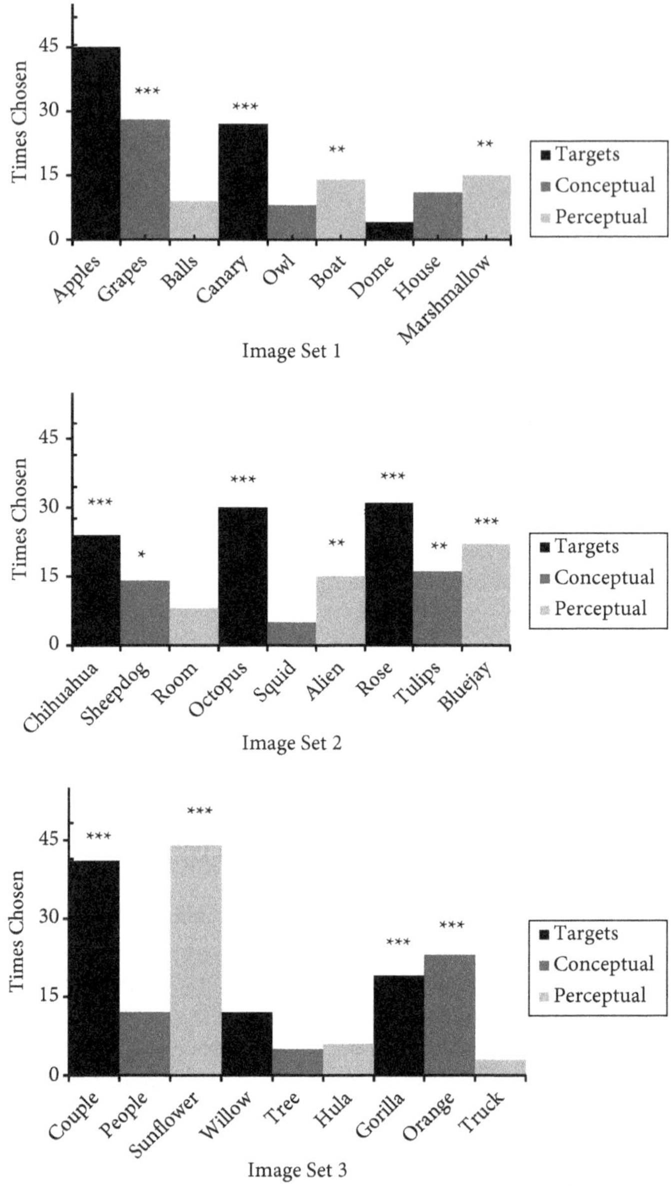

Figure 14.4 Number of choices for particular items within each test set in Experiment 1. *** = p <.001, ** = p <.01, * = p <.05.

suggests that two-dimensional stimuli are meaningful to nonhuman primates (see also Davenport & Rogers, 1971; Fagot, Martin-Malivel, & Dépy, 1999).

It appeared that Joe's choices were driven more by idiosyncratic preferences for particular stimuli rather than for memory for studied items in the experimental procedure. In addition, it is possible that low motivation to work for reinforcement affected Joe's performance. Furthermore, a better experimental procedure might have been to present Joe with longer lists for brief durations and to present completely novel lists on each experimental session. However, given the difficulty composing lists of pictorial stimuli that satisfied our requirements for matches that were categorical but not perceptual and vice versa, and were independently related to each other, it was not possible to create enough unique sets of items until Joe reached criterion on the experimental procedure, or to create multiple test sets. It also might have been advantageous to present Joe with a single choice between a studied item and a distractor item on each test trial, as was done by Beran (2011). Alternatively, it might have been interesting to omit the option of choosing the target to see if Joe was more likely to choose perceptual versus conceptual distractors when the target was not an option, but this procedure might confuse Joe in that he might not have perceived the options as a memory test without the presentation of previously "studied" items. It is likely that, given such a procedure, his choices would be random. Furthermore, in this study, we wished to compare the rates of choices of studied items and distractors of different types when all items were equally available. We were also interested in whether Joe would organize in memory target items with items categorically or perceptually similar and touch all of these items, belonging to a set based on similarity, rather than selecting the three unrelated targets. However, in future, a better procedure might be to present each item with only its own conceptual and perceptual distractor on each of three test trials to more carefully control the selections and probabilities of choice of each item. The downside of such a procedure would be that one would never be able to determine if the subject spontaneously used such strategies as selecting all items from one grouping, such as target A and both its conceptual and perceptual distractor, as we sometimes observed here. Given that so little is known about any nonhuman's use of semantic organization in memory (or lack thereof), we wished to allow the greatest flexibility for determining any use of strategies in Joe's recall. This method is similar to that used by Swartz & Himmanen, 2001, 2007) for investigating use of strategies in orangutan's recall.

EXPERIMENT 2: ODDITY

Given that Joe's choices in Experiment 1 seemed driven more by personal preferences rather than by memory or categorization, in Experiment 2 we elected a different method to examine if Joe might organize categorically related items together in memory. If items are organized in memory at an implicit or automatic level, items that do not belong to the category in question might stand out by virtue of being more distinctive, as per the isolation or von Restorff effect (von Restorff, 1933). In addition, items that are even further unrelated to the main category might be more distinctive and thus better remembered. Thus we manipulated category membership and the level of category that items belonged to, and were out-group members of, in a second experiment. By manipulating level of category membership we also manipulated the extent to which the odd item was perceptually versus semantically distinctive, and thus could further determine if Joe might be relying on perceptual rather than categorical similarity to guide his organization of list memory in this task. Thus, this experiment is a more direct test of the influence of category membership/distinctiveness on memory.

Method

Participant

Joe participated again in the current experiment. This experiment was conducted more than 2 years following the beginning of Experiment 1 and after Joe had completed a series of natural category formation experiments (Vonk et al., 2013). However, it began immediately following completion of the third test set in Experiment 1, so Joe was still familiar with the test procedure when he began testing in Experiment 2.

Materials

The same experimental setup and materials were used for Experiment 2. New sets of stimuli were created such that there were 12 sets of stimuli: 4 in which two items of the set belonged to a concrete-level category and the odd item differed in terms of only one concrete-level feature, such as color; 4 in which two items belonged to an intermediate-level category and the odd item belonged within the same superordinate category grouping; and 4 in which two items belonged to a coherent category and the odd item was completely unrelated. Although we did not establish whether Joe actually perceived these items as belonging to coherent categories by having him complete a categorization task with these items simultaneously, to avoid oversaturation with the items and interference with the familiarity aspect that was being tested in the current experiment, we used items that were from categories similar to those that Joe had successfully discriminated in another study (e.g., carnivores, animals, foods, and landscapes; Vonk et al., 2013) and/or that were familiar in his environment (foods, birds, tools, manmade objects). This decision is in line with that of Beran (2011), who similarly wished to avoid interference with repeated presentation of test items in a different context. Oversaturation during a categorization task might mitigate against the likelihood of obtaining an isolation effect. In addition, finding an isolation effect itself implies that the other items are perceived of as categorically similar. None of the items had been used in previous experiments, and the specific categories had not been discriminated previously. However, by using images of only familiar objects we hoped to avoid the issue of idiosyncratic preferences for more familiar items that we encountered in Experiment 1, while also presenting images of items that would be recognizable to Joe.

For concrete sets, for instance, the "bananas" set included two images of standard yellow bananas and the odd item was a cluster of red bananas. In the "broccoli" set, two images were of green broccoli and the odd item was of purple broccoli. In the "bear" set, two images were of black bears and the odd image was of a white bear. In the "tigers" set, two tigers were of traditional pattern, while the odd item was a white tiger. Thus, there were four such concrete-level sets, two including food items and two including animals among the targets. At this concrete level, then, the odd item was as perceptually distinct as it was categorically distinct. However, only the items from the bear and broccoli set passed an analysis of perceptual similarity using a visual similarity calculator. The Visual Similarity Duplicate Image Finder (version 4.2.0.1), available at MindGems.com, was used. This program examines images to determine whether they are duplicates or not and can find them even if the file name, format, and size are different or the images have been rotated. A percentage of image similarity that will be used during the scan to limit or extend the search results can be specified. The program used advanced algorithms to visually compare all photographs and reported any clusters of images that shared greater than 50 percent physical similarity.

At testing, all three studied items were presented, as in Experiment 1, along with six distractor items. As before, the six distractor items came from two homogenous sets. For bananas and broccoli, the distractors were either foods or tools (but were novel to each set). For bears, the distractors were birds and cakes. For tigers, the distractors were birds and berries. In this manner, one set of distractors was more closely related to the studied items conceptually than the other set of distractors. For each new set, all of the images were completely novel.

There were also four "intermediate sets" in which the odd item belonged to a different set within an intermediate or abstract category. So, for instance, within the "fruits" set, there were two fruits of the same type but different colors (a red and a green apple) and a different kind of fruit (blueberries) as the odd item. For the "vegetables" set, there were two different kinds of beans (fava and string beans) and a green pepper for the odd item. The distractors for these sets were also foods and tools (for fruits) and birds (for vegetables). The "cats" set contained two different cat species and the odd item was a non-feline animal (a goat). The distractors for cats were fish and cakes. The foods set contained two fruits and a non-fruit food item (marshmallow). This differed from the "fruits" set because in that set all items were fruits; in this set all items were foods, but not all items were fruits (thus, the overall category membership was superordinate). Joe was familiar with the foods used in this task. The distractors for foods were a different set of food items and animals. At this level of category membership, the odd item was categorically but not distinctly perceptually different from the other members of the list. Visual similarity analysis confirmed that members of these sets were not perceptually similar.

Lastly, there were four sets where two of the items were either from the same concrete category (rats, melons) or more intermediate/abstract category (flowers, tools) and the third item was completely unrelated. In the case of rats, it was a stool and the distractors were tables and toys. In the case of melons, it was also a stool (but a different type of stool) and the distractors were benches and nuts. In the case of flowers, random items were used. In the case of brooms, it was a dust cloth and the distractors were nuts and other tools. Thus, in this set, the odd item was always categorically distinct but varied in the extent to which it was perceptually distinctive from the other list items. Rats, brooms, and flowers were considered perceptually similar to each other and distinct from the odd item. No other items passed the test of visual similarity.

Although the point of the study was to manipulate the level of category membership and the odd items were chosen based on the fact that they were categorically distinct from the homogenous category members, it is also true that the odd items are more perceptually distinct as well. But this is even more the case at the concrete level relative to the intermediate level, in which category members are also perceptually distinct but still belong to the same category (e.g., fava beans and string beans, red apple, green apple). Thus, manipulating category membership was also a way of measuring if distinctiveness pertained to perceptual rather than categorical similarity.

Analysis with the Visual Similarity Duplicate Image Finder confirmed that odd items shared less than 50 percent similarity with other list set items for all sets. In addition, only rats, black bears, broccoli, brooms, and flowers shared greater than 55 percent similarity with each other. In other words, many of the less concrete categories, such as foods, cats, and vegetables, did not share a high degree of perceptual similarity. However, some concrete-level categories, such as tigers and bananas, were also considered to share less than 50 percent similarity.

PROCEDURE

The basic procedure was identical to that of Experiment 1. Joe was first tested on the bananas, fruits, and flowers sets. When he had completed eight sessions of each of those sets, he moved on to the broccoli, vegetables, and tools sets. When he had completed eight sessions of each of those sets, he moved on to the last six sets. He was tested on one session of each of those sets each day for 8 days. Thus, Joe completed six sessions a day, one with each of the different sets of stimuli in semi-random order such that he began at least one testing day with each of the different sets of oddity stimuli. Across testing, order of testing at each level of category membership was balanced. Joe was not required to reach any kind of criterion level of responding on one set before moving on to another.

Results and Discussion

There were no significant differences between Joe's choice of studied items between concrete, intermediate, or outside sets (Friedman's two-way analysis of ranks, $p = .25$). There were also no significant differences in Joe's choice of odd items between concrete, intermediate, or outside sets (Friedman's two-way analysis of ranks, $p = .93$). Figure 14.5 (top) depicts the percentage of choices of odd items as a function of category, and Figure 14.5 (bottom) depicts the percentage of choices of correct items as a function of category. Joe did not choose studied items at levels significantly greater than chance in any of the categories, according to one-sample t-tests (all $ps > .20$). Chance was set at a level of .33 for choosing studied items, given that this was the average probability across all the possibilities for choosing a studied item rather than an item from one of the other two homogenous distractor sets given all the other options on a trial, based on prior choices.

However, Joe did choose odd items in the concrete categories and "outside" categories at levels greater than chance; concrete, $t(3) = 4.88$, $p = .02$, $CI = .09–.41$, outside, $t(3) = 3.56$, $p = .04$, $CI = .03–.53$. For odd items, chance was determined as .126 given that there was only one odd item available on each test screen and there were three options to choose first among nine options on each test screen ($p = .11$), then among eight options ($p = .125$), then among seven options ($p = .14$) The average of these three probabilities is .126.

Given Joe's tendency to choose particular images regardless of category membership, it is likely that he was more motivated to choose images that he liked rather than those that were technically "correct" from an experimental point of view. Further evidence for this is the finding that Joe's choice of targets was not significantly above chance in Experiment 2, although he had clearly already mastered the procedure at

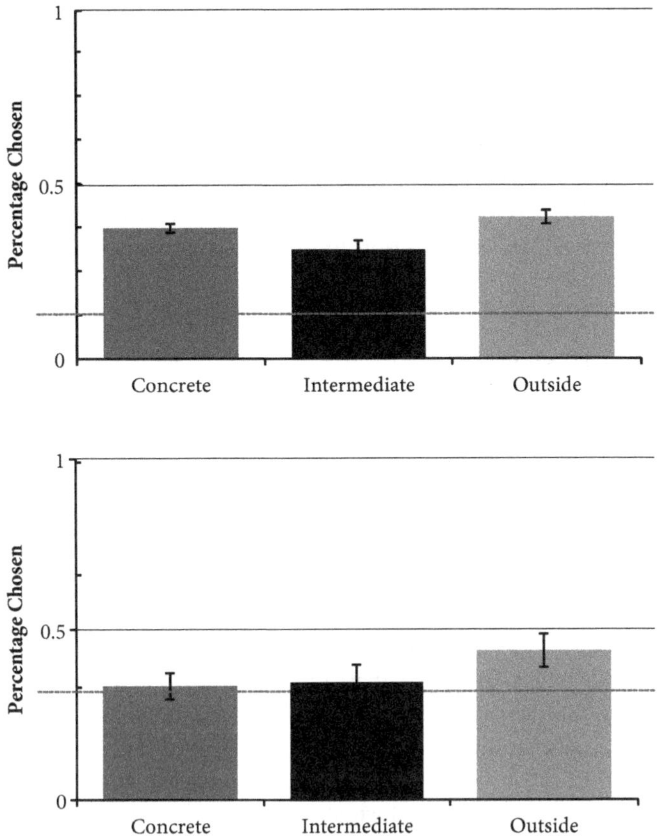

Figure 14.5 (Top) Percentage of choices of odd items according to set category in Experiment 2. Chance is indicated by a dotted line at .126. (Bottom) Percentage of choices of studied items according to set category in Experiment 2. Chance is indicated by a dotted line at .33.

this point. Despite this likelihood, we still obtained evidence for the isolation effect in that Joe was more likely to select odd items from two of the categories. Thus, although he did not choose targets above chance overall, when he did choose them, he was most likely to choose the targets that were odd items of the set. He was more likely to select odd items that were more perceptually distinct, given that these items differed from members of a concrete category in terms of color or some other obvious property, or were members of a completely unrelated category from the other two members of the list, as in the "outside" sets. This finding is similar to that of Beran (2011), who found that two of four chimpanzees were more likely to remember odd items than items from the same position in a homogenous list that did not contain odd items. It is therefore possible that this procedure can be applied usefully to other subjects who are more motivated to earn food rewards.

Of course there are also many other methodological limitations. With more novel lists rather than repeated presentations of the same items, and with a larger list, it is possible that odd items would have been even more distinctive and better remembered (Overman & Doty, 1980; Roberts, Mazmanian, & Kraemer, 1984; Sands &

Wright, 1980a, 1980b). Although Beran (2011) found an effect of isolation using only four item lists with two chimpanzees, two other chimpanzees did not show the effect. It is likely that larger list sets would facilitate the isolation effect and allow better determination of the use of strategies for memory organization (Treichler & Van Tillburg, 1996). Future work should modify the existing procedure with a larger set size and a larger sample size of participants. In addition, perceptual similarity and conceptual relatedness (from the participant's perspective) need to be determined using more objective measures *a priori* to increase the validity of the experimental procedure.

GENERAL DISCUSSION

These experiments are an important first step toward bridging the literature on categorization and memory in nonhumans to explore the potential use of semantic or categorical information in memory organization in species without language. Such research has the potential to inform the organization of cognitive structures in both nonhumans and humans with regard to the dependence or independence of language, abstraction, and concept acquisition. With the exception of the recent study by Beran (2011), no other published reports have explicitly focused on semantic rather than perceptual organization in memory in nonhumans. As noted by Beran, some of the chimpanzees that participated in that study had a history of training to attend to semantic information, so replication with a chimpanzee with no history of symbolic or "language" training is important.

Furthermore, no prior reports have attempted to tease apart nonhumans' reliance on conceptual or semantic information over perceptual information, or vice versa. The current results offer only tentative implications in this regard, given the chimpanzee's idiosyncratic preferences for non-target images in both experiments, despite evidence that he had learned the experimental procedure. Despite this result, the findings of Experiment 1 suggested that the chimpanzee might be slightly more likely to select perceptual, rather than conceptual, matches for target items. In line with this finding, in Experiment 2, odd items were more likely to be selected at test if they were perceptually distinct from other list items; that is, if they belonged to the concrete or outside category sets, rather than the intermediate category sets, in which the odd item was distinctive more by virtue of category membership rather than a single concrete perceptual feature such as color.

It is important to note that presentations of visual stimuli may never produce comparable results to presentation of word lists, which can be presented only to verbal organisms such as humans. Research has shown that conceptual processing of visual information may differ from conceptual processing of verbal information (Hanna & Loftus, 1993) in that typically robust serial position effects are not obtained. Thus, one must be cautioned in proclaiming species differences rather than methodological differences. Future studies should test both human adults and children in completely analogous paradigms. Unfortunately it will not be possible to test nonhumans with word lists, although it is possible to test symbol-trained chimpanzees with lexigram lists. Hampton and Schwartz (2004) outlined a program of research for demonstrating a correspondence between human and nonhuman memory with regard to episodic memory, but many more strides need to be taken to achieve that general goal. However, as one step in the direction of achieving better correspondence with memory tests, Basile and Hampton (2011) recently devised a clever recall test for use with nonhuman primates. Such advances will open the door for investigations of parallel processes in both human and nonhuman species.

The findings of Experiment 1, in which the chimpanzee was most likely to select both targets and items that were either conceptual or perceptual matches for a target that was also chosen on that session (rather than distractor items), suggests the intriguing possibility that he may have categorized items together that both appeared similar to and belonged to the same category as items that he recalled from the study set. It might make sense that he would respond to the test in such a way given that he was simultaneously working on categorization experiments (Vonk et al., 2013). It is important when designing such experiments to allow for the possibility that an animal might use a strategy not conceived of by a researcher, so as not to limit the possibilities for interesting responses. We think that these findings encourage the pursuit of this and other related methodologies with both other chimpanzee subjects and different species, as this is an untapped research question with much to inform us about the evolution of categorization and language in our own and both closely and distantly related species.

Two prolific areas of research in nonhuman memory deal with the possibility of episodic-like memory (Crystal, 2010; Dere, Kart-Teke, Huston, & De, 2006; Schwartz et al., 2005) and metacognition (Hampton, 2001, 2009; Terrace & Metcalfe, 2005) in nonhumans. Hampton (2001, 2003) distinguishes between direct and indirect metacognition. To date, evidence for memory awareness in monkeys using tests that have provided clear evidence in humans and apes has been mixed, with better evidence for memory awareness in rhesus macaques than capuchins (capuchins; Basile, Hampton, Suomi, & Murray, 2009; rhesus macaques; Hampton & Hampstead; 2006; Hampton, Zivin, & Murray, 2004; Templer & Hampton, 2012). Similar studies have yet to be conducted with a wide range of non-primate species. Furthermore, few researchers have applied the concept of conscious or explicit memory processes as defined in work with humans to the study of consciousness in nonhumans despite its potential to elucidate possible continuities in conscious processing between nonhumans and humans. Given that deeper processing (along with other variables) had long been assumed to affect conscious but not unconscious use of memory in humans (Graf & Mandler, 1984; Jacoby & Dallas, 1981; Roediger, & McDermott, 1993), the effects of conceptual processing on nonhuman memory has been grossly understudied. Rob Hampton's work has most closely approached this topic with one particular study aimed at dissociating automatic and conscious memory in rhesus monkeys (Tu, Hampton, & Murray, 2011). In this study, the authors applied the process dissociation procedure developed by Jacoby (1991) to provide evidence for two memory systems (assessed by manipulating habit and one-trial memory on visual matching-to-sample performance) in monkeys. In subsequent tests, two of the same authors manipulated memory delay intervals and other encoding conditions and found that certain factors affected one trial memory but not habit scores and vice versa (Tu & Hampton, 2012), providing empirical support for the research strategy proposed here. The additional development of a recall test for nonhumans (Basile & Hampton, 2011) should provide rich opportunities to investigate similarities and differences in the factors that affect recognition and recall in nonhumans and humans.

Such studies, along with the type of experiments presented here, yield great promise in demonstrating the use of conscious memory strategies in nonhumans, and it is hoped that preliminary findings such as these will promote further attempts to investigate the topic using procedures pioneered for the study of memory systems in humans. In addition, given that methodologies such as computerized testing have now been successfully applied to a variety of more distantly related species, such as dogs (Range, Aust, Steurer, & Huber, 2008) and bears (Vonk & Beran, 2012; Vonk et al., 2012), it is hoped that such tests will be applied more broadly in the animal kingdom in the near future.

Applying these studies to the topic of consciousness extends the relevance of the research to the question of adaptive memory processes. If animals exhibit conscious control over memory retrieval, they could utilize metacognitive processes to aid in memory retrieval or assist with planning for the future. Shettleworth (2012) has suggested that animals are unlikely to use episodic-like memory to reflect on past experiences in an autobiographical sense but may very well use such processes to assist in planning future foraging (as just one example). The study of semantic or conceptual processing and its effects on memory is just one tool to contribute to the study of conscious rather than automatic memory in nonhumans. Although it is important to note that the idea that conceptual processing affects only explicit but not implicit memory is controversial (Brown & Mitchell, 1994; Challis & Brodbeck, 1992; Thapar & Greene, 1994), the study of such factors on different memory tasks as applied to human memory is a practical general strategy for implicating different memory systems or processes in nonhumans as well. Only when researchers apply similar constructs and methodologies to the study of human and nonhuman cognition, taking into account necessary modifications to adapt to morphological, perceptual, and behavioral differences, will they be better able to draw inferences regarding cognitive continuities or discontinuities.

ACKNOWLEDGMENTS

We are indebted to the Mobile Zoo, especially its director, John Hightower. Without his support and assistance these experiments could not have been conducted. In addition, special thanks to Dr. Joan Sinnott for her support, as well as the assistance of Stephanie E. Jett, Tamra Cater, and Valeria Zamisch during data collection. We also thank three anonymous reviewers for their comments on an earlier draft.

REFERENCES

Basile, B. M., & Hampton, R. R. (2011). Monkeys recall and reproduce simple shapes from memory. *Current Biology, 21*, 774–778.

Basile, B. M., & Hampton, R. R. (2012). Monkeys show recognition without priming in a classification task. *Behavioural Processes*, doi: http://dx.doi.org/10.1016/j.beproc.2012.08.005

Basile, B. M., Hampton, R. R., Suomi, S. J., & Murray, E. A. (2009). An assessment of memory awareness in tufted capuchin monkeys (*Cebus apella*). *Animal Cognition, 12*, 169–180. doi: http://dx.doi.org/10.1007/s10071-008-0180-1

Beran, M. J. (2011). Chimpanzees (*Pan troglodytes*) show the isolation effect during serial list recognition memory tests. *Animal Cognition, 14*, 637–645.

Brodbeck, D. R. (1997). Picture fragment completion: Priming in the pigeon. *Journal of Experimental Psychology: Animal Behavior Processes, 23*, 461–468.

Brown, A. S., & Mitchell, D. B. (1994). A reevaluation of semantic versus nonsemantic processing in implicit memory. *Memory & Cognition, 22*, 533–541.

Brunel, N., & Lavigne, F. (2009). Semantic priming in a cortical network model. *Journal of Cognitive Neuroscience, 21*, 2300–2319. doi:10.1162/jocn.2008.21156

Calfee, R. C., & Peterson, R. E. (1968). Effect of list organization on short-term probe recall. *Journal of Experimental Psychology, 78*, 468–474. doi: http://dx.doi.org/10.1037/h0026464

Challis, B. H., & Brodbeck, D. R. (1992). Level of processing affects priming in word fragment completion. *Journal of Experimental Psychology: Learning, Memory & Cognition, 18*, 595–607.

Clayton, N. S., Yu, K. S., & Dickinson, A. (2001). Scrub jays (*Aphelocoma coerulescens*) form integrated memories of the multiple features of caching episodes. *Journal of Experimental Psychology: Animal Behavior Processes, 27*, 17–29. doi: http://dx.doi.org/10.1037/0097-7403.27.1.17

Cofer, C. N., Bruce, D. R., & Reicher, G. M. (1966). Clustering in free recall as a function of certain methodological variations. *Journal of Experimental Psychology, 71*, 858–866. doi:10.1037/h0023217

Cohen, B. H. (1963). Recall of categorized words lists. *Journal of Experimental Psychology, 66*, 227–234. doi:10.1037/h0048846

Cook, R. G., Wright, A. A., & Drachman, E. E. (2012). Categorization of birds, mammals, and chimeras by pigeons. *Behavioral Processes*. http://dx.doi.org/10.1016/j.beproc.2012.11.006

Craik, F. I., & Lockhart, R. S. (1972). Levels of processing: A framework for memory research. *Journal of Verbal Learning & Verbal Behavior, 11*, 671–684. doi:10.1016/S0022- 5371(72)80001-X

Crystal, J. D. (2010). Episodic-like memory in animals. *Behavioural Brain Research, 215*, 235–243. doi: http://dx.doi.org/10.1016/j.bbr.2010.03.005

Davenport, R. K., & Menzel, E. W. (1960). Oddity preference in the chimpanzee. *Psychological Reports, 7*, 523–526. doi:10.2466/PR0.7.7.523-526

Davenport, R. K., & Rogers, C. M. (1971). Perception of photographs by apes. *Behaviour, 39*(2-4), 318–320. doi:10.1163/156853971X00285

Davis, R. T., Leary, R. W., Stevens, D. A., & Thompson, R. F. (1967). Learning and perception of oddity problems by lemurs and seven species of monkey. *Primates, 8*, 311–322. doi:10.1007/BF01792016

Deese, J. (1959). On the prediction of occurrence of particular verbal intrusions in immediate recall. *Journal of Experimental Psychology, 58*, 17–22. doi:10.1037/h0046671

Dere, E., Kart-Teke, E., Huston, J. P., & De, S. S. (2006). The case for episodic memory in animals. *Neuroscience and Biobehavioral Reviews, 30*, 1206–1224. doi: http://dx.doi.org/10.1016/j.neubiorev.2006.09.005

Erickson, C. A., & Desimone, R. (1999). Responses of macaque perirhinal neurons during and after visual stimulus association learning. *The Journal of Neuroscience, 19*, 10404–10416.

Fagot, J., Martin-Malivel, J., & Dépy, D. (1999). What is the evidence for an equivalence between objects and pictures in birds and nonhuman primates? *Cahiers de Psychologie Cognitive/Current Psychology of Cognition, 18*(5-6), 923–949.

Fize, D., Cauchoix, M., & Fabre-Thorpe, M. (2011). Humans and monkeys share visual representations. *Proceedings of the National Academy of Sciences of the United States of America, 108*(18), 7635–7640. doi:10.1073/pnas.1016213108

Gisiner, R., & Schusterman, R. J. (1992). Sequence, syntax, and semantics: Responses of a language-trained sea lion (*Zalophus californianus*) to novel sign combinations. *Journal of Comparative Psychology, 106*, 78–91. doi: http://dx.doi.org/10.1037/0735-7036.106.1.78

Graf, P., & Mandler, G. (1984). Activation makes words more accessible, but not necessarily more retrievable. *Journal of Verbal Learning and Verbal Behaviour, 23*,553–568.

Guerin, S. A., & Miller, M. B. (2008). Semantic organization of study materials has opposite effects on recognition and recall. *Psychonomic Bulletin & Review, 15*, 302–308. doi:10.3758/PBR.15.2.302

Hampton, R. R. (2001). Rhesus monkeys know when they remember, *Proceedings of the National Academy of Sciences USA, 98*, 5359–5362.

Hampton, R. R. (2003). Metacognition as evidence for explicit representation in nonhumans. *Behavioral and Brain Sciences, 26*, 346–347.

Hampton, R. R. (2009). Multiple demonstrations of metacognition in nonhumans: Converging evidence or multiple mechanisms? *Comparative Cognition & Behavior Reviews, 4*, 17–28. doi: http://dx.doi.org/10.3819/ccbr.2009.40002

Hampton, R. R., & Hampstead, B. M. (2006). Spontaneous behavior of a rhesus monkey (*Macaca mulatta*) during memory tests suggests memory awareness. *Behavioural Processes, 72*, 184–189. doi: http://dx.doi.org/10.1016/j.beproc.2006.01.007

Hampton, R. R., & Schwartz, B. L. (2004). Episodic memory in nonhumans: What, and where, is when? *Current Opinion in Neurobiology, 14*, 192–197. doi: http://dx.doi.org/10.1016/j.conb.2004.03.006

Hampton, R. R., Zivin, A., & Murray, E. A. (2004). Rhesus monkeys (*Macaca mulatta*) discriminate between knowing and not knowing and collect information as needed before acting. *Animal Cognition, 7*, 239–246. doi: http://dx.doi.org/10.1007/s10071-004-0215-1

Hanna, A., & Loftus, G. (1993). A model for conceptual processing of naturalistic scenes. *Canadian Journal of Experimental Psychology/Revue Canadienne de Psychologie Expérimentale, 47*(3), 548–569. doi:10.1037/h0078851

Herman, L. M., Kuczaj, S. A., & Holder, M. D. (1993). Responses to anomalous gestural sequences by a language-trained dolphin: Evidence for processing of semantic relations and syntactic information. *Journal of Experimental Psychology: General, 122*, 184–194. doi: http://dx.doi.org/10.1037/0096-3445.122.2.184

Intraub, H. (1984). Conceptual masking: The effects of subsequent visual events on memory for pictures. *Journal of Experimental Psychology: Learning, Memory, and Cognition, 10*, 115–125. doi:10.1037/0278-7393.10.1.115

Jacoby, L. L. (1991). A process dissociation framework—separating automatic from intentional use of memory. *Journal of Memory and Language, 30*, 513–541.

Jacoby, L. L., & Dallas, M. (1981). On the relationship between autobiographical memory and perceptual learning. *Journal of Experimental Psychology General, 110*, 306Ð340.

Lockhart, R. S., & Craik, F. I. (1990). Levels of processing: A retrospective commentary on a framework for memory research. *Canadian Journal of Psychology/Revue Canadienne de Psychologie, 44*, 87–112. doi:10.1037/h0084237

Loftus, G. R., & Ginn, M. (1984). Perceptual and conceptual masking of pictures. *Journal of Experimental Psychology: Learning, Memory, and Cognition, 10*, 435–441. doi:10.1037/0278-7393.10.3.435

Naya, Y., Sakai, K., & Miyashita, Y. (1996). Activity of primate inferotemporal neurons related to a sought target in pair-association task. *Proceedings of the National Academy of Sciences USA, 93*, 2664–2669.

Overman, W. H., & Doty, R.W. (1980). Prolonged visual memory in macaques and man. *Neuroscience, 5*, 1825–1831.

Overman, W. H., McLain, C., Ormsby, G. E., & Brooks, V. (1983). Visual recognition memory in squirrel monkeys. *Animal Learning & Behavior, 11*, 483–488.

Paris, S. G. (1978). Memory organization during children's repeated recall. *Developmental Psychology, 14*, 99–106. doi: http://dx.doi.org/10.1037/0012-1649.14.1.99

Parker, A., Wilding, E., & Akerman, C. (1998). The von Restorff effect in visual object recognition memory in humans and monkeys: The role of frontal/perirhinal interaction. *Journal of Cognitive Neuroscience, 10*, 691–703. doi:10.1162/089892998563103

Pepperberg, I. M. (2005). An avian perspective on language evolution: Implications of simultaneous development of vocal and physical object combinations by a grey parrot (*Psittacus erithacus*). In I. M. Pepperberg, *Language origins: Perspectives on evolution* (pp. 239–261). New York: Oxford University Press.

Platt, M. L., Brannon, E. M., Briese, T. L., & French, J. A. (1996). Differences in feeding ecology predict differences in performance between golden lion tamarins (*Leontopithecus rosalia*) and Wied's marmosets (*Callithrix kuhli*) on spatial and visual memory tasks. *Animal Learning & Behavior, 24*, 384–393. doi: http://dx.doi.org/10.3758/BF03199010

Potter, M. C. (1976). Short-term conceptual memory for pictures. *Journal of Experimental Psychology: Human Learning and Memory, 2*, 509–522. doi:10.1037/0278-7393.2.5.509

Rainer, G., Rao, S., & Miller, E. K. (1999). Prospective coding for objects in primate prefrontal cortex. *Journal of Neuroscience, 19*(13), 5493–5505.

Range, F., Aust, U., Steurer, M., & Huber, L. (2008). Visual categorization of natural stimuli by domestic dogs. *Animal Cognition, 11*, 339–347. doi: http://dx.doi.org/10.1007/s10071-007-0123-2

Reed, P., Chih-Ta, T., Aggleton, J. P., & Rawlins, J. N. (1991). Primacy, recency, and the von Restorff effect in rats' nonspatial recognition memory. *Journal of Experimental Psychology: Animal Behavior Processes, 17*, 36–44. doi:10.1037/0097-7403.17.1.36

Reed, P., & Richards, A. (1996). The von Restorff effect in rats (*Rattus norvegicus*). *Journal of Comparative Psychology, 110*, 193–198. doi:10.1037/0735-7036.110.2.193

Richards, D. G., Wolz, J. P., & Herman, L. M. (1984). Vocal mimicry of computer-generated sounds and vocal labeling of objects by a bottlenosed dolphin, *Tursiops truncatus*. *Journal of Comparative Psychology, 98*, 10–28. doi: http://dx.doi.org/10.1037/0735-7036.98.1.10

Roberts, W. A., & Mazmanian, D. S. (1988). Concept learning at different levels of abstraction by pigeons, monkeys, and people. *Journal of Experimental Psychology: Animal Behaviour Processes, 14*, 247–260.

Roberts, W. A., Mazmanian, D. S., & Kraemer, P. J. (1984). Directed forgetting in monkeys. *Animal Learning & Behavior, 12*, 29–40.

Roediger, H. L., III, & McDermott, K. B. (1993). Implicit memory in normal human participants. In F. Boller &J. Grafman (Eds.), *Handbook of neuropsychology,Vol. 8* (pp. 63–131). Amsterdam: Elsevier Science.

Roediger, H. L., & McDermott, K. B. (1995). Creating false memories: Remembering words not presented in lists. *Journal of Experimental Psychology: Learning, Memory, and Cognition, 21*(4), 803–814. doi:10.1037/0278-7393.21.4.803

Sakai, K., & Miyashita, Y. (1991). Neural organization for the long-term memory of paired associates. *Nature, 354*, 152–155.

Sands, S. F., & Wright, A. A. (1980a). Primate memory: Retention of serial list items by a rhesus monkey. *Science, 209*(4459), 938–940. doi:10.1126/science.6773143

Sands, S. F., & Wright, A. A. (1980b). Serial probe recognition performance by a rhesus monkey and a human with 10- and 20-item lists. *Journal of Experimental Psychology: Animal Behavior Processes, 6*, 386–396.

Savage-Rumbaugh, S., Sevcik, R. A., Brakke, K. E., & Rumbaugh, D. M. (1990). Symbols: their communicative use, comprehension, and combination by bonobos (*Pan paniscus*). *Advances in Infancy Research, 6*, 221–278.

Savage-Rumbaugh, S., Sevcik, R. A., & Hopkins, W. D. (1988). Symbolic cross-modal transfer in two species of chimpanzees. *Child Development, 59*, 617–625. doi: http://dx.doi.org/10.2307/1130561

Schacter, D. L. (1994). In D. L. Schacter & E. Tulving (Eds.), *Priming and multiple memory system: Perceptual mechanisms of implicit memory* (pp. 233–268). Cambridge, MA: MIT Press.

Schwartz, B. L., Hoffman, M. L., & Evans, S. (2005). Episodic-like memory in a gorilla: A review and new findings. *Learning and Motivation, 36*, 226–244. doi: http://dx.doi.org/10.1016/j.lmot.2005.02.012

Sevcik, R. A., & Savage-Rumbaugh, E. (1994). Language comprehension and use by great apes. *Language & Communication, 14*, 37–58. doi: http://dx.doi.org/10.1016/0271-5309 90019-1

Shettleworth, S. J. (2012). Darwin, Tinbergen, and the evolution of comparative cognition. In J. Vonk & T. K. Shackelford (Eds.), *The Oxford handbook of comparative evolutionary psychology* (pp. 529–546). New York: Oxford University Press.

Swartz, K. B., Chen, S., & Terrace, H. S. (1991). Serial learning by rhesus monkeys: I. Acquisition and retention of multiple four-item lists. *Journal of Experimental Psychology: Animal Behavior Processes, 17*, 396–410.

Swartz, K. B., & Himmanen, S. A. (2001). Serial learning by rhesus monkeys: Strategies for list learning and execution. In S. B. Fountain, M. D. Bunsey, J. H. Danks, & M. K. McBeath (Eds.), *Animal cognition and sequential behavior* (pp. 57–67). Boston: Kluwer Academic.

Swartz, K. B., Himmanen, S. A., & Shumaker, R. W. (2007). Response strategies in list learning by orangutans (*Pongo pygmaeus P. abelii*). *Journal of Comparative Psychology, 121*, 260–269.

Templer, V. L., & Hampton, R. R. (2012). Rhesus monkeys (*Macaca mulatta*) show robust evidence for memory awareness across multiple generalization tests. *Animal Cognition, 15*, 409–419. doi: http://dx.doi.org/10.1007/s10071-011-0468-4.

Terrace, H. S., & Metcalfe, J. (2005). *The missing link in cognition: Origins of self-reflective consciousness.* New York: Oxford University Press.

Terrace, H. S., Son, L. K., & Brannon, E. M. (2003). Serial expertise of rhesus macaques. *Psychological Science, 14*, 66–73.

Thapar, A., & Greene, R. L. (1994). Effects of level of processing on implicit and explicit tasks. *Journal of Experimental Psychology: Learning, Memory and Cognition, 20*, 671–679.

Treichler, F., & Van Tilburg, D. (1996). Concurrent conditional discrimination tests of transitive inference by macaque monkeys: List linking. *Journal of Experimental Psychology: Animal Behavior Processes, 22*, 105–117. doi:10.1037/0097-7403.22.1.105

Tu, H.-W., & Hampton, R. R. (2012). One-trial memory and habit contribute independently to matching-to-sample performance in rhesus monkeys (*Macaca mulatta*). *Journal of Comparative Psychology*. Advance online publication. doi:10.1037/a0030496

Tu, H., Hampton, R. R., & Murray, E. A. (2011). Perirhinal cortex removal dissociates two memory systems in matching-to-sample performance in rhesus monkeys. *Journal of Neuroscience, 31*, 16336–16343. doi:http://dx.doi.org/10.1523/JNEUROSCI.2338- 11.2011

Tulving, E. (1999). On the uniqueness of episodic memory. In L. Nilsson & H. J. Markowitsch (Eds.), *Cognitive neuroscience of memory* (pp. 11–42). Ashland, OH: Hogrefe & Huber Publishers.

Tulving, E., & Lepage, M. (2000). Where in the brain is the awareness of one's past? In E. Tulving & M. Lepage, *Memory, brain, and belief* (pp. 208–228). Cambridge, MA: Harvard University Press.

Tulving, E., & Schacter, D. L. (1990). Priming and human memory systems. *Science, 247*(4940), 301–306. doi:10.1126/science.2296719

von Restorff, H. (1933) Uber die Wirkung von Bereichsbildungen im Spurenfeld. *Psychologische Forschung, 18*, 299–342.

Vonk, J. (2003). Gorilla (Gorilla gorilla gorilla) and Orangutan (Pongo abelii) understanding of first and second order relations. *Animal Cognition*, *6*, 77–86.

Vonk, J. (in review). Social and non-social category discriminations in chimpanzees (*Pan troglodytes*) and American black bears (*Ursus americanus*).

Vonk, J., & Beran, M.J. (2012). Bears "count" too: Quantity estimation and comparison in black bears (*Ursus americanus*). *Animal Behaviour, 84*, 231–238.

Vonk, J., Jett, S.E., & Mosteller, K.W. (*2012*). Concept formation in American black bears (*Ursus americanus*). *Animal Behaviour, 84*, 953–964.

Vonk, J., Jett, S. E., Mosteller, K. W., & Galvan, M. (2013). Natural category discrimination in chimpanzees (*Pan troglodytes*) at three levels of abstraction. *Learning & Behavior, 41*, 271–284, doi:http://dx.doi.org/10.3758/s13420-013-0103-0

Vonk, J., & MacDonald, S. E. (2002). Natural concept formation in a juvenile gorilla (*Gorilla gorilla gorilla*) at 3 levels of abstraction. *Journal of the Experimental Analysis of Behaviour, 78*, 315–332.

Vonk, J., & MacDonald, S. E. (2004). Levels of abstraction in orangutan (*Pongo abelii*) categorization. *Journal of Comparative Psychology, 118*, 3–13.

Vonk, J., & Vedder, C. (in press) at International Journal of Comparative Cognition. A zoo-housed chimpanzee's (*Pan troglodytes*) responses to potentially arousing stimuli.

Zentall, T. R., Wasserman, E. A., Lazareva, O. F., Thompson, R. R., & Rattermann, M. (2008). Concept learning in animals. *Comparative Cognition & Behavior Review, 3* 13–45. doi:10.3819/ccbr.2008.30002

15

The Development of Adaptive Memory

PATRICK DOUGLAS SELLERS II AND DAVID F. BJORKLUND

There can be no debate that a functioning memory system is adaptive to an organism. All animals with a nervous system benefit, at the least, from recognizing previously noxious versus innocuous events (reflected, for example, in the sea slug *aplysia*'s ability to habituate to repeated harmless stimuli), and, for long-lived animals living in diverse environments or complicated social communities, remembering aspects of one's physical surroundings and one's conspecifics can mean the difference between life and death. There can be little argument that cognitive abilities in general, and memory abilities specifically, have long evolutionary histories and have gone through the sieve of natural selection. The ecological and historical conditions for each species are unique, of course, and thus so should be their cognitive systems. For example, Platt, Brannon, Briese, and French (1996) found differences in memory performance between marmosets and tamarinds consistent with differences in their foraging habits and ecology. Humans are no exception, and much of what is unique about *Homo sapiens* is our cognition, particularly our ability to form memory representations of objects, people, and events in the past and to generate imaginary and possible events in the future. From this perspective, much of what is special about human memory is related to what is special about human cognition in general, particularly our ability for symbolic representation. However, *what* an animal can easily remember and its ability to retain ecologically important information is also an adaptation (likely several of them), and, as most of the chapters in this volume reflect, humans' memories are well adapted for the ancient environments in which they evolved.

However, what is often overlooked in evolutionary accounts of adaptive behavior is that most adaptations *develop*—they do not arise fully formed in adulthood, but rather have an ontogenetic history (see Bjorklund & Sellers, in press; Howe, 2011, in press). For instance, human adults' abilities to mentally encode, store, and retrieve information and to "remember" the future all have their origins in childhood, as do many of the more specific adaptations, such as showing better recall when items are related to a survival theme than to nonsurvival themes (see Otgaar & Smeets, 2010). Also, there are specific memory adaptations of infancy and childhood that serve not so much as "preparation" for later adult functioning, but rather to adapt the young person to the niche of childhood, termed *ontogenetic adaptations* (Bjorklund & Green, 1992).

In this chapter, we first look briefly at the phylogenetic and ontogenetic origins of *Homo sapiens*' impressive and perhaps unique memory representational abilities. This includes discussions of semantic and episodic memory, following the theory of Tulving (1983, 2002), the advent of autobiographical memory during the preschool years, and the development of "mental time travel," the ability to mentally transport oneself in time, from remembering to pick up your dry cleaning this afternoon to imagining a future self and the various roles and actions in which that self might engage. We then examine some specific memory adaptations that are observed in children as well as in adults, such as the tendency to remember better negative as opposed to positive events and memory for words encoded in a "survival" context. We then look at some proposed ontogenetic adaptations of memory that seem well suited to the niche of infancy or childhood that may, or may not, have counterparts later in life.

THE EVOLUTION AND DEVELOPMENT OF HUMAN MEMORY

Memory permits an organism to encode its experience with the world as a mental representation, existing as an abstract trace of its past. Carrying this representation with it allows an animal to adjust future behavior based upon past contingencies, a vital ability of profound evolutionary significance. Without this capacity, an organism perpetually lives in a state of immediate experience, unable to draw upon more than what it finds currently before its senses. All animals able to represent experience or knowledge share this basic common ability; however, the nature of these representations has changed with evolutionary pressures on cognition (e.g., Nelson, 2005; Schacter, Wagner, & Buckner, 2000), with the memory systems of *Homo sapiens* being perhaps unique among the animal world.

Explicit and Implicit Memory

Tulving's conceptualization of these various representational abilities into three major categories, *episodic*, *semantic*, and *procedural*, provides a basis for investigating phylogenetic as well as ontogenetic change (Tulving, 1983, 2002, 2005). Together episodic and semantic memory form the *declarative*, or *explicit*, memory system. Explicit memory allows for the active retrieval of information into consciousness. Colloquial references to memory typically refer to such explicit forms of memory. Episodic memory is memory for autobiographical moments (e.g., what you wore to work yesterday, your first day of high school, etc.). It allows for the transport through time of the *autonoetic* (self-knowing) self. When remembering an episodic memory, we are transporting our conscious self back to a specific time and place. In contrast, *semantic* memory is memory for pieces of information (e.g., facts or dates), including language terms. Although semantic memory is still conscious and within awareness, it is the domain of temporarily and contextually unspecific pieces of information. Tulving referred to such knowledge as *noetic*. For example, knowing that the Atlanta Braves won the 1995 World Series is factual information that falls under the domain of semantic memory. Knowing that fact is independent of any self-experience. Remembering that when the Atlanta Braves won the 1995 World Series I (PS) was jumping up and down on my grandmother's couch in suburban Atlanta (and was subsequently scolded for doing so) is episodic because it involves the experience and memory of a specific autobiographical event from my life.

In contrast to explicit memory, *implicit*, or *procedural*, memory is an evolutionarily older memory system, one that falls below consciousness and does not require self-awareness. Tulving referred to such knowledge as *anoetic*. Associative or reinforcing contingencies (e.g., taste aversion and reward), as described by classical and operant conditioning, are prime examples of procedural memory as they require no conscious awareness but represent the enactment of a mental representation created from previous experience. Such memories do not require the engagement of the self, only the application of a previously experienced, and then encoded, connection between two stimuli. Automaticity of complex behavioral scripts through practice and repetition, for example in learning a sport or musical instrument, represent another type of common procedural memory.

The distinction between self-knowing episodic memory and anoetic implicit memory is supported by much comparative work (see Suddendorf & Corballis, 2010; Suddendorf, Corballis, & Collier-Baker, 2009) and the investigation of amnesiac patients, such as H.M, who are able to display improvements on simple procedural tasks over time yet lack the ability to convert short-term memory into long-term memory (Milner, 1964).

Donald (1991) proposed that the achievement of truly autonoetic, episodic memory is a critical variable setting hominins[1] apart from our ape ancestors. Explicitly self-referenced memory allows for the recursion of time and the self, which in turn allows for the explicit representation of ideas through symbolic language and the transmission of complex culture from generation to generation. The exponential compounding effect on the sophistication of technology opened the door to the rapid cultural and technological developments that have so quickly set humans apart from the world of other animals over the last 200,000 years. Although there is some evidence that great apes, and perhaps even corvids, engage in complex behaviors with episodic-like components, humans are widely considered to be unique in their ability to engage in episodic memory (see Suddendorf & Corballis, 2010; Suddendorf et al., 2009). However, a minimalist definition of episodic memory (see Clayton & Russell, 2009) opens the door for inclusion of some animals into the exclusive club of episodic rememberers. This is accomplished through the omission of Tulving's autonoetic component as a requirement and demands simply the recall of "what," "where," and "when." However, even if we grant some degree of episodic memory to nonhuman animals, their use of it is greatly limited in comparison to that of humans. An analogy with tool use is appropriate here, we believe. Humans are not the only species to use and even manufacture tools, but no other species displays the facility and broad extension of tool use that humans do. Such differences are not simply matters of degree, but are of such a magnitude to reflect qualitative differences. A similar claim, we argue, can be made for species differences in episodic memory.

The Development of Autobiographical Memory

Perhaps the classic example of explicit, autonoetic, memory is *autobiographical memory*, self-aware memory for events that happened to the individual. Almost by definition, to have autobiographical memory one must have a sense of self, and this develops gradually over the preschool years, first marked by the ability to recognize oneself in a mirror. Children are able to identify themselves in a mirror by around 18 months (e.g., Lewis & Brooks-Gunn, 1979; Nielsen, Suddendorf, & Slaughter, 2006), although such self-recognition is slower to develop when children see themselves in photographs or videos (Povinelli, Landau, & Perilloux, 1996; Skouteris, Spartaro, & Lazarid, 2006). (Other species seem to have at least rudimentary self-awareness similar to that observed in human 2-year-olds, as reflected by mirror self-recognition, including some apes, elephants, dolphins, and birds [see Gallup, 1979; Plotnik, de Waal, & Reiss, 2006; Prior, Schwarz, & Güntürkün, 2008; Reiss & Marino, 2001].)

However, self-recognition is only a rudimentary measure of a sense of self, and young children have a difficult time recounting personal events (things that happened to "me"), typically forming incoherent narratives about their past experiences (Nelson, 1993). In fact, most researchers believe that true autobiographical memory coincides with the offset of *infantile amnesia*, the seeming inability to recall information from infancy and early childhood. Indeed, few people have verifiable memories much before the age of 3.5 or 4 years (Rubin, 2000; West & Bauer, 1999). Although there have been many hypotheses for the cause of infantile amnesia, most are based on differences in how infants and young children versus older children and adults represent experience (e.g., see Bauer, 2007; Brainerd & Reyna, 2005; Howe, Courage, & Rooksby, 2009). For example, some have proposed that children's sense of self is not adequately developed to form true autobiographical memories until about 4 years of age (e.g., Bogdan, 2010; Fivush, 1988; Howe et al., 2009), and unless events can be related to the self, they cannot be retrieved later.

Other research emphasizes the importance of language in forming autobiographical memories, proposing that as children learn to tell stories via language they are learning to share memories of their experiences with other people (e.g., Fivush & Hamond, 1990; Nelson, 1993, 1996). Evidence of the centrality of language in autobiographical memory comes from a longitudinal study in which 27- to 39-month-old children experienced an event and were interviewed 6 or 12 months later for their recollection of the experience (Simcock & Hayne, 2002). The researchers reported that children's subsequent recall of the event was a function of their level of vocabulary *at the time of the experience*, not at the time of recall. According to Simcock and Hayne (2002, p. 229), "children's verbal reports were frozen in time, reflecting their verbal skill at the time of encoding, rather than at the time of test."

Note that we are *not* saying that human infants are incapable of remembering individual events before their fourth birthdays. In fact, infants and toddlers show evidence of long-term episodic-like memory, expressed both verbally and through deferred-imitation tasks (see Bauer, 2007; Fivush, Kuebli, & Clubb, 1992). However, because the representation of information at the time of encoding for infants and young children differs from the representation system used later in life, self-conscious (autonoetic) retrieval is not possible. From this perspective, 2- and 3-year-old children may not understand their lives as stretching back in the past and extending into the future, as older children and adults presumably do, but "Instead, the memories, images, and thoughts pop in and out of consciousness as they are cued by present events, or by other memories, images, and thoughts" (Gopnik, 2009, p. 154).

The Development of Mental Time Travel

To this point we have been discussing children's abilities to remember events in the past, or their *retrospective memory*. However, there is a related form of memory that is seemingly highly dependent upon advanced representational abilities, and that is remembering to do something in the future. This is referred to as *prospective memory* (e.g., Einstein & McDaniels, 2005) and can involve what is sometimes referred to as *mental time travel*, or anticipating the future and planning for it (see Suddendorf & Corballis, 2007, 2010; Tulving, 2005).

Seamlessly transporting oneself (mentally) to not only different places but also different times is a hallmark of human cognition. One can see and feel his Olympic-themed fifth birthday party or imagine his future as a world-famous academic, as one can perceive his current reality. Similar to Tulving's conceptualization of episodic memory, this ability requires an autonoetic (self-knowing) component, reaching beyond simple procedural memory, permitting individuals to represent the self in the future. This has resulted in some people referring to this type of memory as *episodic future thought* (Quon & Atance 2010; Szpunar, 2010). Episodic future thought is distinguished from prospective memory by the inclusion of an autonoetic component (much like the minimalist approach to episodic memory discussed earlier). Indeed, remembering to engage in a future action does not necessarily require the transport of the self into a hypothetical future, although such transport is indeed possible.

The ability to move beyond immediate environmental cues for decision-making opens a whole world of possibilities for planned, coordinated, and complicated sequences of action leading to the satisfying of a goal state many steps removed from the immediate moment. Many cognitive abilities are required for such prospective considerations, and furthermore, these abilities have emerged across evolutionary time and develop within a lifetime.

Mental time travel is easily documented in human adults, and the results are enormously impressive and obvious, as evidenced by modern human culture. Advanced cultural behavior (e.g., art, government, science) requires forethought in the ability to conceptualize a future end state and engage in action in the form of a complex, multistep solution toward that end. In addition, these cultural accomplishments all require consideration of past states and engaging with memories of the past to direct the future behavior of contributing participants.

In comparison to the development of retrospective memory, less attention has been paid to the development of prospective memory, although this is beginning to change (see Kvavilashvili, Kyle, & Messer, 2008 for a review). Early research showed that even 2-year-old children display prospective memory in some contexts, particularly when the delay is short (e.g., minutes) and the thing-to-be-remembered is of high interest (e.g., buying candy at the store: Somerville, Wellman, & Cultice, 1983), although more recent research has questioned the prospective memory abilities of 2-year-olds after failing to find evidence of such young children carrying out future intentions (e.g., Kliegel & Jager, 2007). Similar to retrospective memory, children's prospective memory performance improves over the preschool (e.g., Quon & Atance, 2010) and school years (e.g., Atance, 2008; Kerns, 2000; Smith, Bayen, & Martin, 2010). For example, Kerns (2000) had 7- to 12-year-old children play a computer game called CyberCruiser in which they used a joystick to maneuver around obstacles. They also had to occasionally check the fuel gauge to make sure they did not run out of gas and to fill up when they had less than a quarter of a tank. Perhaps not surprisingly, younger children ran out of gas more often than older children.

The benefits to a species of mental time travel seem self-evident. The ability to recall episodes from the past, as well as seeing oneself in some future time or place, is critical for creative problem-solving. Of course, the behaviors learned by classical and operant conditioning, presumably forms of procedural memory, can also be "carried forward" and used in environments in which they were not acquired. But the procedural memories underlying such learning do not allow individuals to reflect upon past experience, but merely to apply behavioral acquisitions in new (likely similar) contexts. Prospective memory permits someone to consciously anticipate and plan for the future, a skill that may have been crucial to our large-brained but physically puny ancestors (at least in comparison to many of the predators of the African savannah: Suddendorf, 2006) and has its origins in early childhood.

However, humans are not the only animals who display cognitions similar to mental time travel. The ability of nonhuman animals to anticipate and plan for the future has been a point of contention among comparative psychologists and biologists. For example, the Bischof-Köhler hypothesis (Bischof, 1978; Bischof-Köhler, 1985) contends that the ability to anticipate and act on future needs or drive states sets hominin cognition apart from those of other species. Yet, observations in both birds and great apes have questioned this contention. For example, scrub jays have demonstrated future-directed behaviors in the domain of gathering and storing food for subsequent consumption, reportedly in accordance with the anticipation of future needs (Clayton & Dickinson, 1999; Correia, Dickinson, & Clayton, 2007). Although this may appear to meet the criterion for mental time travel, Suddendorf and Corballis (2010) dispute the results and maintain that no evidence yet exists necessitating an assignment of autonoetic future mental time travel in scrub jays. Additionally, Salwiczek, Watanabe, and Clayton (2010), in a review of evidence for episodic-like memory in multiple avian species (scrub jays, magpies, chickadees, and pigeons), suggest that acceptable evidence exists to conclude that many avian species may possess episodic-like memory, but experimental evidence of a truly autonoetic component is ultimately lacking.

Another challenge to the Bischof-Köhler hypothesis comes from research with primates. Apes have shown some limited planning abilities in tool-use situations (e.g., Dufour & Sterck, 2008; Mulcahy & Call, 2006), including keeping a tool for later use for up to an hour (Osvath & Osvath, 2008; see also Tulving, 2005). Additionally, at least one gorilla has been shown to be capable of episodic-like memory for "what" and "who" memory tests (Schwartz, Colon, Sanchez, Rodriquez, & Evans, 2002). Once again, such behavior appears to indicate the inhibition of current need states in favor of planning actions directed at anticipated future needs, and thus would necessitate rejection of the Bischof-Köhler hypothesis. However, these conclusions are again tempered by possible associative explanations of the observed behavior (Suddendorf, Corballis, & Collier-Baker, 2009) and the absence of simultaneous "who," "what," "where," and "when" memory.

Apes certainly display more sophisticated versions of future-directed behavior than other nonhuman animals, and it is possible that they are capable of some limited degree of autonoetic, representational foresight. In any respect, degrees of foresight are likely to exist. Even if the scientific community unequivocally documents limited foresight to chimpanzees at some future date, their abilities will still pale in comparison to the human capacity for planning and anticipation. Sheer timescale differences (for apes) and flexibility of application across domains (for scrub jays) are enough for human foresight abilities to outstrip anything apes (or scrub jays) may one day be given credit for. It appears the Bischof-Köhler hypothesis is safe for now, although its future refutation appears possible as comparative research methodologies continue to improve in sophistication.

ADAPTED MEMORY PROCESSES IN CHILDREN AND ADULTS

Although the human representational system, including the ability to "mentally travel through time," is likely an adaptation, there are more mundane aspects of adapted memory that have been proposed in adults and that have their origins in infancy and childhood. Among these are biases to remember emotional (especially negative) stimuli and enhanced memory for information encoded in a "survival" context, and we examine each of these briefly here.

Memory for Emotional Stimuli

What people remember is influenced by a host of factors, one of which is the emotional valence and arousal of a stimulus or event. For example, adults are more attentive to and generally remember better emotionally arousing events relative to neutral ones (e.g., Calvo & Lang, 2005; Kensinger & Corkin, 2003, 2004). Although this pattern generally holds true regardless of the valence of the stimuli, from an evolutionary perspective it may be especially relevant for negative stimuli, which are signs of potential threat or harm. According to Baumeister and his colleagues, "it is evolutionarily adaptive for bad to be stronger than good. We believe that throughout our evolutionary history, organisms that were better attuned to bad things would have been more likely to survive threats and, consequently, would have increased probability of passing along their genes" (Baumeister, Bratslavsky, Finkenauer, & Vohs, 2001, p. 325; see also Calvo & Lang, 2004; Vaish, Grossman, & Woodward, 2008). Children also display such a negativity bias, giving special attention and processing privilege to negative over neutral or positive stimuli (see Chae, Goodman, & Edelstein, 2011; Vaish et al., 2008).

Children's negativity bias can be observed in their memory. For example, in one series of studies, preschool children were shown photos of faces described variously as "mean" or "nice" (Kinzler & Shutts, 2008). Children were later shown some of these target faces paired with novel faces and asked to select those that they had seen previously. Children were significantly better at recognizing faces described as "mean" rather than "nice" (see also Baltazar, Shutts, & Kinzler, 2012).

This bias for remembering negative information extends to older children's memory, with children generally remembering better highly emotional versus less emotional events (e.g., Davidson, Lou, & Burden, 2001; see Quas & Fivush, 2009). In fact, whereas memory performance improves substantially with age for the recall of non-emotional stimuli (see Courage & Cowan, 2009), age-related increases in memory performance are substantially smaller for emotional stimuli (Davidson et al., 2001). In general, children's memory performance is better for negative than for neutral, and sometimes positive, information. For example, in a naturalistic study, the narratives of children living in violent communities for negative events were more coherent and included more discussion of their thoughts and emotions than those for positive events, although their narratives of positive events included more descriptive details than for negative events (Fivush, Hazzard, Sales, Sarfati, & Brown, 2003). In other research, both (nonclinically) depressed and nondepressed 5- to 11-year-old children displayed higher recall for emotionally negative than positive stories (Bishop, Dalgleish, & Yule, 2004), and 7- to 12-year-old children, as well as adults, were better able to recognize photos of aversive images than neutral images (Cordon, Melinder, Goodman, & Edelstein, in press). Furthermore, despite usually producing higher levels of performance, in some studies, negative stimuli produce higher levels of false recall in both adults (e.g., Brainerd, Stein, Silveira, Rohenkohl, & Reyna, 2008) and children (e.g., Howe, 2007), suggesting that negatively connotative words, pictures, or events are less discriminable than more neutral stimuli.

What we remember depends on what we attend to and how we encode, store, and retrieve information; and some information is simply more important to attend to and remember than others. From an evolutionary perspective, it is adaptive to be attentive to stimuli that can cause harm, whether it be in the form of potentially dangerous animals such as spiders or snakes (e.g., LoBue & DeLoache, 2008) or a "mean" person, as reflected by attributed past deeds (e.g., Kinzler & Shutts, 2008) or an unpleasant expression (Mumme, Fernald, & Herrera, 1996). As we noted, these biases, characteristic of adults, are first observed in infancy and early childhood, suggesting that they are part of an early developing system adapted to alert children (and later adults) to potentially negative and dangerous situations, permitting them to make efforts to avoid or prepare for them. One mechanism for this phenomenon in infancy may be familiarity. For instance, although typically developing infants are more apt to be attentive to negative than positive faces, infants of depressed mothers show the opposite pattern (Striano, Brennan, & Vanman, 2002). Unlike infants of nondepressed mothers, infants of depressed mothers are more familiar with sad, angry, or neutral expressions as opposed to happy and cheerful ones. The novelty of the positive facial expressions for infants of depressed mothers thus provides more useful information than the more typically experienced negative faces. This situation is of course reversed for typically developing children of nondepressed mothers.

Memory in "Survival" Contexts

Another much-studied topic in adapted memory in adult humans is memory in "survival" contexts (see chapters in this volume by Altarriba & Kazanas [Chapter 7]; Burns & Hart [Chapter 4]; Howe & Derbish [Chapter 5]; Schwartz & Brothers [Chapter 9]; Soderstrom & Cleary [Chapter 6]; Toglia et al.). Initial research by Nairne and his colleagues (e.g., Nairne, Pandeirada, & Thompson, 2008; Nairne, Thompson, & Pandeirada, 2007; see Nairne, 2010, for a review) reported that adults who were asked to think about sets of words in terms of survival (e.g., "you are stranded in the grasslands of a foreign land, without any basic survival materials") versus other nonsurvival contexts (e.g., "you are planning to move") showed higher subsequent incidental recall of the target items. Nairne and his colleagues argued that natural selection tuned human memory systems to process information related to fitness, specifically to facilitate retention of information in a survival context.

Given that natural selection has acted as substantially on the early stages as the later stages of life, if not more so, and that the likelihood of surviving to adolescence was likely no greater than 50 percent for our ancient ancestors (e.g., Volk & Atkinson, 2008), children might also display a "survival" memory bias, suggesting that this is an early developing, evolved, adaptive strategy. To our knowledge, there are only two published studies using the survival processing task with children. In the first such study, Otgaar and Smeets (2010) used the Deese-Roediger-McDermott (DRM) procedure in a series of experiments with 8- to 11-year-old children and adults. The DRM procedure involves presenting participants with a list of words all related to a central concept, although that concept itself is not among the items on the word list (e.g., *baker, flour, dough*; all related to *bread*). After making some decisions about each word in the list, participants are given either a recall or recognition test for the words. Falsely recalling or recognizing the lure item from each category (in this example, *bread*) is used as an indication that the category prototype was encoded and incorrectly remembered.

In Otgaar and Smeets' (2010) study, children and adults were asked to make different judgments about sets of DRM lists. Participants were told to rate words according to one of three conditions, each taken from Nairne et al. (2007):

> *Survival*: In this task, we would like you to imagine that you are stranded in the grasslands of a foreign land, without any basic materials. Over the next few months, you'll need to find steady supplies of food and water and protect yourself from predators. We are going to show you some words, and we would like you to rate how relevant each word would be in this survival condition. Some of the words may be relevant and others not—it's up to you to decide.
>
> *Moving*: In this task, we would like you to imagine that you are planning to move to a foreign land with your parents. Over the next few months, you'll need to make new friends and learn a new language. We are going to show you some words, and we would like you to rate how relevant each word would be in this moving condition. Some of the words may be relevant and others not—it's up to you to decide.
>
> *Pleasantness:* In this task, we are going to show you some words, and we would like you to rate the pleasantness of each word. Some of the words may be pleasant and others may not—it's up to you to decide.

After a brief distractor task, participants were unexpectedly asked to write down all the words they had rated. Otgaar and Smeets reported that both the adults (Experiment 1) and the 8- and 11-year-old children (Experiment 2) recalled more words correctly in the survival context than in the moving or pleasantness contexts, replicating earlier research with adults. However, each age group also had higher levels of false recall in the survival condition than in the other two conditions, such that overall accuracy [correct recall/(correct + false recall)] did not differ among the three conditions (see also Howe & Derbish, 2010, for similar results with adults).

In the second published developmental study of adaptive memory we are aware of, Aslan and Bäuml (2012) used age-appropriate survival and pleasantness conditions, similar to those used in research with adults and by Otgaar and Smeets, as well as a condition in which 4- to 10-year-old children rated whether words sounded short or long (word-length condition) in a recognition-memory task (Experiment 1). They reported that overall correct recognition improved with age (9/10-year-olds > 7/8-year-olds > 4/6-year-olds) but that at each age, performance was highest for the survival condition and worst for the word-length condition (survival > pleasant > word-length). A second experiment included a nonsurvival condition that involved relatively "deep" levels of processing (cf., Craik & Lockhart, 1972) (staying overnight at a friend's house), as well as a survival-related scenario (parents forgetting to pick a child up from school), and yielded similar results: correct recognition increased with age and was greater in the survival ("forgotten at school") than the nonsurvival ("staying overnight at a friend's house") condition at each age level. The authors argued that their results could not be attributed solely to memory processes, such as enhanced schematic or self-referencing mechanisms, and, given that the youngest children do not generally engage in explicit mnemonic activities (see Bjorklund, Dukes, & Brown, 2009), are not likely a byproduct of strategic mechanisms. Rather, they suggested that even young children have been adapted by natural selection to be sensitive to threatening stimuli and, like adults, give those stimuli processing priority.

Although there are too few developmental studies following the logic of survival processing tasks for us to come to any definitive conclusions, the few experiments that have been conducted indicate that children as young as 4 years of age perform comparably to adults. Moreover, the pattern of results is in many ways similar to that we described earlier for children's and adults' memories for negative stimuli and events: memory performance is typically greater for negative information, but so is false memory, consistent with the proposition that highly emotional, negative stimuli are less discriminable than less emotional stimuli. In rating the words in the survival conditions of these experiments, participants likely perceived them as more emotionally arousing with potentially negative consequences than when rating words in the nonsurvival conditions. Thus, despite sometimes comparable overall accuracy when survival themes are activated, the high false-memory levels of survival processing may still be adaptive in that such processing "primes related information in memory, information that may be subsequently used to guide attention to other survival-related materials" (Howe & Derbish, 2010, p. 264). That such processing is apparent during the preschool years suggests that it is an early developing phenomenon, but one likely based on experience dealing with the consequences of negative emotion, and perhaps the ability to anticipate what the consequences might be should one be placed in a life-threatening situation (i.e., mental time travel).

We should note that two of the three developmental experiments assessing children's processing of survival-related themes (Otgaar & Smeets, Experiment 2; Aslan & Bäuml, Experiment 1) used scenarios similar to those used by adults (i.e., stranded in the grasslands), which may be more appropriate for adults than for children. Scenarios that reflect greater fitness-enhancing benefits for ancestral children may be more related to seeking out strong adults, hiding, and generally not putting oneself at risk, more similar to the scenario used by Aslan and Bäuml (2012) in their second experiment. Thus, survival processing for children may be more about protection and hiding, whereas for adults it may be more about evaluating risk and seeking survival-relevant materials. Although speculative, this suggests that use of different scenarios may produce enhanced effects of survival-relevant stimuli, especially for young children.[2]

ONTOGENETIC ADAPTATIONS IN CHILDREN'S MEMORY

Developmental investigations of memory have provided the psychological community with an elaborate understanding of children's representational and memory abilities, from infancy through adolescence (see Bauer & Fivush, in press; Courage & Cowan, 2009). Description and cataloguing of these capabilities has elucidated the nature of the child's mind and greatly affected education and treatment of particular developmental irregularities (e.g., autism, fetal alcohol syndrome, intellectual impairment). However, most of this work has, either implicitly or explicitly, viewed developmentally as a means to an end. Abilities of a child are assessed in reference to the proportion of adult functioning achieved. In contrast, an evolutionary account of memory development partially shifts the focus of investigation toward the child's current abilities as ends themselves, in the form of *ontogenetic adaptations*.

Ontogenetic adaptations refer to adaptations that are specific to a particular time in life and "disappear" when they are no longer needed (see Bjorklund, 1997; Bjorklund & Green, 1992; Bjorklund, Periss, & Causey, 2009; Oppenheim, 1981). Thus, they serve to adapt infants/children to their current environment and not to prepare them for a later one. Some ontogenetic adaptations of the prenatal and neonatal period are easily recognized. For instance, the umbilical cord of the fetus serves a life-sustaining function but becomes useless and is discarded at birth when a radical switch point in terms of the infant's physiology takes place. Similarly, many neonatal reflexes serve a specific function shortly after birth (e.g., the sucking reflex) but are lost (and, in fact, are a sign of neurologic impairment if they persist) once infants gain greater intentional control of their behavior.[3]

Humans are immature for such a large portion of their lives that reaching sexual maturity requires overcoming significant obstacles. Surviving to adulthood is taken for granted in the developed world; however, a significant proportion of our hominin predecessors did not achieve this feat. Infancy and childhood, ripe with danger and uncertainty, can be considered a survival bottleneck, a time of particularly strong selection pressures that would have often favored adaptations of immediate benefit over long-term payoffs. Thus, childhood can be thought of as the crucible of evolution (Periss & Bjorklund, 2011; Volk & Atkinson, 2008). Adaptations related to attaining and maintaining a mate or to successful parenting, for example, would be useless if one did not survive infancy and childhood, and adaptations specific to surviving the early years of life would have been favored by natural selection.

However, we must be careful in selecting those behaviors we wish to classify as ontogenetic adaptations. First, not all universally observed behaviors associated with memory development can be said to be adaptive. Their universal occurrence alone is not sufficient grounds upon which to confer such a status. In fact, many memory abilities in children can be considered byproducts of the process of normal development. For example, infantile amnesia is most likely a reflection of the relative immaturity of explicit memory combined with a lack of language abilities with which to encode new memories. Given that older children and adults retrieve memories using language cues, memories encoded prior to sufficient language acquisition languish in the doldrums of the adult mind. Infantile amnesia does not likely serve any grand purpose. Rather, it is a species-typical phenomenon attributable to the interaction of numerous other processes having no direct selection pressures of its own (although autobiographical memory itself surely does).

A second point must be made regarding the criteria for appropriate selection of ontogenetic adaptations. The fact that an ability occurs both in childhood and adulthood does not preclude it from being classified as an ontogenetic adaptation. If one observes an *immature* version of Behavior A in childhood and a *mature* version of a similar behavior in adulthood, it certainly seems logical to conclude that the immature behavior in childhood served as practice or preparation for the adult function. However, this is not necessarily the case. Provided an adaptive function can be identified for children, there is no reason why a behavior cannot contain both ontogenetic and deferred adaptive elements. Childhood play provides a wonderful example. Play behaviors prepare children for adult roles and allow children to practice and understand common interactions in a low-stakes environment, but also help to establish a social hierarchy among the children that confers direct and immediate advantages and disadvantages (e.g., Pellegrini, in press; Pellegrini & Bjorklund, 2004). A developmental memory example to which we will return in greater detail is the special treatment afforded to faces by both children and adults. Although both adults and children treat faces as a special category of stimuli, it would be foolish to conclude that children reap no development-specific advantage because adults exhibit similar behavior.

Using these basic criteria, we have identified a number of processes in infancy and childhood that may be considered ontogenetic adaptations within memory development—in other words, adaptive developmental memory systems. What follows is a brief account of those phenomena we find most critical to an understanding of (a) infants' and (b) children's memory, along with evidence and reasoning placing them into an evolutionary context.

Ontogenetic Memory Adaptations in Infancy

There is little in life that does not involve memory at some level, and this begins in infancy. Memory changes in both qualitative and quantitative ways between infancy and childhood (see Bauer & Fivush, in press), and some characteristics of infant memory can be described as ontogenetic adaptations—well suited for the niche of infancy and changing along with the life conditions of the rememberer (see Bjorklund & Sellers, in press; Nelson, 2005). For example, Nelson (2005) proposed that because of infants' prolonged and nearly complete dependence on adults, they should be especially attentive to recurring events in their social world. According to Nelson, young infants' major cognitive goal is to build up memory representations of the important people, places, and things in their environments, tasks that can be adequately accomplished by implicit memory systems. Thus, for Nelson, it is not surprising that an explicit/declarative memory system, characteristic of older children and adults, would be of little use to infants, and that "building in complex cognitive goals, and the mechanisms such as explicit or declarative memory for achieving them, appears burdensome and likely to interfere with the primary requirements of this period of life" (Nelson, 2005, p. 369).

Young infants' memory also seems tied to context. For example, Rovee-Collier and her colleagues (Rovee-Collier, Schechter, Shyi, & Shields, 1992) showed that 2-month-old infants, who learned that kicking their legs while lying in a crib caused a mobile overhead to move, retained this knowledge 24 hours later, but only when the crib liner was the same during both acquisition and testing. Rovee-Collier and Shyi (1992) suggested that this somewhat extreme dependency on context prevents young infants from retrieving memories in "inappropriate" situations. Given infants' poor inhibitory abilities (e.g., Baird, Kagan, Gaudette, Walz, Hershlag, & Boas, 2002; Diamond, 1985; Holmboe, Pasco Fearon, Csibra, Tucker, & Johnson, 2008), such dependency on context may prevent infants from retrieving previously learned memories (actions) in inappropriate situations.

Infants become less dependent on context for remembering their actions over the course of the first year, accompanied by increased mobility and contact with their physical (and social) environment (Learmonth, Lambert, & Rovee-Collier, 2004; Rovee-Collier & Cuevas, 2009), and this also seems of benefit to developing infants. Although all individuals must be cautious in overgeneralizing learned responses, as infants' experiences become more varied, memories acquired in one context can be usefully applied in similar contexts. According to Hartshorn and her colleagues (2004, p. 76), "As the physical world of the developing infant progressively expands and the infant's niche also changes, the behavioral solutions to problems that characterized the relatively static habitat and niche of the younger infant must also change or lose their adaptive utility."

One class of stimuli that are particularly important to infants (and to older children and adults as well) is faces (see Maguinness & Newell, Chapter 13 in this volume, for a review of face memory). Neurologic evidence clearly shows that there is specialized neurologic processing of faces (McCarthy, Puce, Gore, & Allison, 1997) and neurologic differentiation between dynamic and static facial information (Pitcher, Dilks, Saxe, Triantafyllou, & Kanwisher, 2011). Additionally, the existence of two particular neurologic disorders further establishes faces as specially treated stimuli. *Prosopagnosia*, deficits in facial processing with normal object processing (Duchaine & Nakayama, 2005), and its newly discovered inverse, *object recognition deficits* with normal face processing (Germine, Cashdollar, Duzel, & Duchaine, 2011), extend findings from normal functioning to the domain of neurologic abnormality.

In fact, specialized face processing functions early in life as newborns attend to faces more readily than they do other stimuli (Goren, Sarty, & Wu, 1975; Johnson, Dziurawiec, Ellis, & Morton, 1991; Mondloch, Lewis, Budreau, Maurer, Dannemiller, Stephens, & Kleiner-Gathercoal, 1999). Other research indicates that newborns can discriminate among human faces and look longer at the faces of their mothers than of other women (Bushnell, Sai, & Mullin, 1989) and will modify their rate of sucking to see a picture of their mother's face compared with that of another woman (Walton, Bower, & Bower, 1992). Additionally, children aged 14 to 42 months diagnosed with autism spectrum disorders (a class of disorders characterized by profound social-cognitive deficits) prefer to look at geometric figures rather than pictures of faces, suggesting their deficits in social-cognitive development may be partially due to a lack of attention to faces (Pierce, Conant, Hazin, Stoner, & Desmond, 2011; see also Sanna et al., 2011).

As mentioned earlier, infants treat faces as special stimuli. However, some faces are more special than others. At 3 months of age, infants equally recognize faces with which they have been familiarized regardless of race, but at 6 months they show a bias for recognizing faces of their own race (Kelly, Quinn, Slater, Lee, Ge, & Pascalis, 2007). Furthermore, at 3 months, infants use featural processing when shown both upright and inverted faces; however, at 7 months they have shifted to processing only upright faces holistically (Cashon, 2011). These differences reflect a common nature of change across infancy known as *perceptual narrowing*, in which infants respond to a wider array of stimuli, but through experience hone their attention to more environmentally relevant features (e.g., upright and own-race faces vs. inverted and other-race faces). Narrowing even occurs in cross-species discrimination of faces (Pascalis, de Haan, & Nelson, 2002). At 6 months of age, infants' ability to discriminate between upright faces is comparable whether the faces being discriminated are two monkeys or two humans. But adults and 9-month-old infants are superior at discriminating between two human faces than two monkey faces. Memory benefits are observed in adults for these more relevant classes of faces and when engaging in holistic processing (Cashon, 2011; Richler, Cheung, & Gauthier, 2011). Therefore, infants narrow their preferential perception, discrimination, and processing of faces in such a way that prioritizes environmentally relevant stimuli and is related to better recognition memory in adulthood.

Infancy is a perilous time in development. The extreme helplessness of infants demands that they establish social relations with caretakers and learn (and remember) appropriate behaviors in specific contexts. Their memory seems well suited to these tasks, focusing on repeated events in familiar contexts and being especially attentive to faces, to the point that infants only hours old can recognize their own mother's face.

It is no surprise to evolutionary psychologists that humans respond differentially to faces than other stimuli, and that particular mechanisms appear to be dedicated to the process. A core tenet of the phylogenetic history of hominins is the special affordance placed on social functioning and behavior, and thus social-cognitive processing. The human mind can be conceptualized as a tool for navigating complex and varied social arrangements (Flinn & Ward, 2005), as the rapid rate of encephalization from great ape to modern human is mirrored by increased social complexity (Dunbar, 1998, 2010; Joffe, 1997). What makes humans special is our ability to abstractly represent concepts, both mentally and physically, and then communicate these ideas in a strategic manner to important others (Donald, 1991). Such complexity arose from the importance of social relationships for survival and reproduction, which drove still greater abstraction abilities (and perhaps all memory abilities beyond the most basic implicit memories). Therefore, stimuli providing critical social information (e.g., faces) are likely to be afforded special mechanisms of adaptation.

Ontogenetic Memory Adaptations in Childhood

The transition from infancy to childhood brings with it the emergence and sophistication of explicit memory (Nelson, 2005). As such, children can now engage with their memories in the world or, perhaps more accurately, engage the world with their memories. The possibility for a bidirectional interaction between explicit memory representations and the outside world adds to a child's memories, most notably, the influence of social interactions. Given the pervasive influence of social relationships on human development, and the phylogenetic importance of human social interaction (see Dunbar, 1998, for discussion of the *social brain hypothesis*), searching for ontogenetic adaptations within this framework is likely to yield substantial results. Two possible adaptations, typically viewed as "deficiencies" associated with young children's thinking, are egocentricity and suggestibility, each of which we examine briefly here.

Egocentricity

Preschool-aged children's egocentric, or self-centered, worldview is well established and is considered a hallmark of early childhood cognition (Piaget, 1983). Young children, more so than older children and adults, are more likely to spontaneously reference events and objects to themselves, across domains. Such a self-referencing bias affects many cognitive processes, of which memory and source monitoring are the most relevant to the discussion here. Although typically viewed as a reflection of immature cognition, young children's persistent egocentricity can also provide some cognitive advantages (see Bjorklund & Green, 1992), among them elaborated representations for increased memory.

Source monitoring refers to the ability to determine the source of one's information, action, or memory. Such referential decisions are critical to the establishment of autobiographical memory (self-sourcing) and additionally for judgments of information validity based upon the source's trustworthiness (Johnson, Hashtroudi, & Lindsay, 1993). Source-monitoring abilities improve with age into adulthood, although the types of errors children make on such tasks are not random. After engaging in a joint task with an adult (e.g., putting objects on a collage) and then asked who had taken which actions (themselves or the adult), preschool children are more likely to make self-sourcing errors, remembering the adult's actions as theirs, than other-sourcing errors, remembering their own actions as those of the adult (e.g., Foley & Ratner, 1998; Foley, Ratner, & Gentes, 2010; Foley, Ratner, & Passalacqua, 1993). Additionally, such "I did it" errors are associated with greater task learning (Ratner et al., 2002; Sommerville & Hammond, 2007), and later memory is enhanced after performing a joint task with an adult, even in the presence of "I did it" errors (Ratner, Foley, & Gilbert, 2002).

Self-referencing of memories leads to representational elaboration, providing a richer catalog of experiences to draw upon during memory retrieval. The self is likely the most elaborated of representations; indeed, we have more contact and interaction with the self than any other physical entity or psychological concept. Therefore, if one can reference the self at the encoding of information, recall becomes more likely due to the increased number of possible cues. In line with this position is the fact that self-referencing is known to increase recall in adults (Lord, 1980; see also Pratkanis & Greenwald, 1985) and in children (Mood, 1979; Ross, Anderson, & Campbell, 2011). For example, in studies by Ross and her colleagues (2011), 3- and 4-year-old children demonstrated enhanced recognition memory for self-referenced items, including "shopping" items owned by the children and actions attributed to a cartoon character with the child's name.

In general, young children's explicit memory is relatively poor, and thus any cognitive advantage that may enhance their memory would a plus. An egocentric bias seems to provide such an advantage as it prompts children to reference the self more frequently, leading to more richly elaborated representations. Even given the inaccuracies associated with young children's "promiscuous self-sourcing," this tendency is likely to be adaptive as it leads to greater memory for the relevant actions or ideas.

Suggestibility

A very real problem plaguing the legal system, and also providing a ripe source of theoretical investigation, is the susceptibility of children's memory (and subsequent testimony) to the influence of others. In general, children provide less elaborate and less accurate responses when questioned about a recent event than adults. Moreover, preschool and early school-age children are more likely than older children and adults to be influenced by misleading questions and to change their original answers when repeatedly asked a direct question by an adult interviewer (see Brainerd & Reyna, 2005; Ceci & Bruck, 1993, for reviews).

Understandably, such heightened suggestibility is usually viewed as a problem in legal settings and as a cognitive shortcoming. However, evolutionary developmental theory leads to a reinterpretation of these phenomena as phylogenetically and ontogenetically potentially useful. Deference to an adult is likely to have produced recurrent benefits for children across human history. Encountering an unfamiliar, ambiguous, or difficult circumstance is a common occurrence for children. Given that most of young children's day-to-day interactions, both in contemporary and ancient environments, were with adult kin who were likely to have the child's best interest in mind, it may be advantageous for a child to adjust his or her opinion or memory to align with that of an adult. Attentiveness to even subtle suggestion from high-status members of a social group is likely to pay dividends for children in their attempts to gain important information about the world. Such a tendency will sometimes result in "false memories," or at least false reporting, but, on average, changing one's mind (or profession of belief) in line with that of a more knowledgeable adult was (and remains) adaptive for young children.

Related to young children's enhanced suggestibility with respect to memory is their tendency to "overimitate" the actions of others. Recent research consistently indicates that children beginning at the age of 3 will imitate indiscriminately the actions of an adult, repeating even those actions that are unnecessary for task completion, provided the action of the adult is seemingly intentional (e.g., Gardiner, Greif, & Bjorklund, 2011; Horner & Whiten, 2005; Lyons, Young, & Keil, 2007). Such precise imitation undoubtedly meets the goal of executing the relevant actions of a task but leads to a child repeating some actions that serve no purpose toward the desired goal. (Chimpanzees, in comparison, do not copy irrelevant actions; they copy only those necessary to achieve the goal, engaging in emulation as opposed to imitation [see Whiten, 2010].) "Better safe than sorry" appears to be the default operation when children imitate adults, and this is indeed a common theme among adaptations (e.g., fight or flight, paternity uncertainty). Thus, children often engage in unnecessary actions by following suit with observed adult behavior, just as they are apt to adjust their recollections at the behest of adults' suggestion.

CONCLUSION

As humans change and develop across our lives our memory systems follow suit, building from implicit, unaware memory to systems that are explicit and self-referent. In the same fashion, memory itself has *evolved*. These changes have led to the development of adaptive memory, both in children and adults. Some, perhaps all, of the adaptive functions of adult memory have their origins in childhood; they do not arise *de novo*, fully formed with maturity. Others, which we have referred to as ontogenetic adaptations (adaptations specific to the niches of infancy and childhood), have evolved to deal with the recurrent problems faced by our evolutionary forechildren. These make children's memory not simply an immature form of adult functioning, but an independently useful and necessary entity all its own.

To consider children's developing abilities as evolutionarily relevant behaviors, we must view the process of development not simply as a means to an end (i.e., necessary steps to achieving adult functioning). Rather, development and children's cognitions, emotions, and behaviors must be considered fodder for evolutionary change in the same way as adult sexual or survival behaviors are. Children carry with them through ontogeny not simply a history of their own past behaviors, decisions, contingencies, and learning, but also the cumulative history of hominin children whose experiences of childhood provided the framework for the development of adaptive childhood and memory. Consideration of this shared history and its active interaction over ontogeny can reframe development as an important entity in and of itself.

NOTES

1. Here we use the term *hominin* to refer to humans and human ancestors at the exclusion of the other great apes (chimpanzees, gorillas, and orangutans). Although there is debate regarding taxonomic divisions between these species (and subsequent differential use of the terms *hominin* and *hominid*), we believe the present distinction is appropriately aligned with current anthropologic opinions and findings (e.g., Gee, 2001; Henneberg & Miguel, 2004).
2. Thanks to Bennett Schwartz for suggesting this to us.
3. There certainly are some adaptations that appear early in development and serve to prepare children for later life, termed *deferred adaptations* (Hernández Blasi & Bjorklund, 2003). We do not mean to imply any teleological implications here. Deferred adaptations, such as sex differences in play styles and dealing with dominance hierarchies (e.g., Hawley, 1999; Pellegrini & Smith, 1998), likely also provide some immediate benefits to children as well as preparing them for life as adults (cf., the memory adaptations discussed in the previous section).

REFERENCES

Aslan, A., & Bäuml, K.-H. T. (2012). Adaptive memory: Young children show enhanced retention of fitness-related information. *Cognition*, *122*, 118–122.

Atance, C. M. (2008). Future thinking in young children. *Current Directions in Psychological Science*, *17*, 295–298.

Baird, A. A., Kagan, J., Gaudette, T., Walz, K. A., Hershlag, N., & Boas, D. A. (2002). Frontal lobe activation during object permanence: Data from near-infrared spectroscopy. *NeuroImage, 16*, 120–126.

Baltazar, N. C., Shutts, K., & Kinzler, K. D. (2012). Children show heightened memory for threatening social actions. *Journal of Experimental Child Psychology, 112*, 102–110.

Bauer, P. J. (2007). *Remembering the times of our lives: Memory in infancy and beyond*. Mahwah, NJ: Erlbaum.

Bauer, P. J., & Fivush, R. (Eds.) (in press). *Wiley-Blackwell handbook on the development of children's memory*. New York: Wiley-Blackwell.

Baumeister, R. F., Bratslavsky, E., Finkenauer, C., & Vohs, K. D. (2001). Bad is stronger than good. *Review of General Psychology, 5*, 323–370.

Bischof, N. (1978). On the phylogeny of human morality. In G. S. Stent (Ed.), *Morality as a biological phenomenon* (pp. 53–74). Berlin: Dahlem Konferenzen.

Bischof-Köhler, D. (1985). Zur Phylogenese menschlicher Motivation. In L. H. Eckensberger & E. D. Lantermann (Eds.), *Emotion und Reflexivitat* (pp. 3–47). Vienna: Urban & Schwarzenberg.

Bishop, S. J., Dalgleish, T., & Yule, W. (2004). Memory for emotional stories in high and low depressed children. *Memory, 12*, 214–230.

Bjorklund, D. F. (1997). The role of immaturity in human development. *Psychological Bulletin, 122*, 153–169.

Bjorklund, D. F., Dukes, C., & Brown, R. D. (2009). The development of memory strategies. In M. Courage & N. Cowan (Eds.), *The development of memory in infancy and childhood* (pp. 145–175). New York: Psychology Press.

Bjorklund, D. F., & Green, B. L. (1992). The adaptive nature of cognitive immaturity. *American Psychologist, 47*, 46–54.

Bjorklund, D. F., Periss, V., & Causey, K. (2009). The benefits of youth. *European Journal of Developmental Psychology, 6*, 120–137.

Bjorklund, F. F., & Sellers, P. D. II. (in press). Memory development in evolutionary perspective. In P. Bauer & R. Fivush (Eds.), *Wiley-Blackwell handbook on the development of children's memory*. New York: Wiley-Blackwell.

Bogdan, R. J. (2010). *Our own minds: Sociocultural grounds for self-consciousness*. Cambridge, MA: MIT Press.

Brainerd, C. J., & Reyna, V. F. (2005). *The science of false memory*. New York: Oxford University Press.

Brainerd, C. J., Stein, L. M., Silveira, R. A., Rohenkohl, G., & Reyna, V. F. (2008). How does negative emotion cause false memories? *Psychological Science, 19*, 919–925.

Bushnell, I. W. R., Sai, F., & Mullin, J. T. (1989). Neonatal recognition of the mother's face. *British Journal of Developmental Psychology, 7*, 3–15.

Calvo, M. G., & Lang, P. J. (2005). Parafoveal semantic processing of emotional visual scenes. *Journal of Experimental Psychology: Human Perception and Performance, 31*, 502–519.

Cashon, C. H. (2011). Development of specialized face perception in infants: An information processing perspective. In L. M. Oakes, C. H. Cashon, M. Casasola, & D. H. Rakison (Eds.), *Infant perception and cognition: Recent advances, emerging theories, and future direction* (pp. 69–84). New York: Oxford University Press.

Ceci, S. J., & Bruck, M. (1993). Suggestibility of the child witness: A historical review and synthesis. *Psychological Bulletin, 113*, 403–439.

Chae, Y., Goodman, G. S., & Edelstein, R. S. (2011). Autobiographical memory development from an attachment perspective: The special role of negative events. In J. Benson (Ed.), *Advances in Child Development and Behavior, 40*, 1–49.

Clayton, N. S., & Dickinson, A. (1999). Scrub jays (*Aphelocoma coerulescens*) remember the relative time of caching as well as the location and content of their caches. *Journal of Comparative Psychology, 113*, 403–416.

Clayton, N. S., & Russell, J. (2009). Looking for episodic memory in animals and young children: Prospects for a new minimalism. *Neuropsychologia, 47*, 2330–2340.

Cordon, I. M., Melinder, A. M., Goodman, G. S., & Edelstein, R. S. (2013). Children's and adults' memory for emotional pictures: Examining age-related patterns using the Developmental Affective Photo System (DAPS). *Journal of Experimental Child Psychology, 114*, 339–356.

Correia, S. P. C., Dickinson, A., & Clayton, N. S. (2007). Western scrub-jays anticipate future needs independently of their current motivational state. *Current Biology, 17*, 1–6.

Courage, M., & Cowan, N. (Eds.) (2009). *The development of memory in infancy and childhood*. New York: Psychology Press.

Craik, F. I. M., & Lockhart, R. S. (1972). Levels of processing: A framework for memory research. *Journal of Verbal Learning and Verbal behavior, 11*, 671–684.

Davidson, D., Luo, Z., & Burden, M. J. (2001). Children's recall of emotional behaviors, emotional labels, and nonemotional behaviors: Does emotion enhance memory? *Cognition and Emotion, 15*, 1–26.

Diamond, A. (1985). Development of the ability to use recall to guide action as indicated by infants' performance on AB. *Child Development, 56*, 868–883.

Donald, M. (1991). *Origins of the modern mind: Three stages in the evolution of culture and cognition*. Cambridge, MA: Harvard University Press.

Duchaine, B., & Nakayama, K. (2005). Dissociations of face and object recognition in developmental prosopagnosia. *Journal of Cognitive Neuroscience, 17*, 1–13.

Dufour, V., & Sterck, E. H. M. (2008). Chimpanzees fail to plan in an exchange task but succeed in a tool-using procedure. *Behavioural Processes, 79*, 19–27.

Dunbar, R. I. M. (1998). The social brain hypothesis. *Evolutionary Anthropology, 6*, 178–190.

Dunbar, R. I. M. (2010). Brain and behaviour in primate evolution. In P. M. Kappler & J. B. Silk (Eds.), *Mind the gap: Tracing the origins of human universals* (pp. 315–330). New York: Springer.

Einstein, G. O., & McDaniel, M. A. (2005). Prospective memory: Multiple retrieval processes. *Current Directions in Psychological Science, 14*, 286–290.

Fivush, R. (1988). The functions of event memory: Some comments on Nelson and Barsalou. In U. Neisser & E. Winograd (Eds.), *Remembering reconsidered: Ecological and traditional approaches to the study of memory* (pp. 277–282). New York: Cambridge University Press.

Fivush, R., & Hamond, N. R. (1990). Autobiographical memory across the preschool years: Toward reconceptualizing childhood amnesia. In R. Fivush & J. A. Hudson (Eds.), *Knowing and remembering in young children* (pp. 223–248). Cambridge: Cambridge University Press.

Fivush, R., Hazzard, A., Sales, J. M., Sarfati, D., & Brown, T. (2003). Creating coherence out of chaos? Children's narratives of emotionally positive and negative events. *Applied Cognitive Psychology, 17*, 1–19.

Fivush, R., Kuebli, J., & Clubb, P. A. (1992). The structure of events and event representations: A developmental analysis. *Child Development, 63*, 188–201.

Flinn, M., & Ward, C. (2005). Evolution of the human child. In B. J. Ellis & D. F. Bjorklund (Eds.), *Origins of the social mind: Evolutionary psychology and child development* (pp. 19–44). New York: Guilford.

Foley, M. A., & Ratner, H. H. (1998). Distinguishing between memories for thoughts and deeds: The role of prospective processing in children's source monitoring. *British Journal of Developmental Psychology, 16*, 465–484.

Foley, M. A., Ratner, H. H., & Gentes, E. (2010). Helping children enter into another's experiences: The look and feel of it. *Journal of Cognition and Development, 11*, 217–239.

Foley, M. A., Ratner, H. H., & Passalacqua, C. (1993). Appropriating the actions of another: Implications for children's memory and learning. *Cognitive Development, 8*, 373–401.

Gallup, G. G.,Jr. (1979). Self-recognition in chimpanzees and man: A developmental and comparative perspective. In M. Lewis & L. A. Rosenblum (Eds.), *Genesis of behavior: Vol. 2. The child and its family* (pp. 107–126). New York: Plenum.

Gardiner, A., Greif, M., & Bjorklund, D. F. (2011). Guided by intention: Preschoolers' imitation reflects inferences of causation. *Journal of Cognition and Development, 12*, 355–373.

Gee, H. (2001). Return to the planet of the apes. *Nature, 412*, 131–132.

Germine, L., Cashdollar, N., Duzel, E., & Duchaine, B. (2011). A new selective developmental deficit: Impaired object recognition with normal face recognition. *Cortex, 47*, 598–607.

Gopnik, A. (2009). *The philosophical baby: What children's minds tell us about truth, love, and the meaning of life*. New York: Farber, Straus, and Giroux.

Goren, C. C., Sarty, M., & Wu, P. Y. K. (1975). Visual following and pattern discrimination of face-like stimuli by newborn infants. *Pediatrics, 56*, 544–549.

Hartshorn, K., Rovee-Collier, C., Gerhardstein, P., Bhatt, R. S., Klein, P. J., Aaron, F., Wondoloski, T. L., & Wurtzel, N. (2004). Developmental changes in the specificity of memory over the first year of life. *Developmental Psychobiology, 32*, 69–89.

Hawley, P. H. (1999). The ontogenesis of social dominance: A strategy-based evolutionary perspective. *Developmental Review, 19*, 97–132.

Henneberg, M., & de Miguel, C. (2004). Hominins are a single lineage: brain and body size variability does not reflect postulated taxonomic diversity of hominins. *Journal of Comparative Human Biology, 55*, 21–37.

Hernández Blasi, C., & Bjorklund, D. F. (2003). Evolutionary developmental psychology: A new tool for better understanding human ontogeny. *Human Development, 46*, 259–281.

Holmboe, K., Pasco Fearon, R. M., Csibra, G., Tucker, L., & Johnson, M. H. (2008). "Freeze-Frame": A new infant inhibition task and its relation to frontal cortex tasks in infancy and early childhood. *Journal of Experimental Child Psychology, 100*, 89–114.

Horner, V., & Whiten, A. (2005). Causal knowledge and imitation/emulation switching in chimpanzees (*Pan troglodytes*) and children (*Homo sapiens*). *Animal Cognition, 8*, 164–181.

Howe, M. L. (2007). Children's emotional false memories. *Psychological Science, 18*, 856–860.

Howe, M. L. (2011). *The nature of early memory: An adaptive theory of the genesis and development of memory*. New York: Oxford University Press.

Howe, M. L. (in press). The co-emergence of the self and autobiographical memory: An adaptive view of early memory. In P. Bauer & R. Fivush (Eds.). *Wiley-Blackwell handbook on the development of children's memory*. New York: Wiley-Blackwell.

Howe, M. L., Courage, M. L., & Rooksby, M. (2009). The genesis and development of autobiographical memory. In M. L. Courage & N. Cowan (Eds.), *The development of memory in infancy and childhood* (pp. 178–196). New York: Psychology Press.

Howe, M. L., & Derbish, M. H. (2010). On the susceptibility of adaptive memory to false memory illusions. *Cognition, 115*, 252–267.

Joffe, T. H. (1997). Social pressures have selected for an extended juvenile period in primates. *Journal of Human Evolution, 32*, 593–605.

Johnson, M. H., Dziurawiec, S., Ellis, H. D., & Morton, J. (1991). Newborns' preferential tracking of faces and its subsequent decline. *Cognition, 40*, 1–19.

Johnson, M. K., Hashtroudi, S., & Lindsay, D. S. (1993). Source monitoring. *Psychological Bulletin, 114*, 3–28.

Kelly, D. J., Quinn, P. C., Slater, A. M., Lee, K., Ge, L., & Pascalis, O. (2007). The other-race effect develops during infancy. *Psychological Science, 18*, 1084–1089.

Kensinger, E. A., & Corkin, S. (2003). Memory enhancement for emotional words: Are emotional words more vividly remembered than neutral words? *Memory & Cognition, 31*, 1169–1180.

Kensinger, E., & Corkin, S. (2004). Two routes to emotional memory: Distinct neural processes for valence and arousal. *Proceedings of the National Academy of Sciences USA, 101*, 3310–3315.

Kerns, K. A. (2000). The CyberCruiser: An investigation of development of prospective memory in children. *Journal of the International Neuropsychological Society, 6*, 62–70.

Kinzler, K. D., & Shutts, K. (2008). Memory for "mean" over "nice": The influence of threat on children's face memory. *Cognition, 107*, 775–783.

Kliegel, M., & Jäger, T. (2007). The effects of age and cue-action reminders on event-based prospective memory performance in preschoolers. *Cognitive Development, 22*, 33–46.

Kvavilashvili, L., Kyle, F., & Messer, D. J. (2008). The development of prospective memory in children: Methodological issues, empirical findings and future directions. In M. Kliegel, M. A. McDaniel, & G. O. Einstein (Eds.), *Prospective memory: Cognitive, neuroscience, developmental, and applied perspectives* (pp. 115–140). Mahwah, NJ: Erlbaum.

Learmonth, A. E., Lambert, R., & Rovee-Collier, C. (2004). Generalization of deferred imitation during the first year of life. *Journal of Experimental Child Psychology, 88*, 297–318.

Lewis, M., & Brooks-Gunn, J. (1979). *Social cognition and the acquisition of self*. New York: Plenum Press.

LoBue, V., & DeLoache, J. S. (2008). Detecting the snake in the grass. Attention to fear-relevant stimuli by adults and young children. *Psychological Science, 19*, 284–289.

Lord, C. G. (1980). Schemas and images as memory aids: Two modes of processing social information. *Journal of Personality and Social Psychology, 38*, 257–269.

Lyons, D. E., Young, A. G., & Keil, F. C. (2007). The hidden structure of overimitation. *Proceedings of the National Academy of Sciences USA, 104*, 19751–19756.

McCarthy, G., Puce, A., Gore, J. C., & Allison, T. (1997). Face-specific processing in the human fusiform gyrus. *Journal of Cognitive Neuroscience, 9*, 605–610.

Milner, B. (1964). Some effects of frontal lobectomy in man. In J. M. Warren & K. Akert (Eds.), *The frontal granular cortex and behavior* (pp. 314–334). New York: McGraw-Hill.

Mondloch, C. J., Lewis, T. L., Burdeau, D. R., Maurer, D., Dannemiller, J. L., Stephens, B., R., & Kleiner-Gathercoal, K. A. (1999). Face perception during early infancy. *Psychological Science, 10*, 419–422.

Mood, D. W. (1979). Sentence comprehension in preschool children: Testing an adaptive egocentrism hypothesis. *Child Development, 50*, 247–250.

Mulcahy, N. J., & Call, J. (2006). How great apes perform on a modified trap-tube task. *Animal Cognition, 9*, 193–199.

Mumme, D. L., Fernald, A., & Herrera, C. (1996). Infants' responses to facial and vocal emotional signals in a social referencing paradigm. *Child Development, 67*, 3219–3237.

Nairne, J. S. (2010). Adaptive memory: Evolutionary constraints on remembering. In B. H. Ross (Ed.), *The psychology of learning and motivation: Advances in research and theory* (Vol. 53, pp. 1–32). San Diego, CA: Elsevier.

Nairne, J. S., Pandeirada, J. N. S., & Thompson, S. R. (2008). Adaptive memory: The comparative value of survival processing. *Psychological Science, 19*, 176–180.

Nairne, J. S., Thompson, S. R., & Pandeirada, J. N. S. (2007). Adaptive memory: Survival processing enhances retention. *Journal of Experimental Psychology: Learning, Memory, and Cognition, 33*, 263–273.

Nelson, K. (1993). The psychological and social origins of autobiographical memory. *Psychological Science, 4*, 7–14.

Nelson, K. (1996). *Language in cognitive development: The emergence of the mediated mind*. New York: Cambridge University Press.

Nelson, K. (2005). Evolution and development of human memory systems. In B. J. Ellis & D. F. Bjorklund (Eds.), *Origins of the social mind: Evolutionary psychology and child development* (pp. 354–382). New York: Guilford.

Nielsen, M., Suddendorf, T., & Slaughter, V. (2006). Self-recognition beyond the face. *Child Development, 77*, 176–185.

Oppenheim, R. W. (1981). Ontogenetic adaptations and retrogressive processes in the development of the nervous system and behavior. In K. J. Connolly & H. F. R. Prechtl (Eds.), *Maturation and development: Biological and psychological perspectives* (pp. 73–108). Philadelphia: International Medical Publications.

Osvath, M., & Osvath, H. (2008). Chimpanzee (*Pan troglodytes*) and orangutan (*Pongo abelii*) forethought: self-control and pre-experience in the face of future tool use. *Animal Cognition, 11*, 661–674.

Otgaar, H., & Smeets, T. (2010). Adaptive memory: Survival processing increases both true and false memory in adults and children. *Journal of Experimental Psychology: Learning, Memory, and Cognition, 36*, 1010–1016.

Pascalis, O., de Haan, M., & Nelson, C. A. (2002). Is face processing species-specific during the first year of life? *Science, 296* (17 May), 1321–1323.

Pellegrini, A. D. (in press). Play. In P. Zelazo (Ed.), *Oxford handbook of developmental psychology*. New York: Oxford University Press.

Pellegrini, A. D., & Bjorklund, D. F. (2004). The ontogeny and phylogeny of children's object and fantasy play. *Human Nature, 15*, 23–43.

Pellegrini, A. D., & Smith, P. K. (1998). Physical activity play: The nature and function of neglected aspect of play. *Child Development, 69*, 577–598.

Periss, V., & Bjorklund, D. F. (2011) Trials and tribulations of childhood: An evolutionary perspective. In C. Salmon & T. Shackelford (Eds.) *Oxford handbook of evolutionary family psychology* (pp. 149–168). Oxford: Oxford University Press.

Piaget, J. (1983). Piaget's theory. In P. Mussen (ed.). *Handbook of child psychology, 4th ed., Vol. 1* (pp. 703–732). New York: Wiley.

Pierce, K., Conant, D., Hazin, R., Stoner, R., & Desmond, J. (2011). Preference for geometirc patterns early in life as a risk factor for austism. *Archive of General Psychiatry, 68*, 101–109.

Pitcher, D., Dilks, D. D., Saxe, R., Triantafyllou, C., & Kanwisher, N. (2011). Differential selectivity for dynamic versus static information in face-selective cortical regions. *NeuroImage, 56*, 2356–2363.

Platt, M. L., Brannon, E. M., Briese, T. L., & French, J. A. (1996). Differences in feeding ecology predict differences in performance between golden lion tamarins (*Leontopithecus rosalia*) and Wied's marmosets (*Callithrix kuhli*) on spatial and visual memory tasks. *Learning and Behavior, 24*, 384–393.

Plotnik, J. M., de Waal, F. B. M., & Reiss, D. (2006). Self-recognition in an Asian elephant. *Proceedings of the National Academy of Sciences USA, 103*, 17053–17057.

Povinelli, D. J., Landau, K. R., & Perilloux, H. K. (1996). Self-recognition in young children using delayed versus live feedback: Evidence of a developmental asynchrony. *Child Development, 67*, 1540–1554.

Pratkanis, A. R., & Greenwald, A. B. (1985). How shall the self be conceived? *Journal for the Theory of Social Behavior, 15*, 311–328.

Prior, H., Schwarz, A., & Güntürkün, O. (2008). Mirror-induced behavior in the magpie (*Pica pica*): Evidence of self-recognition. *PLoS Biol 6*(8): e202 doi:10.1371/journal.pbio.0060202.

Quas, J. A., & Fivush, R. (2009). *Emotion in memory and development: Biological, cognitive, and social considerations*. New York: Oxford University Press.

Quon, E., & Atance, C. M. (2010). A comparison of preschoolers' memory, knowledge, and anticipation of events. *Journal of Cognition and Development, 11*, 37–60.

Ratner, H. H., Foley, M. A., & Gimpert, N. (2002). The role of collaborative planning in children's source-monitoring errors and learning. *Journal of Experimental Child Psychology, 81*, 44–73.

Reiss, D.,& Marino, L. (2001). Mirror self-recognition in the bottlenose dolphin: A case of cognitive convergence. *Proceedings of the National Academy of Sciences USA, 98*, 5937–5942.

Richler, J. J., Cheung, O. S., & Gauthier, I. (2011). Holistic processing predicts face recognition. *Psychological Science, 22*, 464–471.

Ross, J., Anderson, J. R., & Campbell, R. N. (2011). I remember me: Mnemonic self-reference effects in preschool children. *Monographs of the Society for Research in Child Development, 76* (Number 3, Serial No. 300).

Rovee-Collier, C., & Cuevas, K. (2009). The development of infant memory. In M. L. Courage & N. Cowan (Eds.), *The development of memory in infancy and childhood* (pp. 11–41). New York: Psychology Press.

Rovee-Collier, C., Schechter, A., Shyi, C-W. G. & Shields, P. (1992). Perceptual identification of contextual attributes and infant memory retrieval. *Developmental Psychology, 28*, 307–318.

Rovee-Collier, C., & Shyi, C-W. G. (1992). A functional and cognitive analysis of infant long-term retention (pp. 3–55). In M. L. Howe, C. J. Brainerd, & V. F. Reyna (Eds.), *Development of long-term retention*. New York: Springer-Verlag.

Rubin, D. C. (2000). The distribution of early childhood memories. *Memory, 8*, 265–269.

Schacter, D. L., Wagner, A. D., & Buckner, R. L. (2000). Memory systems of 1999. In E. Tulving & F. I. M. Craik (Eds.), *Handbook of memory* (pp. 627–644). New York: Oxford University Press.

Salwiczek, L. H., Watanabe, A., & Clayton, N. S. (2010). Ten years of research into avian models of episodic-like memory and its implications for developmental and comparative cognition. *Behavioural Brain Research, 215*, 221–234.

Sanna, K., Eira, J., Alice, C., Rachel, P., Katja, J., Marja-Leena, M., Jukka, R., Hana, E., David, P., & Irma, M. (2011). Face memory and object recognition in children with high functioning autism or Asperger syndrome and in their parents. *Research in Autism Spectrum Disorders, 5*, 622–628.

Schwartz, B. L., Colon, M. R., Sanchez, I. C., Rodriquez, I. A., & Evans, S. (2002). Single-trial learning of "what" and "who" information in a gorilla (*Gorilla gorilla gorilla*): Implications for episodic memory. *Animal Cognition, 5*, 85–90.

Simcock, G., & Hayne, H. (2002). Breaking the barrier? Children fail to translate their preverbal memories into language. *Psychological Science, 13*, 225–231.

Skouteris, H., Spataro, J., & Lazarids, M. (2006). Young children's use of a delayed video representation to solve a retrieval problem pertaining to self. *Developmental Science, 9*, 505–517.

Smith, R. E., Bayen, U. J., & Martin, C. (2010). The cognitive processes underlying event-based prospective memory in school-age children and young adults: A formal model-based study. *Developmental Psychology, 46*, 230–244.

Somerville, S. C., Wellman, H. M., & Cultice, J. C. (1983). Young children's deliberate reminding. *Journal of Genetic Psychology, 143*, 87–96.

Sommerville, J. A., & Hammond, A. J. (2007). Treating another's actions as one's own: Children's memory of and learning from joint activity. *Developmental Psychology, 43*, 1003–1018.

Striano, T., Brennan, P. A., & Vanman, E. J. (2002). Maternal depressive symptoms and 6-month-old infants' sensitivity to facial expressions. *Infancy, 3*, 115–126.

Suddendorf, T. (2006). Foresight and evolution of the human mind. *Science, 312*, 1006–1007.

Suddendorf, T., & Corballis, M. C. (2010). Behavioural evidence for mental time travel in nonhuman animals. *Behavioral Brain Research, 215*, 292–298.

Suddendorf, T., Corballis, M. C., & Collier-Baker, E. (2009). How great is great ape foresight? *Animal Cognition, 12*, 751–754.

Szpunar, K. K. (2010). Episodic future thought: An emerging concept. *Perspectives on Psychological Science, 5*, 142–162.

Tulving, E. (1983). *Elements of episodic memory*. New York: Oxford University Press.

Tulving, E. (2002). Episodic memory: From mind to brain. *Annual Review of Psychology, 53*, 1–25.

Tulving, E. (2005). Episodic memory and autonoesis: Uniquely human? In H. S. Terrace & J. Metcalfe (Eds.), *The missing link in cognition: Origins of self-reflective consciousness* (pp. 3–56). New York: Oxford University Press.

Vaish, A., Grossman, T., & Woodward, A. (2008). Not all emotions are created equal: The negativity bias in social-emotional development. *Psychological Bulletin, 134*, 383–403.

Volk, T., & Atkinson, J. (2008). Is child death the crucible of human evolution? *Journal of Social, Cultural, and Evolutionary Psychology, 2*, 247–260.

Walton, G. E., Bower, N. J. A., & Bower, T. G. R. (1992). Recognition of familiar faces by newborns. *Infant Behavior and Development, 15*, 265–269

West, T. A., & Bauer, P. J. (1999). Assumptions of infantile amnesia: Are there differences between early and later memories? *Memory, 7*, 257–278.

Whiten, A. (2010). Ape behavior and the origins of human culture. In P. M. Kappler & J. B. Silk (Eds.), *Mind the gap: Tracing the origins of human universals* (pp. 429–450). New York: Springer.

16

Adaptive Memory: Controversies and Future Directions

JAMES S. NAIRNE

Human memory is adaptive. Our capacity to remember and forget helps us solve problems, everything from remembering where the car is parked to recognizing the person who owes us money. Remembering is the product of an evolutionary process as well. At some point in our ancestral past, memory systems developed because their presence helped us solve problems related to survival and reproduction. An organism with the capacity to remember where the food is located, or categories of potential predators, is more likely to survive than an organism lacking this capacity. There is currently little or no debate on this point.

More controversial, however, is the notion that nature's criterion—the enhancement of inclusive fitness—has relevance to modern memory functioning. The visual system evolved, as did the organs of the body, but it is not necessary to consider evolutionary pressures to gain insight into the structural and functional properties of these systems. Evolutionary theory does not demand that evolved systems maximize current fitness, or even work most efficiently when asked to solve the problems that led to their development (Symons, 1992). Yet those systems might well bear the imprint of those selection pressures. In the case of the visual system, for example, the specific tunings of cone photopigments appear to dovetail nicely with the adaptive problem of identifying ripe fruit in natural environments (e.g., Nathans, 1999).

Research on adaptive memory originated from the claim that our memory systems are functionally specialized to solve adaptive, or fitness-based, problems. Included are problems such as remembering sources of nourishment, threats to survival, activities of kin, potential mating partners, cheaters and free-riders, and so on (Nairne & Pandeirada, 2008). Most of the things we remember are unrelated to fitness, such as where I left my pencil or the movie I saw last night, but the fact that people are capable of general remembering does not mean that our memory systems were *designed* (by nature) to remember generally. On the contrary, from an evolutionary perspective content and specificity are likely to matter (Klein, Cosmides, Tooby, & Chance, 2002).

This brief chapter is divided into three sections. First, I comment on the nature of evolutionary arguments, particularly the distinction between ultimate and proximate explanations of cognitive traits. This distinction can be a confusing one (Scott-Phillips, Dickens, & West, 2011) and has led to some misunderstandings in the literature on the evolutionary determinants of remembering. Second, I discuss two important ancillary issues—domain-specificity and the role of ancestral priorities in empirical accounts of remembering. Third, and finally, I offer suggestions on future directions for adaptive memory research and evolutionary functional analysis.

ULTIMATE VERSUS PROXIMATE: ON THE NATURE OF EVOLUTIONARY ARGUMENTS

Functional accounts of memory do not require an evolutionary stance. One can approach the study of remembering by investigating the problems that memory systems solve—such as remembering the face of a cheater—without worrying about when and why those systems evolved. As I have discussed elsewhere, there are significant advantages to adopting a functional perspective, including the fact that it gives one concrete criteria against which to measure progress (see Nairne, 2005). However, if one chooses an evolutionary stance, which is a special form of functionalism, it is useful to consider the nature of evolutionary arguments—especially the distinction between ultimate and proximate questions about evolved traits (Mayr, 1963).

Ultimate explanations are statements about the functions of a trait and "why" it would have been selected by nature during an evolutionary process. Natural selection is governed by a criterion, the enhancement of inclusive fitness, so ultimate explanations appeal to (and are judged by) this criterion. *Proximate explanations* focus on the mechanisms that produce the trait—that is, they are statements about "how" the trait works and the conditions under which the trait is likely to be expressed. To cite a common example, crying is a behavior that increases an infant's chances of gaining the care and comfort needed for survival; an ultimate explanation of crying focuses on its fitness effects and why such a trait likely gained traction in the population over generations. Alternatively, and perhaps concurrently, one can choose to study the biological mechanisms that induce and control crying, such how glands secrete lachrymal fluid, along with the external triggering conditions that produce the behavior (e.g., separation from the caretaker). The latter would constitute an investigation of the proximate mechanisms that underlie the behavior.

Evolutionary theorists generally agree that both kinds of explanations are necessary for a full accounting of an evolved trait, but it is important not to confuse the two. Proximate mechanisms do not generally provide answers to ultimate questions, which deal with how traits affect inclusive fitness. In the case of adaptive memory, our lab originally proposed an ultimate hypothesis: If nature "tuned" our memory systems to process and remember fitness-relevant information, then our ancestors likely possessed an improved ability to survive and propagate their genetic record (Nairne, Thompson, & Pandeirada, 2007, p. 263). This is an evolutionary argument, one that generated an *a priori* empirical prediction: People should be particularly good at remembering information that is processed in a fitness context. The survival processing effect, among other phenomena, supports this prediction (see Erdfelder & Kroneisen, Chapter 10 in this volume; Nairne, 2010).

Notice that our hypothesis focuses on "why" our memory systems might show tunings and not on "how" those characteristics are actually implemented. Enhanced fitness is the engine driving natural selection, so it seemed sensible to propose that memory evolved because it solved specific problems related to fitness (e.g., remembering information processed in a fitness context). But we were careful not to specify the proximate mechanisms that drive the retention advantages. We speculated that survival processing might tap a special cognitive adaptation—a kind of evolved fitness-based memory module—but questioned the viability of such a mechanism (e.g., "survival" is too broad a construct). We suggested as well that survival processing advantages might be explained by other processes such as rehearsal, elaboration, distinctive processing, or self-referent processing (Nairne et al., 2007, p. 270). Fitness-based tunings seemed certain to produce fitness advantages, we argued, but nature could have achieved those tunings in a variety of ways.

One possibility is that our retention systems are simply reactive to novel or unusual events in the environment (e.g., Sokolov, 1963). It is well known that organisms orient to novel events, such as the sudden appearance of a predator. Orienting responses, in turn, could activate elaborative machinery that promotes good long-term recall of the attended item. From an evolutionary perspective the adaptive problem is solved—we are likely to show excellent retention of predators in our environment. Framed in this way, however, novelty and elaboration are proximate mechanisms that produce a stable mnemonic trait: enhanced retention of new and potentially dangerous things. The question of "why" such a memory tuning developed—the ultimate question—is answered by considering its effects on inclusive fitness. One would need to make the case that novelty tunings increase inclusive fitness; that is, organisms with retention systems tuned uniquely to novelty are associated with greater reproductive success. For reasons discussed below, I believe the case for novelty-based mnemonic tunings, although valid as an ultimate hypothesis, is not compelling.

The ultimate/proximate distinction is important here because proximate analyses are sometimes used to argue against Nairne et al.'s (2007) evolutionary account. Howe and Otgaar (2013) used evidence from scenario-based manipulations of elaboration to conclude that "the mnemonic value associated with survival processing may not lie in the adaptive significance of such processing, but in the fact that such processing recruits other well-known memory processes that enhance retention" (p. 19; see also Howe & Derbish, 2010, and Chapter 5 in this volume); Similar arguments have been made by others: "it is not the evolutionary significance of survival per se that explains the survival processing effect. Rather, the degree to which survival processing invites elaborative, distinctive forms of encoding would predict the mnemonic benefit of survival processing" (Kroneisen & Erdfelder, 2011, p. 1554). Such statements confuse the ultimate hypothesis—nature evolved memory systems sensitive to fitness—with hypotheses about possible proximate mechanisms (e.g., novelty-induced elaboration). The fact that survival situations might naturally invite elaborate, distinctive forms of encoding is consistent with the Nairne et al. (2007) hypothesis, not evidence against it.

To determine what proximate mechanisms actually evolved, of course, the supreme court of appeal is inclusive fitness. Following Howe and Otgaar (2013), one could propose evolved memory systems that show no special sensitivity to fitness-relevant information or contexts (see also Bell & Buchner, Chapter 3 in this volume). As noted above, our memory systems might simply be "tuned" to novelty or to information that has been processed meaningfully (Craik & Lockhart, 1972); one could also subscribe to the conventional notion that we retrieve information that is matched by cues in the retrieval environment (Tulving & Thomson, 1973). However, in these cases, as ultimate hypotheses, one would need to establish that such designs were likely to have become the targets of natural selection. As I have argued elsewhere (Nairne, 2010), viewed through the lens of nature's criterion—the enhancement of inclusive fitness—memory mechanisms that fail to differentiate among the adaptive consequences of an event seem unlikely to have evolved. Our memory systems need "crib sheets," or ways to differentiate between important and unimportant events (e.g., Ermer, Cosmides, & Tooby, 2007). Remembering everything that has been processed meaningfully, or every novel event, leads to combinatorial explosion of storage; similarly, a system that is merely driven by an encoding–retrieval match remembers continuously. Selection advantages will "accrue to memory systems that remember appropriately—that is, to systems that remember information pertinent to improving survival and reproduction" (Nairne, 2010, p. 14).

For this reason, our laboratory has maintained that we evolved domain-specific memory mechanisms that are inherently sensitive to content—specifically, information that is processed for its fitness consequences (either naturally or through experimental manipulations). Such "tunings" are likely to have evolved, as opposed to domain-general systems, because, again, the engine driving natural selection is the enhancement of inclusive fitness. Such reasoning does not rule out a role for elaboration—in fact, elaboration or "richness of encoding" may be the natural consequence of fitness-relevant processing, as argued in the original Nairne et al. (2007) report. But to claim that a domain-general process such as elaboration was the main target of evolution, rather than fitness-relevant processing *per se*, conflicts with what we know about the nature of physical and cognitive adaptations.

DOMAIN-SPECIFICITY AND ANCESTRAL PRIORITIES

The study of proximate mechanisms can be relevant to ultimate hypotheses. Evolved traits, for example, need to arise from proximate mechanisms that are heritable. Others have suggested that adaptations show evidence of special design (e.g., Williams, 1966)—that is, they are engineered to solve specific problems, such as pumping blood (the heart), filtering impurities (kidneys), or transducing environmental messages into the electrochemical language of the brain (the retina). A related characteristic, just mentioned, is domain-specificity, which means that evolved mechanisms are engaged selectively for some kinds of input and not others.

As defined by Barrett and Kurzban (2006), domain-specificity "refers to the idea that a given system accepts or is specialized to operate on only specific classes of information ('domains') for processing" (p. 631). For example, one could propose a face-recognition system that operates or is engaged only by faces, or at least by stimuli that share the formal statistical properties of faces (Kanwisher, 2000). Other possibilities include systems that are designed to handle living things, such as animals, and particular sorts of manmade objects, such as tools (Boyer & Barrett, 2005). Neuropsychological evidence, such as a selective impairment in the ability to process faces or animals, and neuroimaging data have helped build the case for such domain-specific systems (e.g., Caramazza & Shelton, 1998).

Within evolutionary cognitive psychology, Tooby and Cosmides (1992, 2005) have suggested that humans evolved systems specialized to detect cheaters or, more specifically, neurocognitive adaptations designed for social exchange. People show an enhanced ability to reason about violations of social contracts, especially in general reasoning tasks that often produce poor overall reasoning performance. These neurocognitive systems are assumed to be domain-specific, meaning that they are triggered by the processing of social contracts (much like a face-recognition system is initiated only by face-like stimuli). But, importantly, it is the general structure of a social contract rather than any particular kind of social contract that engages the system. Cosmides and Tooby (2005) have shown that people can reason exceptionally well about "wildly unfamiliar" situations as long as the situation fits the form of a social contract. Similarly, our retinae are domain-specific, in that they only process electromagnetic input, but they are capable of handling countless different patterns of electromagnetic information.

Returning to adaptive memory, the hypothesis that our memory systems are "tuned" to the processing of fitness-relevant information certainly implies domain-specificity. We suggested that "special" mnemonic advantages accrue whenever fitness-relevant processing is engaged, presumably because we evolved proximate mechanisms designed specifically to solve fitness-relevant problems. We placed the emphasis on processing, rather than on inherent content, because fitness relevance is often context-specific. As we noted, "food is survival relevant, but more so at the beginning of a meal than at its completion; a fur coat has high [survival]-value at the North Pole, but low at the Equator" (Nairne & Pandeirada, 2008, p. 240). What engages the memory system(s) is fitness-relevant processing, broadly conceived, rather than any particular environmental event.

This emphasis on processing, rather than specific content, helps to explain a puzzling finding in the survival processing literature. In the typical procedure, people are given lists of unrelated words and are asked to rate the relevance of each word to a survival scenario. Retention performance can then be examined as a function of relevance rating, and memory often increases accordingly. Items deemed highly relevant to the survival scene are often (but not always) remembered better than items given lower ratings (Butler, Kang, & Roediger, 2009; Nairne & Pandeirada, 2011). But strong survival effects are still typically found for items given the lowest ratings—that is, people remember words rated as irrelevant to survival better than words rated comparably in control conditions. This result is puzzling because it is not immediately apparent why we would have evolved memory mechanisms that show enhanced retention of information that lacks survival relevance. The reason, though, is straightforward: The survival processing paradigm *forces* participants to consider the fitness-relevant properties of all presented words, regardless of whether those stimuli are considered relevant or not. In natural settings, only fitness-relevant events are likely to engage fitness-relevant processing and show the enhanced retention.

Once again, what is important is fitness-relevant processing, not the particular event or setting that happens to induce the processing. Artificial laboratory tasks can engage survival processing, as can events in natural settings—both presumably lead to enhanced retention of the processed material. Recently, much has been made of the fact that scenarios using settings discrepant from the environment of evolutionary adaptedness (EEA) can produce retention advantages comparable to (or even exceeding) those found using the traditional Nairne et al. (2007) grasslands scenario. Kostic, McFarlan, and Cleary (2012) found strong survival advantages using a scenario in which people imagined themselves "stranded in a spaceship in space." Even more impressive, Soderstrom and McCabe (2011) discovered that "zombie processing"—rating the relevance of words to a possible attack from zombies—produced better retention than the Nairne et al. (2007) grassland-based survival scenario. At face value, these data seem surprising: there were no zombies in the EEA. But again, it is fitness-relevant processing that drives the retention advantage. Zombies are clearly predators and could activate other self-preservation systems as well (e.g., contamination), so it is not surprising that zombie processing leads to retention advantages. Similar arguments apply to "stranded spaceship" scenarios—again, basic self-preservation systems are likely to be activated, which in turn engender fitness-relevant processing.

More pertinent to the present discussion, though, is the issue of domain-specificity. In describing the zombie and spaceship results, Soderstrom and Cleary (Chapter 6 in this volume) conclude the following: "Taken together, these data suggest that survival processing advantages in memory are domain-general insofar as they are not optimally suited for any one environment or situation in particular, but rather operate effectively in a wide range of scenarios" (p. XXX). Although this conclusion is partly true, it confuses the concept of domain-specificity with another issue, *ancestral priorities*, which I discuss below. Our claim is that we evolved mnemonic mechanisms that are specialized to handle fitness-relevant events, particularly ones related to survival or self-preservation. If there is special mnemonic machinery to handle such events, then the input "domain" would be events that activate self-preservation systems regardless of the particular setting or event that happened to initiate the processing. Just as face-recognition systems respond to any face and "cheater modules" respond generally to violations of social contracts, we evolved memory mechanisms that respond generally to survival threats, or fitness processing in general.

In fact, we have known for some time that it is not the "setting" *per se* that produces survival processing advantages. Survival processing advantages have repeatedly been found when both the fitness-relevant and the control scenarios are set in the same natural (even grasslands) environment. For example, Nairne, Pandeirada, and Thompson (2008) compared the standard survival scenario to one in which people were vacationing at a fancy resort—a strong survival effect was obtained. Nairne, Pandeirada, Gregory, and Van Arsdall (2009) compared survival-based "hunting" and "gathering" scenarios to control scenarios that involved exactly the same activities (hunting for food in a natural environment to win a hunting contest, or searching for food in the grasslands as part of a scavenger hunt)—again, a strong survival effect was obtained. It is now clear that survival advantages can be obtained in virtually any environment, even outer space, as long as the scenarios activate basic predator-avoidance and/or self-preservation systems. This does not mean that the evolved memory "tunings" are domain-general; they are domain-specific, in the sense that they are activated by situations that are survival- or fitness-relevant.

Accepting that we may have evolved domain-specific memory tunings—for example, for events that induce fitness-relevant processing (predators, edible foods, potential mating partners, cheaters, etc.)—does not imply that all domain-specific input will activate those tunings to the same degree. Domain-specific systems, such as the auditory and visual systems, characteristically respond best to certain kinds of input. For example, the auditory system responds well to frequencies of sound that match what is produced by spoken language. From an evolutionary perspective, we might expect cognitive systems to respond most efficiently when faced with the problems (and corresponding input stimuli) that were critical to their development. Given that unique aspects of the human brain were sculpted primarily during the Pleistocene, the EEA favored by evolutionary theorists, optimal cognitive performance might well emerge when the task at hand mimics the foraging problems faced by our hunter-gatherer ancestors.

Note, though, that this is merely a hypothesis, not a strong prediction of evolutionary theory. Evolved systems do not necessarily maximize fitness, even when dealing with the selection pressures that led to their development. It is an interesting empirical question, however, and a number of laboratories have produced data consistent with ancestral priorities. For example, our visual and attentional systems may be tuned to the detection of threatening stimuli (LoBue & DeLoache, 2008) or animates (New, Cosmides, & Tooby, 2007). There is evidence as well for enhanced learning and conditioning for aversive stimuli that were likely to have been relevant during the EEA (e.g., snakes; Öhman & Mineka, 2001). Some of these data are controversial, however, because of item-selection concerns (Nairne, 2010; Quinlan, in press).

With respect to adaptive memory, both Nairne and Pandeirada (2010) and Weinstein, Bugg, and Roediger (2008) found evidence for ancestral priorities in the survival processing paradigm. People were asked to rate the relevance of words to either ancestral or modern scenarios. Both scenarios described survival-relevant situations, but one described escaping from a predator in the grasslands (ancestral) whereas the other involved escaping from an attacker in a city (modern). People using the ancestral scenario subsequently remembered the rated words better than those using the modern scenario. This ancestral advantage was replicated using several different wordings of the scenarios—for example, when searching for medicinal plants versus antibiotics to cure an infection (Nairne & Pandeirada, 2010, Experiment 2) or when gathering edible plants in the grasslands versus searching for and buying food in a city (Experiment 3).

Again, it is not the setting *per se* that is important—grasslands versus the city—but the processing systems that are activated by those settings. As Nairne and Pandeirada (2010) argued, dealing with an infection or an attacker in a city is clearly survival-relevant and should activate self-preservation systems, but cities contain support systems, such as police and hospitals, that are absent in the grasslands. Faced with an attacker in the city, or a raging infection, one probably thinks about calling the police or visiting the local pharmacy to get some antibiotics. You also have lots of social support systems in the city that are absent when you are stranded in the grasslands. At the same time, we argued, asking people to search for edible plants in the grasslands presumably activates processing scenarios that are closer to those employed in the EEA, compared to those activated when one attempts to buy food in a modern market, and the "fit" between the problem and the evolved system might contribute to the obtained retention differences.

The important point here is that the demonstration of ancestral priorities is only tangentially related to the question of domain-specificity. Both the ancestral and modern scenarios used by Nairne and Pandeirada (2010) were designed to induce fitness-relevant processing, thus engaging any special mnemonic systems that may have evolved to handle such situations. Indeed, Nairne and Pandeirada (2010), and several other researchers as well (e.g., Klein, 2013), have found strong survival processing effects using scenarios that do not directly reference the conditions of the EEA. Variations in the efficiency of processing, within the selected domain, are important to theory too but do not speak directly to the issue of domain specificity.

FUTURE DIRECTIONS

The functionalist agenda proposes that cognitive systems are functionally designed. As a product of natural selection, our capacity to remember and forget evolved because it helped us solve adaptive problems consistent with nature's criterion—the enhancement of inclusive fitness. As noted throughout, this conclusion does not guarantee that memory's operating characteristics will maximize current fitness or even lead to adaptive behavior in all situations. At the same time, given its origins, it seems unwise to ignore the selection pressures that presumably led to memory's development. Imagine attempting to understand the machinery in a factory without worrying about the product the machinery is designed to produce.

In evolutionary functional analysis, one attempts to generate empirically testable hypotheses about cognitive systems based on an analysis of the recurrent adaptive problems faced by our ancestors. Although we can never be certain about the specific selection pressures that drove the evolution of a memory system, such as episodic memory, we can make reasonable guesses: It would be important to remember the characteristics of predators or prey, the location of potable water or edible plants, the actions of those who violate social contracts, prospective mates, authority figures, and so on. Organisms with memory systems biased to retain such information likely held adaptive advantages over those who did not. Still, these are merely coarse-grained hypotheses rather than airtight predictions of evolutionary theory. One needs to conduct the relevant empirical tests to confirm (or disconfirm) hypotheses about the adaptive nature of a cognitive system. Much of the work described in this volume is directed toward this end.

It is also worth emphasizing that evolutionary functional analysis, as described here, is specifically designed to avoid *post hoc* reasoning—that is, fanciful adaptive explanations of existing phenomena. Evolutionary accounts are frequently criticized for relying on "just-so" stories (e.g., Gould & Lewontin, 1979), but this kind of criticism does not apply to most research on adaptive memory. The mnemonic advantages found for survival processing were predicted *a priori*, based on a hypothesis about possible fitness advantages that might accrue for remembering fitness-relevant information (Nairne et al., 2007). Much of the work described elsewhere in this volume flows from a similar foundation. For instance, work on memory for cheaters, or their personal attributes, follows naturally from social contract theory and various assumptions about the role of reciprocal altruism in human actions (see Bell &

Buchner, Chapter 3 in this volume, for a review). One of the exciting consequences of thinking functionally, especially in evolutionary terms, is the discovery of new empirical phenomena.

With this in mind, researchers should be wary about identifying the concept of adaptive memory exclusively with survival processing. "Survival" is a very broad concept—too general, in fact, to have been the likely target of selection. In our original work, we suggested that survival processing might tap into multiple modules or adaptations working in concert, each one activated to one degree or another by the survival processing task (see Nairne et al., 2007, p. 270). Although effective retention mechanisms might be activated strongly by any form of fitness-relevant processing, echoing the idea of novelty-based encoding, we almost certainly evolved unique mechanisms to deal with different adaptive problems. Retaining critical information about predators, food sources, possible violations of social contracts, potential mating partners, and so forth is unlikely to be handled successfully by a single mnemonic mechanism (e.g., elaboration). Nature usually evolves problem-specific structures because the "solution" to one problem can conflict with the requirements of another (Sherry & Schacter, 1987).

Our laboratory has recently turned its attention to more specific adaptive dimensions. For example, we have conducted experiments exploring the mnemonic effects of animacy, or the distinction between living and nonliving things (Nairne, VanArsdall, Pandeirada, Cogdill, & LeBreton, in press; VanArsdall, Nairne, Pandeirada, & Blunt, 2013). Animacy is a foundational dimension, appearing early in development, and people seem to prioritize the visual processing of animate objects (e.g., New et al., 2007). From an evolutionary perspective, it is reasonable to propose that our memory systems are selectively tuned to process and remember animate things. Predators are animate beings, as are potential mating partners. Methodological issues loom, however, because animate and inanimate stimuli presumably differ along a number of relevant dimensions (e.g., imageability, frequency, meaningfulness, and so on). In our work, we have attempted to equate or control for these potentially confounding dimensions and have found strong animacy advantages in both free recall and paired-associate learning. This research suggests that animacy may turn out to be one of the most important item dimensions ultimately controlling retention (see Nairne et al., in press).

We have also conducted experiments showing that animacy *processing* produces mnemonic advantages (see VanArsdall et al., 2013). Here, instead of using existing words or pictures of real objects, we asked participants to process novel events—pronounceable nonwords (e.g., FRAV)—as either living or nonliving things. The nonwords were paired during presentation with properties characteristic of either living ("dreams at night") or nonliving ("assembled with screws") stimuli. People were then asked to decide whether each nonword likely represented a living thing or an object prior to a final surprise recall or recognition test. We found that memory was significantly enhanced when the nonwords were associated with animate properties. Methodologically, these experiments are important because across conditions everyone was asked to process and remember exactly the same nonwords, thus effectively eliminating item-selection concerns. Again, this research suggests that our cognitive systems may be prepared or tuned to detect and remember animate things, a conclusion that follows nicely from a functional/evolutionary perspective.

We have also been investigating the mnemonic effects of contamination. There is certainly adaptive value in remembering objects in the environment that have been potentially contaminated by disease. There is a large literature on the social-cognitive effects of contamination and disgust (e.g., Tybur, Lieberman, Kurzban, & DeScioli, 2013), but the mnemonic effects of contamination remain unexplored. Specifically, we asked the following empirical question: Will people remember objects that have been touched by a sick person better than ones touched by a healthy person? To address the question, people were shown pictures of everyday objects and were told that each object had been recently touched by a sick or a healthy individual. Accompanying each picture was a short description that helped to categorize who had touched the object. For example, a picture of a "ball" appeared with the descriptor "person with a constant cough" or the descriptor "person with a straight nose." After every third item, the three preceding items appeared again and the task was to decide whether that object had been touched by a sick or a healthy person. Following a series of these trials, everyone received a surprise free recall test for the pictured objects. The free recall results revealed a strong advantage for the "contaminated" items—proportion correct recall for the items touched by a sick person averaged 0.42 whereas the "healthy" items averaged only 0.34, a statistically significant effect. We think this is another important and novel finding because it demonstrates the functional value of remembering and suggests, once again, that our memory systems may be biased to remember information that is fitness-relevant.

Still, as discussed previously about survival processing, the simple demonstration of memory-based contamination or animacy effects does not tell us much about the proximate mechanisms involved. From an ultimate perspective, one can make a compelling argument for *why* animacy and contamination tunings might have evolved—they enhance inclusive fitness—but the *how* remains unknown. As chapters in the current volume show, the study of proximate mechanisms can be revealing, both in establishing the generality of phenomena (Klein, Chapter 2 in this volume; Altarriba & Kazanas, Chapter 7 in this volume) and in establishing boundary conditions (Otgaar, Howe, Smeets, Raymaekers, & van Beers, Chapter 11 in this volume; Schwartz & Brothers, Chapter 9 in this volume). The study of boundary conditions may be particularly important because boundary conditions encourage one to consider the function of a mnemonic mechanism. As discussed above, it is unlikely that we evolved a single mnemonic mechanism to cover all memory-based adaptive problems. The functional requirements associated with mating, for instance, may differ fundamentally from those connected with finding food or avoiding predators (see Klein, 2013). Thus, memory adaptations that work well in one context might not work well in another. The fact that survival processing produces robust effects across a wide variety of materials and designs, but apparently does not transfer well to implicit tests (Tse & Altarriba, 2010) or paired-associate learning (Schwartz & Brothers, Chapter 9 in this volume), undoubtedly will tell us something important about the functional design of the memory mechanisms involved.

Recent work on the development of adaptive memory provides insight into the functional properties of memory mechanisms as well. As Sellers and Bjorklund (Chapter 15 in this volume) perceptively note, cognitive development undoubtedly reflects the age-specific problems faced by the developing organism. A particular mechanism (or adaptation) appropriate for solving the adaptive problems of a

2-year-old might not be adaptive for an adolescent or an adult. We should also expect to find development differences in the appearance of certain mnemonic traits—for example, survival-relevant traits might appear early and remain relatively constant across the lifespan (see also Howe & Otgaar, 2013), whereas adaptations relevant to mating could appear later. Focusing on development is important for another reason as well: Experience almost certainly plays a critical role in shaping the operating characteristics of many evolved systems. The fact that people might have evolved "tunings" for certain recurrent problems does not mean that experience is unimportant in shaping the final characteristics of those traits.

Memory is adaptive—few, if any, researchers would disagree with this conclusion. Yet the bulk of memory research since Ebbinghaus has focused primarily on structure rather than function. Researchers have defined their mission as one of understanding the contents of the mind, much like a chemist deconstructs a chemical compound into its more elemental components. Although a structural stance can produce significant advances in our understanding of cognitive systems, functionless tinkering often leads to directionless research, or to theoretical frameworks designed to explain tasks rather than psychological processes (Klein et al., 2002; Nairne, 2005). As William James once said, it is difficult to understand a house by focusing on its bricks and mortar—one needs to know what the house is for, what the house is designed to do, and it is only in this functional context that bricks and mortar make sense. Identifying the adaptive problems that our memory systems evolved to solve provides just such a functional context in which structural investigations of remembering potentially make sense.

ACKNOWLEDGMENTS

Josefa N. S. Pandeirada, Joshua VanArsdall, and Mindi Cogdill deserve special thanks for their contributions to the research and ideas discussed in this chapter. Our research, and the writing of this chapter, has been supported by a grant from the National Science Foundation (BCS-0843165).

REFERENCES

Barrett, H. C., & Kurzban, R. (2006). Modularity in cognition: Framing the debate. *Psychological Review, 113*, 628–647.

Boyer, P., & Barrett, H. C. (2005). Domain specificity and intuitive ontology. In D. M. Buss (Ed.), *The handbook of evolutionary psychology* (pp. 96–188). Hoboken, NJ: Wiley.

Butler, A. C., Kang, S. H. K., & Roediger, H. L. (2009). Congruity effects between materials and processing tasks in the survival processing paradigm. *Journal of Experimental Psychology: Learning, Memory, and Cognition, 35*, 1477–1486. doi:10.1037/a0017024

Caramazza, A., & Shelton, J. R. (1998). Domain-specific knowledge systems in the brain: The animate-inanimate distinction. *Journal of Cognitive Neuroscience, 10*(1), 1–34. doi:10.1162/089892998563752

Cosmides, L., & Tooby, J. (2005). Neurocognitive adaptations designed for social exchange. In D. M. Buss (Ed.), *The handbook of evolutionary psychology* (pp. 584–627). Hoboken, NJ: Wiley.

Craik, F. I. M., & Lockhart, R. S. (1972). Levels of processing: A framework for memory research. *Journal of Verbal Learning and Verbal Behavior, 11*, 671–684.

Ermer, E., Cosmides, L., & Tooby, J. (2007). Functional specialization and the adaptationist program. In S. W. Gangestad & J. A. Simpson (Eds.), *The evolution of mind: Fundamental questions and controversies* (pp. 86–94). New York: Guilford Press.

Gould, S. J., & Lewontin, R. C. (1979). The spandrels of San Marco and the Panglossian paradigm: A critique of the adaptationist programme. *Proceedings of the Royal Society of London, Series B, 205*, 581–598. doi:10.1098/rspb.1979.0086

Howe, M. L., & Derbish, M. H. (2010). On the susceptibility of adaptive memory to false memory illusions. *Cognition, 115*, 252–267. doi:10.1016/j.cognition.2009.12.016

Howe, M. L., & Otgaar, H. (2013). Proximate mechanisms and the development of adaptive memory. *Current Directions in Psychological Science, 22*, 16–22. doi:10.1177/0963721412469397

Kanwisher, N. (2000). Domain specificity in face recognition. *Nature Neuroscience, 3*, 759–763. doi:10.1038/77664

Klein, S. B. (2013). Does optimal performance in the adaptive memory paradigm require the encoding context to encourage thoughts about the environment of evolutionary adaptation? *Memory & Cognition, 41*, 49–59.

Klein, S. B., Cosmides, L., Tooby, J., & Chance, S. (2002). Decisions and the evolution of memory: Multiple systems, multiple functions. *Psychological Review, 109*(2), 306–329. doi:10.1037/0033-295X.109.2.306

Kostic, B., McFarlan, C. C., & Cleary, A. M. (2012). Extensions of the survival advantage in memory: Examining the role of ancestral context and implied social isolation. *Journal of Experimental Psychology: Learning, Memory, and Cognition, 38*, 1091–1098. doi:10.1037/a0026974

Kroneisen, M., & Erdfelder, E. (2011). On the plasticity of the survival processing effect. *Journal of Experimental Psychology: Learning, Memory, and Cognition, 37*, 1553–1562. doi:10.1037/a0024493

LoBue, V., & DeLoache, J. S. (2008). Detecting the snake in the grass: Attention to fear-relevant stimuli by adults and young children. *Psychological Science, 19*, 284–289. doi:10.1111/j.1467-9280.2008.02081.x

Mayr, E. (1963). *Animal species and evolution*. Cambridge, MA: Harvard University Press.

Nairne, J. S. (2005). The functionalist agenda in memory research. In A. F. Healy (Ed.), *Experimental cognitive psychology and its applications* (pp. 115–126). Washington, DC: American Psychological Association.

Nairne, J. S. (2010). Adaptive memory: Evolutionary constraints on remembering. In B. H. Ross (Ed.), *The psychology of learning and motivation* (Vol. 53, pp. 1–32). Academic Press. doi:10.1016/S0079-7421(10)53001-9.

Nairne, J. S., & Pandeirada, J. N. S. (2008). Adaptive memory: Remembering with a stone-age brain. *Current Directions in Psychological Science, 17*, 239–243. doi:10.1111/j.1467-8721.2008.00582.x

Nairne, J. S., & Pandeirada, J. N. S. (2010). Adaptive memory: Ancestral priorities and the mnemonic value of survival processing. *Cognitive Psychology, 61*, 1–22. doi:10.1016/j.cogpsych.2010.01.005

Nairne, J. S., & Pandeirada, J. N. S. (2011). Congruity effects in the survival processing paradigm. *Journal of Experimental Psychology: Learning, Memory, and Cognition, 37*, 539–549. doi:10.1037/a0021960.

Nairne, J. S., Pandeirada, J. N. S., Gregory, K. J., & Van Arsdall, J. E. (2009). Adaptive memory: Fitness relevance and the hunter-gatherer mind. *Psychological Science, 20*, 740–746. doi:10.1111/j.1467-9280.2009.02356.x

Nairne, J. S., Pandeirada, J. N. S., & Thompson, S. R. (2008). Adaptive memory: The comparative value of survival processing. *Psychological Science, 19*(2), 176–180. doi:10.1111/j.1467-9280.2008.02064.x

Nairne, J. S., Thompson, S. R., & Pandeirada, J. N. S. (2007). Adaptive memory: Survival processing enhances retention. *Journal of Experimental Psychology: Learning, Memory, and Cognition, 33*(2), 263–273. doi:10.1037/0278-7393.33.2.263

Nairne, J. S., VanArsdall, J. E., Pandeirada, J. N. S., Cogdill, M., & LeBreton, J. M. (in press). Adaptive memory: The mnemonic value of animacy. *Psychological Science.*

Nathans, J. (1999). The evolution and physiology of human color vision: Insights from molecular genetic studies of visual pigments. *Neuron, 24*, 299–312. doi:10.1016/S0896-6273(00)80845-4

New, J., Cosmides, L., & Tooby, J. (2007). Category-specific attention for animals reflects ancestral priorities, not expertise. *Proceedings of the National Academy of Sciences USA, 104*(42), 16598–16603. doi:10.1073/pnas.0703913104

Öhman, A., & Mineka, S. (2001). Fears, phobia, and preparedness: Toward an evolved module of fear and fear learning. *Psychological Review, 108*, 483–522. doi:10.1037/0033-295X.108.3.483

Quinlan, P. T. (in press). The visual detection of threat: A cautionary tale. *Psychonomic Bulletin & Review.*

Scott-Phillips, T. C., Dickens, T. E., & West, S. A. (2011). Evolutionary theory and the ultimate-proximate distinction in the human behavioral sciences. *Perspectives on Psychological Science, 6*, 38–47. doi:10.1177/1745691610393528

Sherry, D. F., & Schacter, D. L. (1987). The evolution of multiple memory systems. *Psychological Review, 94*, 439–454.

Soderstrom, N. C., & McCabe, D. P. (2011). Are survival processing memory advantages based on ancestral priorities? *Psychonomic Bulletin & Review, 18*, 564–569. doi:10.3758/s13423-011-0060-6

Sokolov, E. N. (1963). Higher nervous functions: The orienting reflex. *Annual Review of Physiology, 25*, 545–580.

Symons, D. (1992). On the use and misuse of Darwinism in the study of human behavior. In J. H. Barkow, L. Cosmides, & J. Tooby (Eds.), *The adapted mind: Evolutionary psychology and the generation of culture* (pp. 137–159). New York: Oxford University Press.

Tooby, J., & Cosmides, L. (1992). The psychological foundations of culture. In J. H. Barkow, L. Cosmides, & J. Tooby (Eds.), *The adapted mind: Evolutionary psychology and the generation of culture* (pp. 19–136). New York: Oxford University Press.

Tooby, J., & Cosmides, J. (2005). Conceptual foundations of evolutionary psychology. In D. Buss (Ed.), *The handbook of evolutionary psychology* (pp. 5–67). Hoboken, NJ: Wiley.

Tse, C-S., & Altarriba, J. (2010). Does survival processing enhance implicit memory? *Memory & Cognition, 38*, 1110–1121. doi:10.3758/MC.38.8.1110.

Tulving, E., & Thomson, D. M. (1973). Encoding specificity and retrieval processes in episodic memory. *Psychological Review, 80*, 352–373.

Tybur, J. M., Lieberman, D., Kurzban, R., & DeScioli, P. (2013). Disgust: Evolved function and structure. *Psychological Review, 120*, 65–84. doi:10.1037/a0030778

VanArsdall, J.E., Nairne, J.S., Pandeirada, J.N.S., & Blunt, J.R. (2013). Adaptive memory: Animacy processing produces mnemonic advantages. *Experimental Psychology, 60*, 172–178.

Weinstein, Y., Bugg, J. M., & Roediger, H. L. (2008). Can the survival recall advantage be explained by the basic memory processes? *Memory & Cognition*, *36*, 913–919. doi: 10.3758/MC.36.5.913.

Williams, G. C. (1966). *Adaptation and natural selection*. Princeton: Princeton University Press.

Index